Jared Michael Maruskin

Dynamical Systems and Geometric Mechanics

De Gruyter Studies in Mathematical Physics

Volume 48

Jared Michael Maruskin

Dynamical Systems and Geometric Mechanics

An Introduction

2nd edition

DE GRUYTER

Physics and Astronomy Classification Scheme 2010
35-02, 65-02, 65C30, 65C05, 65N35, 65N75, 65N80

Author
Dr. Jared Michael Maruskin
San Jose CA 95128, USA
jared.maruskin@gmail.com

ISBN 978-3-11-059729-5
e-ISBN (PDF) 978-3-11-059780-6
e-ISBN (EPUB) 978-3-11-059803-2
ISSN 2194-3532

Bibliographic information published by the Deutsche Nationalbibliothek
The Deutsche Nationalbibliothek lists this publication in the Deutsche Nationalbibliografie; detailed bibliographic data are available on the Internet at http://dnb.dnb.de.

Typesetting: VTeX UAB, Lithuania
Printing and binding: CPI books GmbH, Leck

www.degruyter.com

Contents

Preface

This book is based on a set of lecture notes from a graduate course in dynamical systems that I taught during spring 2010. The goal of this book is not to serve as an encyclopedic reference for the fields of dynamical systems and geometric mechanics; many wonderful and advanced books already exist, several of which I strive to direct the reader to along the journey contained within these pages. Rather, my goal is to provide a broad perspective on this beautiful subject in a brief and introductory manner that can be taught within a single semester for advanced undergraduates or beginning graduate students. The field of dynamical systems has grown to encompass many branches of mathematics; this text will use notions from linear algebra, analysis, topology, differential geometry, and Lie group theory. Though I strive to make such a book as self-contained as possible, that magical ingredient referred to as mathematical maturity should be taken as prerequisite. This means that, as an eager student, you do not have to have studied any of the particular fields of mathematics listed above before taking on this text, but you should have a certain understanding of mathematical rigor and language that will facilitate your digestion of powerful ideas that are presented in a concise fashion.

Theme

The theme of this book is systems of first-order differential equations. In the first part, we describe various aspects of the theory of solutions of linear and nonlinear first-order differential equations; in the second part, we take up the topic of differential equations on smooth differentiable manifolds. Generally speaking, we are interested in differential equations of the form $\dot{x} = f(x)$. In Part I of the book, we take $x \in \mathbb{R}^n$. We begin Chapter 1 by examining the case where $f(x) = Ax$, for some real-valued, $n \times n$ matrix A. For the remaining chapters of Part I, we broaden our perspective by regarding $f(x)$ as a smooth vector field on $\mathbb{R}^n$. In Part II of the book, we are again interested in the same equation, only x is regarded as a point on a manifold and f as a vector field on that manifold. The solutions can be described as a smooth class of curves on the manifold whose tangent vectors coincide with the vector field f at every point.

In addition to studying the mathematical and theoretical aspects of such systems, we will pay keen attention to applications throughout the text, with an emphasis on applications to mechanics. The second part of the book is particularly suited to the modern theory of geometric mechanics. Newton's laws of motion were originally written down for systems in Euclidean spaces. However, starting with the works of Lagrange and Hamilton, dynamicists began to realize that a system's configuration could more accurately be viewed as a point on a smooth manifold: the simple pendulum

https://doi.org/10.1515/9783110597806-201

lives on a circle, the double pendulum lives on a torus, and a satellite in orbit about the Earth lives on a certain matrix Lie group.

Map of the Book

To provide the greatest amount of flexibility, I have provided a map of this text in Figure 1. The text is divided into two parts: *Dynamical Systems* and *Geometric Mechanics*, respectively. Chapters 1–3 form the core of Part I of the book and should be taken as prerequisite for Part II. Similarly, Chapters 6 and 7 constitute the core of Part II and should be taken as prerequisite from which one may spring directly into Chapter 8, 9, or 10.

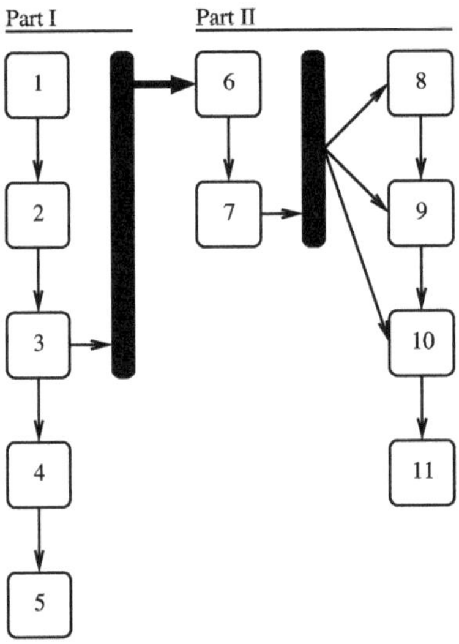

Figure 1: Map of this book.

The purpose of this book is to provide a sophisticated and concise introduction to the fields of dynamical systems and geometric mechanics that can be covered in a single-semester course. For shorter courses that aim to cover either the first or second half of the text, I have included various applications at the end of each chapter that can serve as springboards for further discussion, research, projects, or investigation.

Background and References

I have written this book aimed at graduate students in mathematics, physics, and engineering with previous coursework experience in linear algebra and differential equations and an undergraduate course in dynamical systems, at the level of analyzing two-dimensional systems and phase planes. This book, however, is self-contained, so that astute students without a background in differential equations and planar dynamical systems should still manage to thrive. I kept this flexibility in mind when writing the first part of the book. The first chapter on linear systems, for example, is more advanced than students with an undergraduate background in dynamical systems would have previously encountered, so the material should not bore them, but it is comprehensive enough to provide a complete picture of linear systems to newcomers. A similar statement could be made with regard to the remaining chapters in Part I: material that some students may have previously encountered will be presented in a more sophisticated, yet complete, way, so that experience with the material is not absolutely necessary for a student possessing mathematical maturity.

I should, however, recommend several supplements to brave students who wish to tackle the text with no previous coursework in dynamical systems. Most of the gist can be picked up here; however, for a lighter discussion and reprieve one might consult an elementary book on differential equations such as [34], which also contains a very nice and concise chapter on linear and nonlinear phase plane analysis (some students might want to start with this chapter). A second book on nonlinear phase plane dynamics, such as [115], [138], or [274], should also be referenced so as to form a complete picture of the subject and its applications. The Strogatz text is particularly good in that it abounds with a rich tapestry of applications and conceptual clarity, whereas the other two contain a greater amount of mathematical rigor.

For novice graduate and ambitious undergraduate students taking on this text, it will become paramount, as you begin your careers in research, to learn how to consult a variety of sources on a given topic in order to fully appreciate and comprehend its complexity. Different authors offer different perspectives on any particular topic, and these varying perspectives, taken in aggregate, yield a more holistic understanding of the subject. As such, I have strived to offer a variety of references for additional material throughout the text. I would like to make special mention of several important ones up front. For Part I of the text, [110], [211], [235], and [294] constitute a nice selection of references to complement this text. In those works you will find additional relevant discussions, references, and topics. For Part II, [8], [11], [104, 106], and [222] constitute a set of classic references for Lagrangian, Hamiltonian, and nonholonomic mechanics, whereas [24], [37], and [196] are excellent references that go into greater depth on the geometric formulation of mechanics and that might also form a basis for subsequent studies of this book's subject matter.

Acknowledgments

I am incredibly grateful for and indebted to the many enjoyable and enlightening conversations I have had with Anthony M. Bloch, Daniel J. Scheeres, and Dmitry Zenkov pertaining to many of the topics and discussions contained within these pages, as well as to the suggestions of the anonymous reviewers that led to the formation of a better manuscript and to the copyediting work performed by Glenn Corey, who helped improve the overall readability of the text and helped enforce global consistencies in style.

Finally, I am very pleased that you have chosen to undertake the study of such an exciting and elegant subject. I look forward to embarking on our adventure together in the pages that lie ahead.

San José, California
September 2018

Jared M. Maruskin

Part I: **Dynamical Systems**

1 Linear Systems

Linear systems constitute an important hallmark of dynamical systems theory. Not only are they among the rare class of systems in which exact, analytic solutions can actually be obtained and computed, they have also become a cornerstone even in the analysis of nonlinear systems, including systems on manifolds. We will see later in this book that the flow near a fixed point can be *locally* understood by its associated *linearization*. The different solution modes of a linear system may be analyzed individually using linearly independent solutions obtained from eigenvalues and eigenvectors or matrixwise using operators and operator exponentials. We begin our study of linear systems with a discussion of both approaches and some fundamental consequences.

1.1 Eigenvector Approach

Once the early pioneers of linear dynamical systems realized they could represent these systems in matrix form, it was not long before they realized that, when expressed relative to an eigenbasis of the constant coefficient matrix, a spectacular decoupling results, yielding the solution almost trivially. We begin our study of linear systems with a review of this method, paying particularly close attention to the case of distinct eigenvalues. We will consider the case of repeated eigenvalues later in the chapter.

Linear Systems and the Principle of Superposition

Throughout this chapter, we will consider linear systems of first-order ordinary differential equations of the form

$$\dot{x} = Ax, \tag{1.1}$$

$$x(0) = x_0, \tag{1.2}$$

where $x : \mathbb{R} \to \mathbb{R}^n$ is the unknown *solution* to the above initial value problem, $x_0 \in \mathbb{R}^n$ is the *initial condition*, and $A \in \mathbb{R}^{n\times n}$ is a real-valued matrix called the *coefficient matrix*. The *solution*, or *flow*, of system (1.1) is a smooth function $\varphi : \mathbb{R} \times \mathbb{R}^n \to \mathbb{R}^n$ that satisfies

$$\frac{d\varphi}{dt} = A\varphi(t; x_0), \qquad \text{for all } t \in \mathbb{R}, \tag{1.3}$$

$$\varphi(0; x_0) = x_0, \qquad \text{for all } x_0 \in \mathbb{R}^n. \tag{1.4}$$

The singular curve $x(t) = 0$ trivially satisfies differential equation (1.1) with initial condition $x_0 = 0$ for all $t \in \mathbb{R}$. Hence the point $x = 0$ is referred to as a *fixed point* of the system.

https://doi.org/10.1515/9783110597806-001

We say that a curve $z : \mathbb{R} \to \mathbb{C}^n$ is a *complex-valued solution* to (1.1) if both its real and imaginary parts are solutions to (1.1), i. e., if $z(t) = x(t)+iy(t)$, where $x, y : \mathbb{R} \to \mathbb{R}^n$, then z is a complex-valued solution to (1.1) if and only if both x and y are real-valued solutions to (1.1). To facilitate our analytic understanding of system (1.1), we first state a theorem dealing with linear combinations of particular solutions of the system.

Theorem 1.1 (Principle of Superposition). *Suppose $z_1, \ldots, z_k : \mathbb{R} \to \mathbb{C}^n$ are k distinct (possibly complex-valued) solutions to the system of differential equations* (1.1). *Then the linear combination*

$$z(t) = c_1 z_1(t) + \cdots + c_k z_k(t) \tag{1.5}$$

is also a solution of (1.1) *for arbitrary scalar coefficients $c_1, \ldots, c_k \in \mathbb{C}$.*

Proof. This follows directly from the fact that both A and d/dt are linear operators. First, suppose that $z_1, z_2 : \mathbb{R} \to \mathbb{C}^n$ are *two* complex-valued solutions. To show that an arbitrary linear combination satisfies the system of differential equations (1.1), we simply check that

$$\frac{d}{dt}(c_1 z_1 + c_2 z_2) = c_1 \dot{z}_1 + c_2 \dot{z}_2 = c_1 A z_1 + c_2 A z_2 = A(c_1 z_1 + c_2 z_2).$$

The final result holds by induction. □

Now suppose that $v \in \mathbb{C}^n$ is an eigenvector of coefficient matrix A, with $\lambda \in \mathbb{C}$ being its associated eigenvalue. It is clear that $x(t) = ve^{\lambda t}$ is a particular solution to differential equation (1.1). Assuming the uniqueness of solutions to the preceding linear system, we immediately obtain the following general solution.

Theorem 1.2. *Suppose matrix A has n distinct eigenvalues $\lambda_1, \ldots, \lambda_n \in \mathbb{C}$ with associated eigenvectors $v_1, \ldots, v_n \in \mathbb{C}^n$. Then the general solution of the system of first-order differential equations* (1.1) *is given by*

$$x(t) = c_1 v_1 e^{\lambda_1 t} + \cdots + c_n v_n e^{\lambda_n t}, \tag{1.6}$$

where $c_1, \ldots, c_n \in \mathbb{C}$ are arbitrary constants.

Real, Distinct Eigenvalues

Now let us restrict our attention to systems that have *real, distinct* eigenvalues, so that the general solution (1.6) may be regarded as a real-valued solution, where each of the coefficients $c_1, \ldots, c_n \in \mathbb{R}$.

Example 1.1. Compute the general solution of system (1.1), where

$$A = \begin{bmatrix} 3 & -2 \\ 0 & -1 \end{bmatrix}.$$

The eigenvalues and eigenvectors of A are $\lambda_1 = 3$, $\lambda_2 = -1$, $v_1 = (1,0)$, and $v_2 = (1,2)$, respectively. Therefore, the general solution is given by

$$x(t) = c_1 e^{3t} \begin{bmatrix} 1 \\ 0 \end{bmatrix} + c_2 e^{-t} \begin{bmatrix} 1 \\ 2 \end{bmatrix}.$$

Observe that if the initial condition is $x_0 = \alpha v_1$, then the particular solution of the initial value problem (1.1)–(1.2) is $x(t) = \alpha e^{3t} v_1$. Hence the solution trajectory remains on the line defined by the span of v_1 for all time and the solution grows exponentially in the direction *away* from the origin.

On the other hand, if the initial condition is $x_0 = \alpha v_2$, then the particular solution of the initial value problem is $x(t) = \alpha e^{-t} v_2$. Again, the solution remains on the line defined by the span of v_2 for all time, and the solution decays exponentially *toward* the origin.

Since $\operatorname{span}\{v_1, v_2\} = \mathbb{R}^2$, vectors v_1 and v_2 form a basis for $\mathbb{R}^2$. The constants c_1 and c_2 are simply the components of the initial condition vector with respect to this basis. Hence, the general solution flow will be a linear combination of the two previously mentioned cases. The *phase portrait*, i. e., a graph of representative solution trajectories, is plotted in Figure 1.1. This type of fixed point is called a *saddle point* because of the shape of the surrounding phase flow (it helps to imagine the level sets of the flow in three dimensions). □

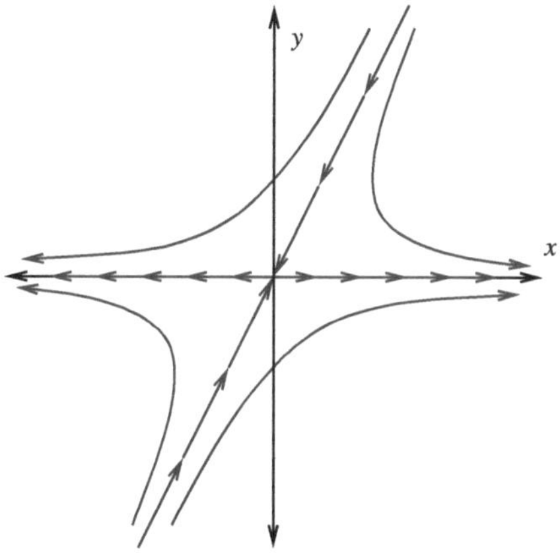

Figure 1.1: Example 1.1: flow around a saddle point.

Example 1.2. Consider linear system (1.1) with the coefficient matrix

$$A = \begin{bmatrix} 1.5 & 0.5 \\ 0.5 & 1.5 \end{bmatrix}.$$

The eigenvalues and eigenvectors of A are $\lambda_1 = 2$, $\lambda_2 = 1$, $v_1 = (1,1)$, and $v_2 = (-1,1)$, respectively. Hence the general solution is given by

$$x(t) = c_1 e^{2t} \begin{bmatrix} 1 \\ 1 \end{bmatrix} + c_2 e^{t} \begin{bmatrix} -1 \\ 1 \end{bmatrix}.$$

Again, if the initial condition lies in the direction of either eigenvector, then the solution flow will remain on the line defined by that eigenvector for all time. For a general initial condition, coefficients c_1 and c_2 will be the components of x_0 with respect to the eigenbasis $\{v_1, v_2\}$. Since both eigenvalues are positive, all nonzero solutions will grow exponentially away from the origin as $t \to \infty$. If x_0 does not lie in the direction defined by either eigenvector, then the solution will veer toward v_1 as $t \to \infty$ since the solution component relative to v_1 has an exponential growth constant λ_1 that is greater than the exponential growth constant λ_2 relative to v_2. A fixed point with two eigenvalues of the same sign is referred to as a *node* or *nodal point*. Since solutions starting near the origin exponentially diverge away from the origin, the fixed point in this example is referred to as an *unstable node*. The phase flow is plotted in Figure 1.2. □

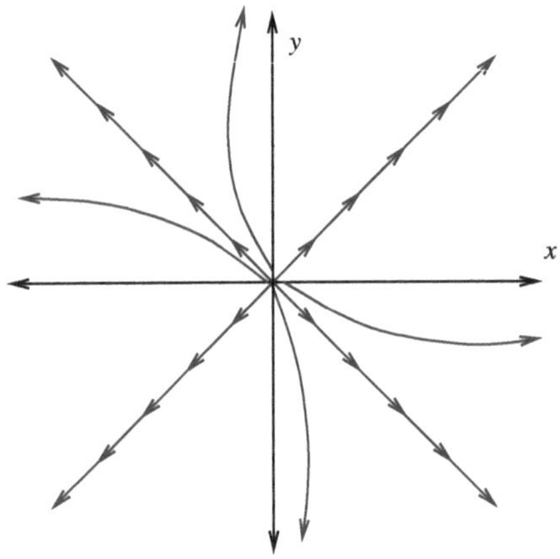

Figure 1.2: Example 1.2: flow around an unstable node.

Exercise 1.1. Compute the general solution of the linear system $\dot{x} = Ax$, where the coefficient matrix is given by

$$A = \begin{bmatrix} 0 & -1 \\ 4 & -5 \end{bmatrix}.$$

Plot the phase portrait of the flow.

Exercise 1.2. Compute the general solution of the linear system $\dot{x} = Ax$, where the coefficient matrix is given by

$$A = \begin{bmatrix} -1 & -1 & 3 \\ 0 & -2 & 3 \\ 0 & 0 & 1 \end{bmatrix}.$$

Plot a sketch of the solution flow in $\mathbb{R}^3$. What happens to trajectories that initially hover above the x_1-x_2 plane as $t \to \infty$?

These examples exhibit a fundamental feature of linear systems: initial conditions in a subspace that is spanned by the eigenvectors associated with all the positive or negative eigenvalues result in trajectories that remain within that subspace. Next we will define these concepts more rigorously and then return to them in §1.4.

Definition 1.1. Suppose $A \in \mathbb{R}^{n\times n}$ has k distinct negative eigenvalues $\lambda_1, \ldots, \lambda_k$ and $(n-k)$ distinct positive eigenvalues $\lambda_{k+1}, \ldots, \lambda_n$. The *stable* and *unstable subspaces* of linear system (1.1) are defined by

$$\begin{aligned} E^s &= \operatorname{span}\{v_1, \ldots, v_k\}, \\ E^u &= \operatorname{span}\{v_{k+1}, \ldots, v_n\}, \end{aligned}$$

respectively.

Definition 1.2. A subspace $W \subset \mathbb{R}^n$ is said to be an *invariant subspace* under the flow of linear system (1.1) if $\varphi(t; x_0) \in W$ for all $t \in \mathbb{R}$ whenever $x_0 \in W$.

Exercise 1.3. Show that if $x_0 \in E^s$, then $\varphi(t; x_0) \in E^s$ for all t and, moreover, that $\lim_{t\to\infty} \varphi(t; x_0) = 0$. Similarly, show that if $x_0 \in E^u$, then $\varphi(t; x_0) \in E^u$ for all t and, moreover, that $\lim_{t\to-\infty} \varphi(t; x_0) = 0$. Conclude that E^s and E^u are *invariant subspaces* of linear system (1.1). The invariant subspaces E^s and E^u are referred to as the *stable* and *unstable subspaces* of the system, respectively.

Complex, Distinct Eigenvalues

We now consider the case in which the coefficient matrix A in (1.1) has $k \leq n$ distinct, real-valued eigenvalues and $(n-k)$ distinct, complex-valued eigenvalues. Since complex eigenvalues and their corresponding eigenvectors always occur in complex conjugate pairs, we may define the integer $s = (n-k)/2$ as the number of such pairs. The eigenvalues may therefore be arranged so that $\lambda_1, \ldots, \lambda_k \in \mathbb{R}$, and

$$\lambda_j = a_j + ib_j, \quad \overline{\lambda_j} = a_j - ib_j, \quad \text{for } j = (k+1), \ldots, (k+s),$$

where $a_j, b_j \in \mathbb{R}$ for $j = (k+1), \ldots, (k+s)$. Similarly, the first k associated eigenvectors may be taken as $w_1, \ldots, w_k \in \mathbb{R}^n$, whereas the associated complex eigenvectors, which also occur in complex conjugate pairs, may be taken as

$$w_j = u_j + iv_j, \quad \overline{w_j} = u_j - iv_j, \quad \text{for } j = (k+1), \ldots, (k+s),$$

where $u_j, v_j \in \mathbb{R}^n$ for $j = (k+1), \ldots, (k+s)$.

In other words, the eigenvalues of A are $\lambda_1, \ldots, \lambda_k, a_{k+1}+ib_{k+1}, a_{k+1}-ib_{k+1}, \ldots, a_{k+s}+ ib_{k+s}, a_{k+s} - ib_{k+s}$. Similarly, the eigenvectors of A are $w_1, \ldots, w_k, u_{k+1} + iv_{k+1}, u_{k+1} - iv_{k+1}, \ldots, u_{k+s} + iv_{k+s}, u_{k+s} - iv_{k+s}$.

Definition 1.3. Given a set of k complex-valued vectors $w_1, \dots, w_k \in \mathbb{C}^n$, we define the *real span* of the vectors as

$$\text{realspan}\{w_1, \dots, w_k\} = \text{span}\{\Re\{w_1\}, \Im\{w_1\}, \dots, \Re\{w_k\}, \Im\{w_k\}\} \subset \mathbb{R}^n,$$

where $\Re\{w\}$ and $\Im\{w\}$ are the real and imaginary parts of w, respectively.

We expect the final solution of (1.1) to be real-valued. Hence the complex-valued coefficients of the general solution (1.6) must be such that their sum is a real-valued vector function in $\mathbb{R}^n$.

We know from (1.6) that $e^{\lambda_j t} w_j$ and $e^{\overline{\lambda_j} t}\overline{w_j}$ are distinct, complex-valued solutions of (1.1) for $j = (k+1), \dots, (k+s)$. From the principle of superposition we know that arbitrary (complex) linear combinations of these solutions are solutions as well. Hence, their real and imaginary parts are solutions. Since $e^{\lambda_j t} w_j$ and $e^{\overline{\lambda_j} t}\overline{w_j}$ are complex conjugates of each other, it follows that

$$\text{realspan}\{e^{\lambda_j t} w_j, e^{\overline{\lambda_j} t}\overline{w_j}\} = \text{span}\{\Re\{e^{\lambda_j t} w_j\}, \Im\{e^{\lambda_j t} w_j\}\},$$

for $j = (k+1), \dots, (k+s)$. We may therefore "trade" the complex solutions $e^{\lambda_j t} w_j$ and $e^{\overline{\lambda_j} t}\overline{w_j}$ for the real solutions $\Re\{e^{\lambda_j t} w_j\}$ and $\Im\{e^{\lambda_j t} w_j\}$ for each $j = (k+1), \dots, (k+s)$. Hence the following theorem.

Theorem 1.3. *Suppose linear system* (1.1) *has* $k \leq n$ *distinct, real-valued eigenvalues* $\lambda_1, \dots, \lambda_k$ *and* $s = (n-k)/2$ *complex conjugate eigenvalue pairs*

$$\lambda_{k+1}, \overline{\lambda_{k+1}}, \dots, \lambda_{k+s}, \overline{\lambda_{k+s}}.$$

Then the general, real-valued solution of (1.1) *is given by*

$$x(t) = \sum_{j=1}^{k} c_j e^{\lambda_j t} w_j + \sum_{j=k+1}^{k+s} (c_j \Re\{e^{\lambda_j t} w_j\} + d_j \Im\{e^{\lambda_j t} w_j\}),$$

where $c_1, \dots, c_{k+s}, d_{k+1}, \dots, d_{k+s} \in \mathbb{R}$.

Notice that the real and imaginary parts of $e^{\lambda_j t} w_j$, up to a sign, are identical to the real and imaginary parts of $e^{\overline{\lambda_j} t}\overline{w_j}$, for $j = (k+1), \dots, (k+s)$. This is why we are allowed to replace complex conjugate pair solutions with their real and imaginary parts.

Example 1.3. Consider the linear system $\dot{x} = Ax$, where coefficient matrix A is given by

$$A = \begin{bmatrix} -2 & 2 \\ -4 & 2 \end{bmatrix}.$$

The eigenvalues and eigenvectors are $\lambda_1 = 2i$, $\overline{\lambda_1} = -2i$, $w_1 = (1, 1+i)$, and $\overline{w_1} = (1, 1-i)$, respectively. Note that the eigenvalues are purely imaginary. By Theorem 1.3, the general solution is given by

$$x(t) = c_1 \mathbb{R}\{e^{\lambda_1 t} w_1\} + d_1 \mathbb{I}\{e^{\lambda_1 t} w_1\},$$

where $c_1, d_1 \in \mathbb{R}$. Using Euler's equation

$$e^{i\theta} = \cos\theta + i\sin\theta, \tag{1.7}$$

we may write

$$\begin{aligned} e^{\lambda_1 t} w_1 &= (\cos(2t) + i\sin(2t)) \begin{bmatrix} 1 \\ 1+i \end{bmatrix} \\ &= \begin{bmatrix} \cos(2t) + i\sin(2t) \\ (\cos(2t) - \sin(2t)) + i(\cos(2t) + \sin(2t)) \end{bmatrix}. \end{aligned}$$

Hence the general solution is given by

$$x(t) = c_1 \begin{bmatrix} \cos(2t) \\ \cos(2t) - \sin(2t) \end{bmatrix} + c_2 \begin{bmatrix} \sin(2t) \\ \cos(2t) + \sin(2t) \end{bmatrix}.$$

The phase portrait is plotted in Figure 1.3. Orbits form concentric, closed periodic loops about the origin. This type of fixed point is referred to as a *center*. □

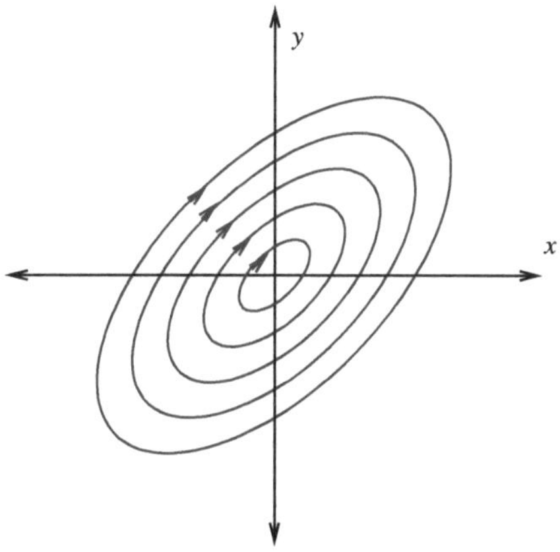

Figure 1.3: Example 1.3: flow around a center.

Example 1.4. Consider the linear system $\dot{x} = Ax$, where coefficient matrix A is given by

$$A = \begin{bmatrix} -1 & 2 \\ -2 & -1 \end{bmatrix}.$$

The eigenvalues and eigenvectors are $\lambda_1 = -1 + 2i$, $\overline{\lambda_1} = -1 - 2i$, $w_1 = (1, i)$, $\overline{w_1} = (1, -i)$. From Euler's equation (1.7) we have

$$e^{(a+bi)} = e^{at}\,(\cos(bt) + i\sin(bt))\,.$$

Hence,

$$e^{\lambda_1 t} w_1 = e^{-t}(\cos(2t) + i\sin(2t)) \begin{bmatrix} 1 \\ i \end{bmatrix} = \begin{bmatrix} e^{-t}\cos(2t) + ie^{-t}\sin(2t) \\ -e^{-t}\sin(2t) + ie^{-t}\cos(2t) \end{bmatrix}.$$

Therefore, by Theorem 1.3, the general solution is given by

$$x(t) = c_1 e^{-t} \begin{bmatrix} \cos(2t) \\ -\sin(2t) \end{bmatrix} + c_2 e^{-t} \begin{bmatrix} \sin(2t) \\ \cos(2t) \end{bmatrix}.$$

Hence the oscillations due to the imaginary eigenvalues decay exponentially so that solutions spiral in toward the origin. In this context, the origin is referred to as a *stable spiral*. The phase portrait for this flow is plotted in Figure 1.4. Notice that the exponential decay is due to the real part of the eigenvalues. Had the real parts been positive, the solutions would have spiraled *away* from the origin, and the origin would have been referred to as an *unstable spiral*. □

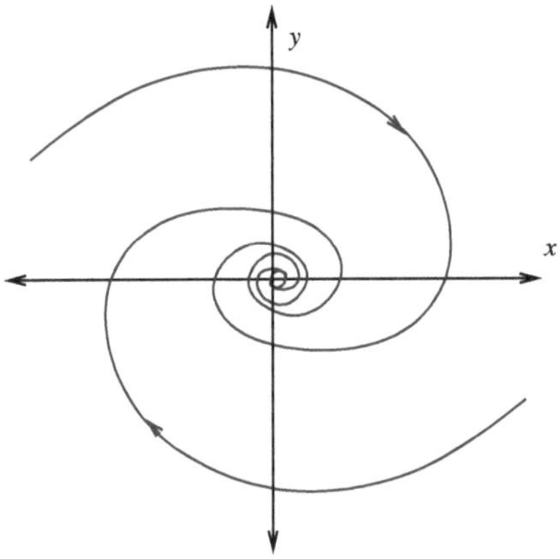

Figure 1.4: Example 1.4: flow around a stable spiral.

Exercise 1.4. Determine the general solution of the system

$$\dot{x} = \begin{bmatrix} -4 & 3 & 2 \\ -6 & 2 & 6 \\ 0 & 0 & -2 \end{bmatrix} x.$$

Sketch the flow in $\mathbb{R}^3$.

1.2 Matrix Exponentials

In a first course on ordinary differential equations, one doubtlessly encounters the fact that the solution to the initial value problem

$$\dot{x} = ax, \quad x(0) = x_0,$$

for $a \in \mathbb{R}$, is $x(t) = x_0 e^{at}$. In §1.3, we will generalize this result to coupled linear systems in $\mathbb{R}^n$, i. e., we will see that if $\dot{x} = Ax$, where $A \in \mathbb{R}^{n\times n}$, then

$$x(t) = e^{At} x_0,$$

given some appropriate understanding of the matrix e^{At}.

Before laying out the precise definition of a *matrix exponential*, we will make a brief interlude to review a few key concepts from functional analysis, so that we may obtain a more sophisticated understanding of this definition. These concepts will also be useful later during our discussion of the graph transform method, which we will use to prove the existence of stable and unstable manifolds of fixed points in nonlinear systems. For more details on functional analysis, see, for example, [107], [171], or [257].

Normed Linear Spaces

We begin with the definition of *normed linear space*. After discussing notions of convergence, we will further introduce the concept of a Banach space, which we will require later in §3.4, preceding our discussion of the contraction mapping theorem.

Definition 1.4. A *real (or complex) vector space* is a set V with the following operations:

1. *Vector addition*: any $x, y \in V$ determines an element $x + y \in V$ that satisfies the following properties:
   (a) $x + y = y + x$ for all $x, y \in V$;
   (b) $x + (y + z) = (x + y) + z$ for all $x, y, z \in V$;
   (c) there is an element of V called 0 such that $0 + x = x$ for all $x \in V$;
   (d) given any element $x \in V$, there is an element called $-x \in V$ such that $x + (-x) = 0$.
2. *Scalar multiplication*: any $x \in V$ and $\alpha \in \mathbb{R}$ (or $\mathbb{C}$) determines an element $\alpha x \in V$ that satisfies the following properties:
   (a) $\alpha(\beta x) = (\alpha\beta)x$ for all $\alpha, \beta \in \mathbb{R}$ (or $\mathbb{C}$) and $x \in V$;
   (b) $1x = x$ for any $x \in V$;
   (c) $(\alpha + \beta)x = \alpha x + \beta x$ for all $\alpha, \beta \in \mathbb{R}$ (or $\mathbb{C}$) and $x \in V$;
   (d) $\alpha(x + y) = \alpha x + \alpha y$ for all $\alpha \in \mathbb{R}$ (or $\mathbb{C}$) and $x, y \in V$.

Example 1.5. As the vectors in $\mathbb{R}^n$ satisfy each of the above axioms, $\mathbb{R}^n$ is clearly a real vector space. □

Exercise 1.5. Show that $C^0([0,1])$, the set of continuous functions on the interval $[0,1]$, is a real vector space.

Exercise 1.6. Show that

$$S_\delta([0,1]) = \{f \in C^0([0,1]) : |f(x) - f(y)| \leq \delta|x-y| \text{ for all } x, y \in [0,1]\},$$

the set of *Lipschitz continuous* functions with *Lipschitz constant* $\delta > 0$, is *not* a real vector space.

Exercise 1.7. Show that $L(\mathbb{R}^n)$, the set of linear transformations A from $\mathbb{R}^n \to \mathbb{R}^n$, is a real vector space.

Definition 1.5. Let V be a real (or complex) vector space. Then a *norm* on V is a mapping $\|\cdot\| : V \to \mathbb{R}$ that satisfies the following conditions:
1. $\|x\| \geq 0$ for all $x \in V$, and $\|x\| = 0$ if and only if $x = 0$;
2. $\|\alpha x\| = |\alpha|\,\|x\|$ for all $x \in V$ and $\alpha \in \mathbb{R}$ (or $\mathbb{C}$);
3. $\|x + y\| \leq \|x\| + \|y\|$ for all $x, y \in V$ (the triangle inequality).

A vector space with a norm is called a *normed linear space.*

Exercise 1.8. Show that

$$\|f\| = \sup_{x \in J} |f(x)| \tag{1.8}$$

defines a norm on $C^0(J)$, where $J \subset \mathbb{R}$ (called the *sup norm*).

Given a normed linear space V, an *ε-ball centered at the point* $x \in V$ is the set $B_\varepsilon(x) = \{y \in V : \|x - y\| < \varepsilon\}$. These ε-balls form a *basis for a topology* on V (see, for example, [216]), i. e., arbitrary *open sets* can be defined as arbitrary unions of these open balls. Moreover, a set is *closed* if its complement in V is open.

An important concept in analysis is that of *convergence* of a sequence of points in a space. Many of the applications in this chapter rely on such a notion. We presume the reader is familiar with convergence of sequences of real numbers, and we introduce the notion of convergence in a normed linear space.

Definition 1.6. Let V be a normed linear space and let $(x_n)_{n=1}^\infty = (x_1, x_2, \ldots)$ be a sequence of points in V. The sequence (x_n) is said to *converge* to the element $x \in V$, denoted

$$\lim_{n \to \infty} x_n = x,$$

if the real numbers $\|x_n - x\| \to 0$ as $n \to \infty$.

Definition 1.7. Let V be a normed linear space and let $(x_n)_{n=1}^\infty$ be a sequence of points in V. The sequence (x_n) is said to be a *Cauchy sequence* if, for every $\varepsilon > 0$, there exists an N such that $\|x_n - x_m\| < \varepsilon$ whenever $n, m > N$.

It is true of the real numbers that every Cauchy sequence converges to some point in $\mathbb{R}$; see, for instance, [252]. However, the same is *not* necessarily true of a generic normed linear space. The situation thus merits the following distinction.

Definition 1.8. A normed linear space V is said to be *complete* if every Cauchy sequence in V converges to some element of V. A *Banach space* is a complete normed linear space.

Exercise 1.9. Show that $C^0([0,1])$ with the sup norm is a Banach space.

Matrix Exponentials

Let $L(\mathbb{R}^n)$ be the space of linear transformations $T : \mathbb{R}^n \to \mathbb{R}^n$. As we saw in Exercise 1.7, $L(\mathbb{R}^n)$ is a real vector space. Before defining the matrix exponential, we must first endow this space with a norm.

Definition 1.9. The *operator norm* of $T \in L(\mathbb{R}^n)$ is defined by

$$\|T\| = \sup\{|T(x)| : x \in \mathbb{R}^n \text{ and } |x| \leq 1\}.$$

Exercise 1.10. Show that the operator norm satisfies the properties that define a norm, i. e., for $S, T \in L(\mathbb{R}^n)$, show that
(a) $\|T\| \geq 0$ and $\|T\| = 0$ if and only if $T = 0$;
(b) $\|kT\| = |k| \cdot \|T\|$ for $k \in \mathbb{R}$;
(c) $\|S + T\| \leq \|S\| + \|T\|$.

Definition 1.10. A sequence of linear operators $T_k \in L(\mathbb{R}^n)$ is said to *converge* to the linear operator $T \in L(\mathbb{R}^n)$ if, for all $\varepsilon > 0$, there exists an $N \in \mathbb{N}$ such that $\|T - T_k\| < \varepsilon$ for all $k > N$.

Definition 1.11 (The Matrix Exponential). Let A be a real-valued, $n \times n$ matrix. For $t \in \mathbb{R}$, we define the *matrix exponential* of At as

$$e^{At} = \sum_{k=0}^{\infty} \frac{A^k t^k}{k!}. \tag{1.9}$$

Our goal is to show that for any $t_0 > 0$, the series defined in (1.9) converges uniformly on the interval $[-t_0, t_0]$. But first we will require the following lemma.

Lemma 1.1. *For $S, T \in L(\mathbb{R}^n)$ and $x \in \mathbb{R}^n$:*
(i) $|T(x)| \leq \|T\| \cdot |x|$;
(ii) $\|TS\| \leq \|T\| \cdot \|S\|$;
(iii) $\|T^k\| \leq \|T\|^k$, *for* $k \in \mathbb{Z}^+$.

Proof. 1. It is obvious that (i) holds for $x = 0$. For $x \neq 0$, define $y = x/|x|$. Then, by the definition of the operator norm, we have

$$\|T\| \geq |T(y)| = \frac{|T(x)|}{|x|},$$

from which (i) follows.

2. From (i) it follows that

$$\begin{aligned}|T(S(x))| &\leq \|T\| \cdot |S(x)| \\ &\leq \|T\| \cdot \|S\| \cdot |x|.\end{aligned}$$

For $|x| \leq 1$, it further follows that $|T(S(x))| \leq \|T\| \cdot \|S\|$. We therefore obtain the inequality

$$\|TS\| = \sup\{|TS(x)| : x \in \mathbb{R}^n \text{ and } |x| \leq 1\} \leq \|T\| \cdot \|S\|.$$

3. The third inequality follows immediately from (ii) using mathematical induction. □

Theorem 1.4. *Given $T \in L(\mathbb{R}^n)$ and $t_0 > 0$, the series*

$$\sum_{k=0}^{\infty} \frac{T^k t^k}{k!}$$

converges uniformly on the interval $-t_0 \leq t \leq t_0$.

Proof. From the lemma we have, for $|t| \leq t_0$,

$$\left\| \frac{T^k t^k}{k!} \right\| \leq \frac{\|T\|^k |t|^k}{k!} \leq \frac{\|T\|^k t_0^k}{k!}.$$

But

$$\sum_{k=0}^{\infty} \frac{\|T\|^k t_0^k}{k!} = e^{\|T\| t_0}.$$

Therefore, the series in question converges by the Weierstrass M-Test. □

Our next theorem, regarding the relation of the matrix exponential of two *similar* matrices, is the key to solving linear systems of differential equations. Recall that two matrices $S, T \in L(\mathbb{R}^n)$ are *similar* if there exists an invertible matrix $P \in L(\mathbb{R}^n)$ such that $S = PTP^{-1}$.

Theorem 1.5. *Let $S, T \in L(\mathbb{R}^n)$ be similar matrices, and let P be an invertible matrix with the property $S = PTP^{-1}$. Then*

$$e^S = Pe^T P^{-1}.$$

Proof. This follows immediately from the definition of the matrix exponential and from the fact that

$$S^k = PT^kP^{-1}$$

for every $k \in \mathbb{N}$. Specifically, we have

$$e^S = \lim_{n\to\infty} \sum_{k=0}^{n} \frac{(PTP^{-1})^k}{k!} = P\left[\lim_{n\to\infty} \sum_{k=0}^{n} \frac{T^k}{k!}\right]P^{-1} = Pe^TP^{-1}. \qquad \square$$

Exercise 1.11. Show that if $S, T \in L(\mathbb{R}^n)$ commute (i. e., $ST = TS$), then $e^{S+T} = e^Se^T$. *Hint:* Use the binomial theorem. Does the same result hold if S and T do not commute?

Exercise 1.12. Show that if $D = \operatorname{diag}\{\lambda_1, \ldots, \lambda_n\}$, where $\lambda_1, \ldots, \lambda_n \in \mathbb{R}$, then $e^{Dt} = \operatorname{diag}\{e^{\lambda_1 t}, \ldots, e^{\lambda_n t}\}$.

Exercise 1.13. Show that

$$e^A = e^a \begin{bmatrix} \cos b & -\sin b \\ \sin b & \cos b \end{bmatrix}, \quad \text{where } A = \begin{bmatrix} a & -b \\ b & a \end{bmatrix}.$$

Exercise 1.14. Show that

$$e^A = e^a \begin{bmatrix} 1 & b \\ 0 & 1 \end{bmatrix}, \quad \text{where } A = \begin{bmatrix} a & b \\ 0 & a \end{bmatrix}.$$

1.3 Matrix Representation of Solutions

We now have the ability to discuss the solutions of linear systems in a more elegant and succinct way than before. Our basic aim is to show how solutions may be represented using matrix exponentials, diagonalization, and block diagonalization and to explore the particular form solutions take for a variety of different types of eigenvalues. For an early reference on such operator representations of solutions, see [193].

Real, Distinct Eigenvalues

Let us first revisit linear systems of the form (1.1), where coefficient matrix A has real, distinct eigenvalues $\lambda_1, \ldots, \lambda_n \in \mathbb{R}$ and associated eigenvectors $v_1, \ldots, v_n$. The general solution is given by

$$x(t) = c_1e^{\lambda_1 t}v_1 + \cdots + c_ne^{\lambda_n t}v_n,$$

where $(c_1, \ldots, c_n)$ are the components of the initial condition x_0 relative to the eigenbasis $v_1, \ldots, v_n$. Before we proceed, let us recall a theorem from linear algebra.

Theorem 1.6 (Diagonalization). *If the eigenvalues $\lambda_1, \dots, \lambda_n \in \mathbb{C}$ of an $n \times n$ matrix $A \in \mathbb{R}^{n\times n}$ are distinct, then the matrix $V \in \mathbb{C}^{n\times n}$ defined by*

$$V = \begin{bmatrix} | & & | \\ v_1 & \cdots & v_n \\ | & & | \end{bmatrix}, \tag{1.10}$$

where $v_1, \dots, v_n \in \mathbb{C}^n$ are the associated eigenvectors, is invertible and

$$V^{-1}AV = \operatorname{diag}\{\lambda_1, \dots, \lambda_n\} = D. \tag{1.11}$$

Proof. Equation (1.11) is equivalent to the matrix equation $AV = VD$. Now, the ith column of this equation is simply $Av_i = \lambda_i v_i$, which is true by definition of the eigenvalues and eigenvectors. The invertibility of V follows from the fact that the eigenvalues are distinct, which implies that their associated eigenvectors must be linearly independent. □

This theorem is equally applicable to any combination of real or complex eigenvalues. For the remainder of this section, we will restrict our attention to the case of *n real*, distinct eigenvalues. Let us define matrix V as in (1.10) and D as the diagonal matrix whose entries consist of the eigenvalues of A, as in (1.11). We then introduce the change of variables $y = V^{-1}x$. The differential equations for vector y will then be given by

$$\dot{y} = V^{-1}\dot{x} = V^{-1}Ax = V^{-1}AVy = Dy.$$

Hence, relative to eigenbasis V, system (1.1) decouples as

$$\begin{aligned} \dot{y}_1 &= \lambda_1 y_1, \\ &\vdots \\ \dot{y}_n &= \lambda_n y_n. \end{aligned}$$

The general solution, in terms of the new variable y, is therefore given by

$$\begin{aligned} y_1(t) &= c_1 e^{\lambda_1 t}, \\ &\vdots \\ y_n(t) &= c_n e^{\lambda_n t}, \end{aligned}$$

or, in matrix form, as

$$y(t) = \begin{bmatrix} e^{\lambda_1 t} & & \\ & \ddots & \\ & & e^{\lambda_n t} \end{bmatrix} y(0) = e^{Dt} y_0,$$

where $e^{Dt} = \text{diag}\{e^{\lambda_1 t}, \ldots, e^{\lambda_n t}\}$; see Exercise 1.12. Since $y_0 = V^{-1}x_0$ and $x(t) = Vy(t)$, we can write the general solution to equations (1.1), as originally posed relative to the x-variables, as

$$x(t) = Ve^{Dt}V^{-1}x_0.$$

This is exactly equivalent to (1.6), derived in Chapter 1, where coefficient vector c is related to the initial conditions by $c = V^{-1}x_0$. Moreover, combining the results of Theorems 1.5 and 1.6, we see the solution may be expressed as

$$x(t) = e^{At}x_0.$$

Does a similar formula hold for general linear systems?

General Solution

Our next task will be to show that the general solution of $\dot{x} = Ax$ can always be represented as $x(t) = e^{At}x_0$, even if we drop the restriction that the eigenvalues must be real.

Theorem 1.7. *Let $A \in \mathbb{R}^{n\times n}$. Then for a given $x_0 \in \mathbb{R}^n$, the initial value problem*

$$\dot{x} = Ax, \quad x(0) = x_0 \tag{1.12}$$

has a unique solution given by

$$x(t) = e^{At}x_0. \tag{1.13}$$

Proof. First we must show that the solution given by (1.13) actually satisfies the differential equation. To see this, let us first compute

$$\begin{aligned}
\frac{d}{dt}e^{At} &= \lim_{h\to 0}\frac{e^{A(t+h)} - e^{at}}{h} \\
&= \lim_{h\to 0} e^{At}\frac{e^{Ah} - I}{h} \\
&= e^{At}\lim_{h\to 0}\left[\lim_{n\to\infty}\left(A + \frac{A^2h}{2!} + \cdots + \frac{A^n h^{n-1}}{n!}\right)\right] \\
&= e^{At}\lim_{n\to\infty}\left[\lim_{h\to 0}\left(A + \frac{A^2h}{2!} + \cdots + \frac{A^n h^{n-1}}{n!}\right)\right] \\
&= e^{At}A \\
&= Ae^{At}.
\end{aligned}$$

Note that the two limits may be interchanged only because the series defining e^{Ah} converges *uniformly* on the interval, say, $-1 \le h \le 1$. In addition, we may rewrite

$e^{At}A = Ae^{At}$ only because A commutes with itself, and hence with e^{At}, since it commutes with each term in the defining power series of e^{At}.

Clearly the proposed solution satisfies the initial condition $x(0) = x_0$.

Finally, to show uniqueness, let $x(t)$ be *any* solution of the initial value problem (1.12). Now set $y(t) = e^{-At}x(t)$. Differentiating, we obtain

$$\dot{y} = -Ae^{-At}x(t) + e^{-At}\dot{x}(t) = -Ae^{-At}x(t) + e^{-At}A\dot{x}(t) = 0.$$

Since A and e^{At} commute, $y(t)$ must be a constant function of time, $y(t) = x_0$. Therefore, it must follow that $x(t) = e^{At}x_0$. □

Example 1.6. Find the solution to the linear system $\dot{x} = Ax$, where coefficient matrix A is given by

$$A = \begin{bmatrix} -2 & -7 \\ 7 & 2 \end{bmatrix}.$$

Recall from Exercise 1.13 that

$$e^{At} = e^{-2t}\begin{bmatrix} \cos(7t) & -\sin(7t) \\ \sin(7t) & \cos(7t) \end{bmatrix}.$$

Hence the general solution is

$$x(t) = e^{-2t}\begin{bmatrix} \cos(7t) & -\sin(7t) \\ \sin(7t) & \cos(7t) \end{bmatrix} x_0 = c_1 e^{-2t}\begin{bmatrix} \cos(7t) \\ \sin(7t) \end{bmatrix} + c_2 e^{-2t}\begin{bmatrix} -\sin(7t) \\ \cos(7t) \end{bmatrix},$$

where $x_0 = (c_1, c_2)$. Hence the origin is a stable spiral point. □

Complex Eigenvalues

In the previous paragraph, we saw how solutions to linear systems may be represented by the matrix exponential of the coefficient matrix. Now we take up the topic of devising a useful form for this exponential for the case of mixed real and complex eigenvalues. We begin with a theorem from linear algebra.

Theorem 1.8. *Let $A \in \mathbb{R}^{n\times n}$ have k real, distinct eigenvalues $\lambda_1, \dots, \lambda_k$ with associated eigenvectors $u_1, \dots, u_k$ and $s = (n-k)/2$ distinct, complex conjugate eigenvalue pairs $\lambda_{k+1}, \overline{\lambda_{k+1}}, \dots, \lambda_{k+s}, \overline{\lambda_{k+s}}$, where $\lambda_j = a_j + ib_j$ with associated eigenvectors $w_{k+1}, \overline{w_{k+1}}, \dots, w_{k+s}, \overline{w_{k+s}}$, where $w_j = u_j + iv_j$. Further, let*

$$P = \begin{bmatrix} | & & | & | & | & & | & | \\ u_1 & \dots & u_k & v_{k+1} & u_{k+1} & \dots & v_{k+s} & u_{k+s} \\ | & & | & | & | & & | & | \end{bmatrix}.$$

Then $P \in \mathbb{R}^{n \times n}$ is invertible, and

$$D = P^{-1}AP = \operatorname{diag}\{\lambda_1, \dots, \lambda_k, B_{k+1}, \dots, B_{k+s}\},$$

where the 2×2 real blocks B_j, for $j > k$, are given by

$$B_j = \begin{bmatrix} a_j & -b_j \\ b_j & a_j \end{bmatrix} \quad \text{for } j = k+1, \dots, k+s. \tag{1.14}$$

Proof. Let $\lambda = a + bi$ and $w = u + iv$ be one particular complex-valued eigenvalue–eigenvector pair of matrix A. Then $\overline{\lambda} = a - bi$ and $\overline{w} = u - iv$ is also an eigenvalue–eigenvector pair of A. Since $u = \frac{1}{2}(w + \overline{w})$, we have

$$Au = \frac{1}{2}(Aw + A\overline{w}) = \frac{1}{2}(\lambda w + \overline{\lambda w}) = \mathbb{R}\{\lambda w\} = au - bv.$$

Similarly, $v = \frac{1}{2i}(w - \overline{w})$, and hence

$$Av = \frac{1}{2i}(Aw - A\overline{w}) = \frac{1}{2i}(\lambda w - \overline{\lambda w}) = \mathbb{J}\{\lambda w\} = bu + av.$$

Combining the two preceding equations into a single matrix equation we obtain

$$A \begin{bmatrix} | & | \\ v & u \\ | & | \end{bmatrix} = \begin{bmatrix} | & | \\ v & u \\ | & | \end{bmatrix} \begin{bmatrix} a & -b \\ b & a \end{bmatrix}.$$

The result follows. □

Combining the results of Theorems 1.5, 1.7, and 1.8, we immediately obtain the following theorem.

Theorem 1.9. *The general solution of the initial value problem* (1.1) *is given by*

$$x(t) = Pe^{Dt}P^{-1}x_0, \tag{1.15}$$

where

$$e^{Dt} = \operatorname{diag}\{e^{\lambda_1 t}, \dots, e^{\lambda_k t}, e^{B_{k+1} t}, \dots, e^{B_{k+s} t}\} \tag{1.16}$$

and

$$e^{B_j t} = e^{a_j t} \begin{bmatrix} \cos(b_j t) & -\sin(b_j t) \\ \sin(b_j t) & \cos(b_j t) \end{bmatrix} \quad \text{for } j = k+1, \dots, k+s. \tag{1.17}$$

Example 1.7. Find the general solution of the first-order system $\dot{x} = Ax$, where coefficient matrix A is given by

$$A = \begin{bmatrix} 3.5 & 0 & 4.5 & 0 & -7 & -1.5 \\ 0 & -2 & 0 & 1 & 0 & 0 \\ 0 & 0 & -14 & 0 & 15 & 0 \\ 0 & -2 & 12 & 0 & -12 & 0 \\ 0 & 0 & -12 & 0 & 13 & 0 \\ 7.5 & 0 & 22.5 & 0 & -30 & 0.5 \end{bmatrix}.$$

The eigenvalues of A are $\lambda_1 = -2$, $\lambda_2 = 1$, $\lambda_3 = -1+i$, $\overline{\lambda_3} = -1-i$, $\lambda_4 = 2+3i$, and $\overline{\lambda_4} = 2-3i$. The associated eigenvectors are

$$w_1 = \begin{bmatrix} 1 \\ 6 \\ 5 \\ 0 \\ 4 \\ 0 \end{bmatrix}, \quad w_2 = \begin{bmatrix} 1 \\ 0 \\ 1 \\ 0 \\ 1 \\ 0 \end{bmatrix}, \quad w_3 = \begin{bmatrix} 0 \\ 1 \\ 0 \\ 1+i \\ 0 \\ 0 \end{bmatrix}, \quad w_4 = \begin{bmatrix} i \\ 0 \\ 0 \\ 0 \\ 0 \\ 2+i \end{bmatrix},$$

$\overline{w_3}$, and $\overline{w_4}$. Therefore, if we define matrix P as

$$P = \begin{bmatrix} 1 & 1 & 0 & 0 & 1 & 0 \\ 6 & 0 & 0 & 1 & 0 & 0 \\ 5 & 1 & 0 & 0 & 0 & 0 \\ 0 & 0 & 1 & 1 & 0 & 0 \\ 4 & 1 & 0 & 0 & 0 & 0 \\ 0 & 0 & 0 & 0 & 1 & 2 \end{bmatrix},$$

then the general solution will be given by (1.15), with

$$e^{Dt} = \begin{bmatrix} e^{-2t} & 0 & 0 & 0 & 0 & 0 \\ 0 & e^{t} & 0 & 0 & 0 & 0 \\ 0 & 0 & e^{-t}\cos(t) & -e^{-t}\sin(t) & 0 & 0 \\ 0 & 0 & e^{-t}\sin(t) & e^{-t}\cos(t) & 0 & 0 \\ 0 & 0 & 0 & 0 & e^{2t}\cos(3t) & -e^{2t}\sin(3t) \\ 0 & 0 & 0 & 0 & e^{2t}\sin(3t) & e^{2t}\cos(3t) \end{bmatrix}.$$

If $x_0 \in \text{span}\{w_1, u_3, v_3\}$, then the solution will exponentially decay toward 0 as $t \to \infty$. Similarly, if $x_0 \in \text{span}\{w_2, u_4, v_4\}$, then the solution will grow exponentially away from 0 as $t \to \infty$. (The block diagonal matrix e^{Dt} in the preceding equation is literally the solution relative to the basis defined by the columns of P, i. e., w_1, w_2, v_3, u_3, v_4, and u_4.) □

Exercise 1.15. Find the general solution of the linear system $\dot{x} = Ax$, where coefficient matrix A is given by

$$A = \begin{bmatrix} -4 & 0 & 3 \\ 0 & -4 & 0 \\ -6 & 0 & 2 \end{bmatrix}.$$

Express the solution in the form specified in Theorem 1.9.

Exercise 1.16. Find the general solution of the linear system $\dot{x} = Ax$, where coefficient matrix A is given by

$$A = \begin{bmatrix} 2 & 0 & 0 & -4 \\ 1.5 & 1 & -0.5 & 1.5 \\ 5 & 2 & 1 & -9 \\ 2 & 0 & 0 & -2 \end{bmatrix}.$$

Express the solution in the form specified in Theorem 1.9.

Repeated Eigenvalues

Up to this point we have restricted our attention to the case where each of the eigenvalues of the coefficient matrix are distinct. In this section, we take up the case of coefficient matrices with repeated eigenvalues. Recall that the eigenvalues of a matrix $A \in \mathbb{R}^{n\times n}$ are the roots of the characteristic equation $\det(A - \lambda I) = 0$, which is an nth-degree polynomial in λ. The *algebraic multiplicity* of an eigenvalue is the number of times the eigenvalue is repeated as a root of the characteristic equation.

Definition 1.12. Let λ be an eigenvalue of the real $n \times n$ matrix A with algebraic multiplicity $m \leq n$. Then, for $k = 1, \ldots, m$, any nonzero solution v of the equation

$$(A - \lambda I)^k v = 0$$

is called a *generalized eigenvector of A*.

Definition 1.13. An $n \times n$ real matrix N is said to be *nilpotent of order k* if $N^{k-1} \neq 0$ and $N^k = 0$.

We now state a theorem from linear algebra without proof.

Theorem 1.10. *Let $A \in \mathbb{R}^n$ have real eigenvalues $\lambda_1, \ldots, \lambda_n$, repeated according to their multiplicities. Then there exists a basis for $\mathbb{R}^n$ consisting of generalized eigenvectors. Moreover, if $\{v_1, \ldots, v_n\}$ is any basis of generalized eigenvectors, where v_i is a generalized eigenvector of λ_i for $i = 1, \ldots, n$, then the matrix*

$$P = \begin{bmatrix} | & & | \\ v_1 & \cdots & v_n \\ | & & | \end{bmatrix}$$

is invertible, A can be written as $A = S + N$, *where*

$$P^{-1}SP = \operatorname{diag}\{\lambda_1, \dots, \lambda_n\} = D,$$

the matrix $N = A - S$ *is nilpotent of order* $p \le n$, *and* S *and* N *commute.*

Corollary 1.1. *Under the hypotheses of Theorem* 1.10, *the general solution of the system* $\dot{x} = Ax$ *is given by*

$$x(t) = Pe^{Dt}P^{-1}\left[I + Nt + \cdots + \frac{N^{p-1}t^{p-1}}{(p-1)!}\right]x_0,$$

where

$$e^{Dt} = \operatorname{diag}\{e^{\lambda_1 t}, \dots, e^{\lambda_n t}\}.$$

Proof. From Theorem 1.7, the general solution is given by $x(t) = e^{At}x_0$. Since S and N commute, we have

$$e^{At} = e^{St+Nt} = e^{St}e^{Nt}.$$

Now $S = P\operatorname{diag}\{\lambda_1, \dots, \lambda_n\}P^{-1}$, hence $e^{St} = P\operatorname{diag}\{e^{\lambda_1 t}, \dots, e^{\lambda_n t}\}P^{-1}$. Furthermore, since N is nilpotent of order $p \le n$, we obtain

$$e^{Nt} = I + Nt + \cdots + \frac{N^{p-1}t^{p-1}}{(p-1)!}.$$

The result follows. □

Example 1.8. Consider the initial value problem

$$\dot{x} = \begin{bmatrix} -1 & 0 & 1 \\ -3 & 0 & 1 \\ -4 & 0 & 3 \end{bmatrix} x, \quad x(0) = x_0.$$

The coefficient matrix A has eigenvalues $\lambda_1 = 1$, $\lambda_2 = 0$, and $\lambda_3 = 1$. The eigenvalue $\lambda = 1$ is a repeated eigenvalue with algebraic multiplicity 2 and geometric multiplicity 1. The eigenvalues λ_1 and λ_2 have eigenvectors

$$v_1 = \begin{bmatrix} 1 \\ -1 \\ 2 \end{bmatrix} \quad \text{and} \quad v_2 = \begin{bmatrix} 0 \\ 1 \\ 0 \end{bmatrix},$$

respectively. Now, a straightforward computation yields

$$(A - \lambda_1 I)^2 = \begin{bmatrix} 0 & 0 & 0 \\ 5 & 1 & -2 \\ 0 & 0 & 0 \end{bmatrix}.$$

The kernel of $(A - \lambda_1 I)^2$ is spanned by the vectors $(1, -5, 0)^T$ and $(0, 2, 1)^T$; we name the latter v_3 and straightaway construct matrices P and D as follows:

$$P = \begin{bmatrix} 1 & 0 & 0 \\ -1 & 1 & 2 \\ 2 & 0 & 1 \end{bmatrix}, \quad D = \begin{bmatrix} 1 & 0 & 0 \\ 0 & 0 & 0 \\ 0 & 0 & 1 \end{bmatrix}.$$

It follows that

$$S = PDP^{-1} = \begin{bmatrix} 1 & 0 & 0 \\ -5 & 0 & 2 \\ 0 & 0 & 1 \end{bmatrix} \quad \text{and} \quad N = A - S = \begin{bmatrix} -2 & 0 & 1 \\ 2 & 0 & -1 \\ -4 & 0 & 2 \end{bmatrix}.$$

Here, N is nilpotent of order 2. The solution is therefore given by

$$x(t) = Pe^{Dt}P^{-1}(I + Nt)\,x_0 = \begin{bmatrix} (1-2t)e^t & 0 & te^t \\ 2te^t - 5e^5 + 5 & 1 & 2e^t - te^t - 2 \\ -4te^t & 0 & (1+2t)e^t \end{bmatrix} x_0,$$

as can be readily verified. □

Exercise 1.17. Solve the initial value problem

$$\dot{x} = \begin{bmatrix} 1 & 1 & 0 \\ 0 & 1 & 1 \\ 0 & 0 & 1 \end{bmatrix}, \quad x(0) = x_0.$$

Theorem 1.11. *Let $A \in \mathbb{R}^{n \times n}$ have $k \le n$ real eigenvalues, $\lambda_1, \ldots, \lambda_k$, repeated as necessary according to their multiplicities, and $n - k$ complex eigenvalues $\lambda_{k+1} = a_{k+1} + ib_{k+1}$, $\overline{\lambda_{k+1}} = a_{k+1} - ib_{k+1}, \ldots, \lambda_{k+s} = a_{k+s} + ib_{k+s}$, and $\overline{\lambda_{k+s}} = a_{k+s} - ib_{k+s}$, where $s = (n-k)/2$ is the number of complex conjugate pairs, taken with their multiplicities. Then there exists a set of generalized, complex eigenvectors $w_j = u_j + iv_j$, where $v_j = 0$ for $j \le k$, such that the vectors*

$$\{u_1, \ldots, u_k, v_{k+1}, u_{k+1}, \ldots, v_{k+s}, u_{k+s}\}$$

form a basis for $\mathbb{R}^n$; hence the matrix

$$P = \begin{bmatrix} | & & | & | & | & & | & | \\ u_1 & \cdots & u_k & v_{k+1} & u_{k+1} & \cdots & v_{k+s} & u_{k+s} \\ | & & | & | & | & & | & | \end{bmatrix}$$

is invertible. Moreover, A can be written as $A = S + N$, where

$$D = P^{-1}SP = \operatorname{diag}\{\lambda_1, \ldots, \lambda_k, B_{k+1}, \ldots, B_{k+s}\},$$

$N = A - S$ is nilpotent of order $p \le n$, S and N commute, and the real blocks B_i, for $k + 1 \le i \le n$, are defined in (1.14).

Corollary 1.2. *Suppose A has repeated eigenvalues, as above. The solution of the initial value problem* $\dot{x} = Ax$, $x(0) = x_0$, *is given by*

$$x(t) = Pe^{Dt}P^{-1}\left[I + Nt + \cdots + \frac{N^{k-1}t^{k-1}}{(k-1)!}\right]x_0,$$

where e^{Dt} *is given by* (1.16) *and* (1.17).

1.4 Stability Theory

In our study of linear systems thus far, we have already seen several examples that exhibited an interesting property: if the initial condition lies within the span of a certain set of eigenvectors, then the solution trajectory remains within this span for all time. This idea turns out to be true for linear systems in general. In this paragraph, we examine the details of this feature and formalize these concepts in a more mathematically rigorous way.

Invariant Subspaces

We begin with a definition.

Definition 1.14. The vector subspace $W \subset \mathbb{R}^n$ is called an *invariant subspace* of the linear system $\dot{x} = Ax$ if, whenever $x_0 \in W$, $\varphi(t; x_0) \in W$ for all t.

For linear systems, the flow may always be represented by $\varphi(t; x_0) = e^{At}x_0$. Hence, one might say that a subspace W is invariant if and only if $e^{At}W \subset W$ for all t. Here, $e^{At}W = \{y \in \mathbb{R}^n : y = e^{At}x_0 \text{ for some } x_0 \in W\}$, i. e., $e^{At}W$ is the image of subspace W under the flow of $\dot{x} = Ax$. Since 0 is a fixed point for the system, it follows that $0 \in e^{At}W$.

As it turns out, all linear systems exhibit invariant subspaces. Suppose that coefficient matrix A has $k \leq n$ real eigenvalues $\lambda_1, \ldots, \lambda_k$ with associated (generalized) eigenvectors $w_1, \ldots, w_k$ and $n-k$ complex eigenvalues $\lambda_{k+1}, \overline{\lambda_{k+1}}, \ldots, \lambda_{k+s}$, and $\overline{\lambda_{k+s}}$ with associated (generalized) eigenvectors $w_{k+1}, \overline{w_{k+1}}, \ldots, w_{k+s}, \overline{w_{k+s}}$, where $s = (n-k)/2$. Let $\lambda_j = a_j + ib_j$ and $w_j = u_j + iv_j$. Note, if $j \leq k$, i. e., if λ_j is real, then $b_j = 0$ and $v_j = 0$. Then one can form a basis

$$\mathcal{B} = \{u_1, \ldots, u_k, v_{k+1}, u_{k+1}, \ldots, v_{k+s}, u_{k+s}\}$$

for $\mathbb{R}^n$. Given these basic definitions, we now establish the following definition.

Definition 1.15. The subspaces defined by

$$E^s = \operatorname{span}\{u_j, v_j : a_j < 0\},$$
$$E^u = \operatorname{span}\{u_j, v_j : a_j > 0\},$$

$$E^c = \text{span}\{u_j, v_j : a_j = 0\}$$

are referred to as the *stable subspace*, *unstable subspace*, and *center subspace*, respectively.

In other words, the stable subspace E^s is the span of the real and imaginary parts of the generalized eigenvectors associated with an eigenvalue with negative real part. Similarly, the unstable subspace E^u is the span of the real and imaginary parts of the generalized eigenvectors associated with an eigenvalue with positive real part. And finally, the center subspace E^c is the span of the real and imaginary parts of the generalized eigenvectors associated with a purely imaginary eigenvalue.

Linear systems without purely imaginary eigenvalues merit the special nomenclature *hyperbolic*. As we will see in §2.5, the behavior of nonlinear systems near a fixed point is locally topologically equivalent to an associated linearized system whenever its linearized counterpart is *hyperbolic*; this statement is the content of the Hartman–Grobman theorem. Because of this distinction, we now present the following definition.

Definition 1.16. If all of the eigenvalues of A have a nonzero real part, then the flow $e^{At} : \mathbb{R}^n \to \mathbb{R}^n$ is called a *hyperbolic flow*. Similarly, in this case, the fixed point $x = 0$ is called a *hyperbolic fixed point*.

Stability Theorem

Our next goal is to show that E^s, E^u, and E^c are invariant subspaces of the linear system $\dot{x} = Ax$.

Lemma 1.2. *Let E be the generalized eigenspace of A corresponding to the eigenvalue λ. Then $AE \subset E$.*

Proof. Let $(w_1, \ldots, w_k)$ be a basis of generalized eigenvectors for E. Given $w \in E$, we may then write

$$w = \sum_{j=1}^{k} c_j w_j$$

for some collection of constants $c_1, \ldots, c_k \in \mathbb{C}$. Hence

$$Aw = \sum_{j=1}^{k} c_j A w_j.$$

By definition, each w_j satisfies the equation

$$(A - \lambda I)^{k_j} w_j = 0$$

for some minimal k_j. Now write

$$(A - \lambda I)w_j = W_j,$$

where

$$W_j \in \ker(A - \lambda I)^{k_j - 1} \subset E.$$

This last line follows since, by definition,

$$(A - \lambda I)^{k_j - 1} W_j = (A - \lambda I)^{k_j} w_j = 0.$$

Hence

$$Aw_j = \lambda w_j + W_j \in E.$$

Therefore,

$$\sum_{j=1}^{k} c_j A_j w_j \in E,$$

i. e., $Aw \in E$ for all $w \in E$. It follows that $AE \subset E$. □

Theorem 1.12. *Let $A \in \mathbb{R}^{n\times n}$. Then*

$$\mathbb{R}^n = E^s \oplus E^u \oplus E^c, \tag{1.18}$$

where E^s, E^u, and E^c are the stable, unstable, and center subspaces of A, respectively. Furthermore, E^s, E^u, and E^c are invariant with respect to the flow e^{At}.

Proof. Since $\mathcal{B} = \{u_1, \ldots, u_k, v_{k+1}, u_{k+1}, \ldots, v_{k+s}, u_{k+s}\}$ form a basis for $\mathbb{R}^n$, (1.18) follows directly from the definitions of E^s, E^u, and E^c.

Let $x_0 \in E^s$. Then we may write

$$x_0 = \sum_{j=1}^{n_s} c_j z_j,$$

where n_s is the dimension of E^s, $c_j \in \mathbb{R}$, and z_j is either a real or imaginary part of one of the eigenvectors whose eigenvalue has a negative real part, so that $\{z_j\}_{j=1}^{n_s}$ is a basis for E^s. Now

$$e^{At} x_0 = \sum_{j=1}^{n_s} c_j e^{At} z_j.$$

But

$$e^{At} z_j = \lim_{r\to\infty} \left[I + At + \cdots + \frac{A^r t^r}{r!} \right] z_j \in E^s$$

since, by the preceding lemma, $A^r z_j \in E^s$, and E^s is complete. Thus, for all $t \in \mathbb{R}$, $e^{At} x_0 \in E^s$. Therefore, $e^{At} E^s \subset E^s$, i. e., E^s is an invariant subspace under the flow e^{At}.

Similar arguments hold for E^u and E^c. □

Example 1.9. Consider again the linear system discussed in Example 1.7. The eigenvalues and eigenvectors of this system were found to be $\lambda_1 = -2$, $\lambda_2 = 1$, $\lambda_3 = -1 + i$, $\overline{\lambda_3} = -1 - i$, $\lambda_4 = 2 + 3i$, and $\overline{\lambda_4} = 2 - 3i$; and

$$w_1 = \begin{bmatrix} 1 \\ 6 \\ 5 \\ 0 \\ 4 \\ 0 \end{bmatrix}, \quad w_2 = \begin{bmatrix} 1 \\ 0 \\ 1 \\ 0 \\ 1 \\ 0 \end{bmatrix}, \quad w_3 = \begin{bmatrix} 0 \\ 1 \\ 0 \\ 1+i \\ 0 \\ 0 \end{bmatrix}, \quad w_4 = \begin{bmatrix} i \\ 0 \\ 0 \\ 0 \\ 0 \\ 2+i \end{bmatrix},$$

$\overline{w_3}$, and $\overline{w_4}$, respectively. The eigenvalues λ_1, λ_3, and $\overline{\lambda_3}$ each have a negative real part; therefore

$$E^s = \operatorname{span}\left\{ \begin{bmatrix} 1 \\ 6 \\ 5 \\ 0 \\ 4 \\ 0 \end{bmatrix}, \begin{bmatrix} 0 \\ 1 \\ 0 \\ 1 \\ 0 \\ 0 \end{bmatrix}, \begin{bmatrix} 0 \\ 0 \\ 0 \\ 1 \\ 0 \\ 0 \end{bmatrix} \right\},$$

i. e., $E^s = \operatorname{span}\{w_1, \Re\{w_3\}, \Im\{w_3\}\}$. Similarly, the eigenvalues λ_2, λ_4, and $\overline{\lambda_4}$ have a positive real part; therefore

$$E^u = \operatorname{span}\left\{ \begin{bmatrix} 1 \\ 0 \\ 1 \\ 0 \\ 1 \\ 0 \end{bmatrix}, \begin{bmatrix} 0 \\ 0 \\ 0 \\ 0 \\ 0 \\ 2 \end{bmatrix}, \begin{bmatrix} 1 \\ 0 \\ 0 \\ 0 \\ 0 \\ 1 \end{bmatrix} \right\},$$

i. e., $E^u = \operatorname{span}\{w_2, \Re\{w_4\}, \Im\{w_4\}\}$. One readily checks that indeed $\mathbb{R}^6 = E^s \oplus E^u$. Moreover, since the system has no purely imaginary eigenvalues, it follows that $E^c = \{0\}$ and the flow is *hyperbolic*. Initial conditions that lie on E^s will spawn trajectories that asymptotically approach the origin as $t \to \infty$. Moreover, components of such an initial condition that lie along the direction w_1 will decay exponentially, whereas components that lie along the subspace $\operatorname{span}\{\Re\{w_3\}, \Im\{w_3\}\}$ will spiral into the origin. Similarly, a trajectory with initial condition in E^u will remain in E^u for all time, but such trajectories will be unbounded and diverge away from the origin as $t \to \infty$. For a general trajectory, with an initial condition *not* contained in either E^s or E^u, the motion will be similar to that of a saddle; the stable components of the initial condition will decay, whereas the unstable components will grow. Hence the ensuing trajectories will squeeze against the unstable subspace as $t \to \infty$. □

Exercise 1.18. Determine the eigenvalues and eigenvectors for the linear system

$$\dot{x} = \begin{bmatrix} -2 & 0 & 0 & 1.5 & 0 \\ 0 & -1 & 0 & 0 & 0 \\ 0 & -1.5 & 0.5 & 0 & 0 \\ 0 & 0 & 0 & 1 & 0 \\ 0 & 0.5 & -0.5 & 0 & 1.5 \end{bmatrix}.$$

Determine a basis for the stable, unstable, and center subspaces, and then show that $\mathbb{R}^5 = E^s \oplus E^u \oplus E^c$. Write out the general solution of the foregoing system and show that E^s, E^u, and E^c are invariant under the flow. Identify the origin as either a hyperbolic or nonhyperbolic fixed point.

Exercise 1.19. Determine the eigenvalues and eigenvectors for the linear system

$$\dot{x} = \begin{bmatrix} -1 & 0 & 1 & 0 & -2 \\ 0 & 1 & 0 & 1 & 0 \\ -1 & 0 & -1 & 0 & 0 \\ 0 & -4 & 0 & 1 & 0 \\ 0 & 0 & 0 & 0 & -1 \end{bmatrix}.$$

Determine a basis for the stable, unstable, and center subspaces, and then show that $\mathbb{R}^5 = E^s \oplus E^u \oplus E^c$. Write out the general solution of the foregoing system and show that E^s, E^u, and E^c are invariant under the flow. Identify the origin as either a hyperbolic or nonhyperbolic fixed point.

Exercise 1.20. Determine the eigenvalues and eigenvectors for the linear system

$$\dot{x} = \begin{bmatrix} 0 & 1 & 0 & 1 & 0 \\ -1 & 0 & 0 & 1 & 0 \\ 0 & 0 & -1 & 0 & 1 \\ 0 & 0 & 0 & 1 & 0 \\ 0 & 0 & -4 & 0 & -1 \end{bmatrix}.$$

Determine a basis for the stable, unstable, and center subspaces, and then show that $\mathbb{R}^5 = E^s \oplus E^u \oplus E^c$. Write out the general solution of the foregoing system and show that E^s, E^u, and E^c are invariant under the flow. Identify the origin as either a hyperbolic or nonhyperbolic fixed point.

1.5 Fundamental Matrix Solutions

The idea of a fundamental matrix solution for a given linear system is important and it is one that we will carry with us during our discussion of nonlinear systems. For now, these concepts have significance in their ability to allow us to represent solutions to nonautonomous, nonhomogeneous equations using a given integral equation that we will derive in Section 1.6. In this paragraph, our goal will be to lay out some fundamental terminology.

Definition 1.17. Consider a (possibly nonautonomous) linear system,

$$\dot{x} = A(t)x, \tag{1.19}$$

where $A : \mathbb{R} \to \mathbb{R}^{n \times n}$ is a matrix-valued function of time. Any nonsingular matrix-valued function $\Psi : \mathbb{R} \to \mathrm{GL}(n; \mathbb{R})$ that satisfies the matrix differential equation

$$\dot{\Psi}(t) = A(t)\Psi(t) \tag{1.20}$$

for all $t \in \mathbb{R}$ is referred to as a *fundamental matrix solution* of differential equation (1.19).

We would like to note that the set $\mathrm{GL}(n; \mathbb{R})$ is known as the *general linear group* and is defined by

$$\mathrm{GL}(n; \mathbb{R}) = \{A \in \mathbb{R}^{n \times n} : \det(A) \neq 0\},$$

i. e., $\mathrm{GL}(n; \mathbb{R})$ is the set of *invertible*, real, $n \times n$ matrices. It is also an important example of what we will later call a *matrix Lie group*, a topic to which we will return in Chapter 9.

An obvious corollary to the preceding definition sheds some light upon the meaning of these fundamental matrix solutions.

Corollary 1.3. *Let $\Psi(t)$ be a fundamental matrix solution of* (1.19) *and let $x_1, \ldots, x_n : \mathbb{R} \to \mathbb{R}^n$ be the individual column vectors of Ψ. Then the vector-valued functions $x_1, \ldots, x_n$ form a set of n linearly independent solutions of differential equation* (1.19).

Proof. Since $\Psi(t)$ is invertible for all $t \in \mathbb{R}$, each of the column vectors must be linearly independent for all $t \in \mathbb{R}$. The condition that $x_i(t)$ must satisfy differential equation (1.19) is equivalent to the condition that the ith row of matrix Ψ must satisfy the matrix differential equation in (1.20). Since Ψ satisfies (1.20), it follows that each of its columns also satisfies (1.19). □

Definition 1.18. The *principal matrix solution* $\Phi(t)$ of (1.19) is a function $\Phi : \mathbb{R} \to \mathrm{GL}(n : \mathbb{R})$ that is a fundamental matrix solution of (1.19) and that satisfies the initial condition

$$\Phi(0) = I.$$

Although system (1.19) has many fundamental matrix solutions, it has only one principal matrix solution. (This statement follows from the theory of uniqueness of ordinary differential equations.)

Proposition 1.1. *Let $\Psi(t)$ be any fundamental matrix solution to* (1.19). *Then the principal matrix solution to* (1.19) *is determined by*

$$\Phi(t) = \Psi(t)\Psi^{-1}(0). \tag{1.21}$$

Proof. Clearly the matrix defined in (1.21) satisfies the initial condition $\Phi(0) = I$. Hence, if $\Phi(t)$ is a fundamental matrix solution to (1.19), then it is the principal matrix solution. To show that it is a fundamental matrix solution, that is, that it satisfies matrix differential equation (1.20), we simply compute

$$\dot{\Phi}(t) = \dot{\Psi}(t)\Psi^{-1}(0) = A\Psi(t)\Psi^{-1}(0) = A\Phi(t).$$

Hence Φ must be the fundamental matrix solution. □

Proposition 1.2. *If A in* (1.19) *is constant, then the principal matrix solution is given by*

$$\Phi(t) = e^{At}.$$

Proof. This follows immediately from Theorem 1.7. □

The principal matrix solution of a linear system is an important tool to have on hand in one's tool chest when processing batches of initial conditions for a single system. Such a situation is relevant, for example, when visualizing the *flow* of a system, as opposed to individual trajectories. Let us illustrate this idea with two examples.

Example 1.10. Determine the principal matrix solution of the linear system

$$\dot{x} = \begin{bmatrix} -1 & -4 \\ 1 & -1 \end{bmatrix} x.$$

Following a simple computation, one obtains the complex conjugate paired eigenvalues and eigenvectors

$$\lambda_{1,2} = -1 \pm 2i, \quad w_{1,2} = \begin{bmatrix} 2 \\ \mp i \end{bmatrix}.$$

From Theorem 1.3 it follows that the general solution is given by

$$x(t) = c_1 \Re\{e^{\lambda_1 t} w_1\} + c_2 \Im\{e^{\lambda_1 t} w_1\}$$

for arbitrary real constants c_1, c_2. This solution may be written more explicitly as

$$\begin{aligned} x(t) &= c_1 e^{-t} \begin{bmatrix} 2\cos(2t) \\ \sin(2t) \end{bmatrix} + c_2 e^{-t} \begin{bmatrix} 2\sin(2t) \\ -\cos(2t) \end{bmatrix} \\ &= \begin{bmatrix} 2e^{-t}\cos(2t) & 2e^{-t}\sin(2t) \\ e^{-t}\sin(2t) & -e^{-t}\cos(2t) \end{bmatrix} \begin{bmatrix} c_1 \\ c_2 \end{bmatrix}. \end{aligned}$$

It follows that

$$\Psi(t) = \begin{bmatrix} 2e^{-t}\cos(2t) & 2e^{-t}\sin(2t) \\ e^{-t}\sin(2t) & -e^{-t}\cos(2t) \end{bmatrix}$$

is a fundamental matrix solution and, consequently, by Proposition 1.1, that

$$\Phi(t) = \Psi(t)\Psi(0)^{-1} = \begin{bmatrix} e^{-t}\cos(2t) & -2e^{-t}\sin(2t) \\ \frac{1}{2}e^{-t}\sin(2t) & e^{-t}\cos(2t) \end{bmatrix}$$

is the principal matrix solution.

The preceding example can be used rather nicely to plot the flow of a specified initial region. Take, for instance, the parametric curve

$$x_0(\theta) = \begin{bmatrix} 0.8 \\ 0.8 \end{bmatrix} + 0.1 \begin{bmatrix} \cos\theta \\ \sin\theta \end{bmatrix},$$

which represents a circle of radius 0.1 centered at $(0.8, 0.8)$. Considering this to be a spectrum of initial conditions, one finds that its flow is given merely by

$$x(t;\theta) = \Phi(t)x_0(\theta).$$

One can easily animate the flow of the region defined by this circle with the aid of MATLAB. For an introduction on MATLAB, see [100]. Consider the following program.

examp1.m

```
clear
u = linspace(0,2*pi);
x = 0.8 + 0.1*cos(u);
y = 0.8 + 0.1*sin(u);
t = linspace(0,2);
for(i=1:100)
    PHI(1,1) = exp(-t(i))*cos(2*t(i));
    PHI(1,2) = -2*exp(-t(i))*sin(2*t(i));
    PHI(2,1) = exp(-t(i))*sin(2*t(i))/2;
    PHI(2,2) = exp(-t(i))*cos(2*t(i));
    R = PHI*[x;y];
    plot(R(1,:),R(2,:))
    axis([-0.8,1,-0.4,1])
    A(i) = getframe;
end
t = linspace(0, 2, 10);
hold on
for(i=1:10)
```

```
    PHI(1,1) = exp(-t(i))*cos(2*t(i));
    PHI(1,2) = -2*exp(-t(i))*sin(2*t(i));
    PHI(2,1) = exp(-t(i))*sin(2*t(i))/2;
    PHI(2,2) = exp(-t(i))*cos(2*t(i));
    R = PHI*[x;y];
    plot(R(1,:),R(2,:))
    axis([-0.8,1,-0.4,1])
end
```

The MATLAB program `examp1.m` animates the flow of the initial condition circle defined above. Upon executing the program, one first sees the animation of the flow followed by a concurrent plot of 11 temporally equidistant snapshots of the flow of the initial region, as plotted in Figure 1.5. We make two observations as our region flows toward the origin along one of its spiral arms. The first is that the initial circle deforms into an ellipse along the phase flow. Secondly, the area enclosed by the ellipse exponentially decays with time. □

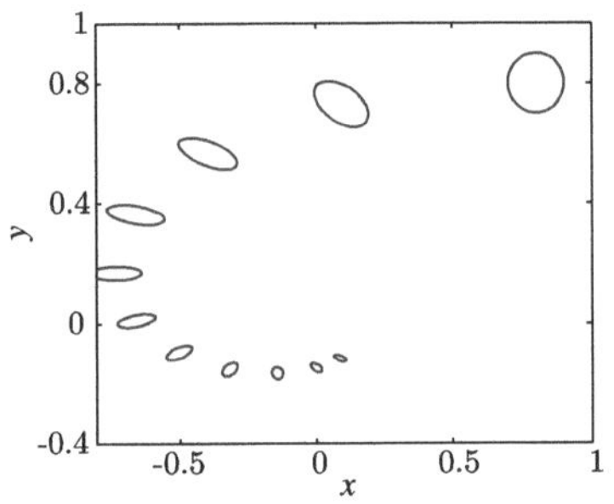

Figure 1.5: Flow near a stable spiral.

Exercise 1.21. Use the code provided by `examp1.m` to reproduce the graph shown in Figure 1.5. Then experiment by defining your own initial region and plotting its flow.

Example 1.11. Determine the principal matrix solution of the linear system

$$\dot{x} = \begin{bmatrix} 1 & 1 \\ 4 & -2 \end{bmatrix} x.$$

One readily checks that the eigenvalues and eigenvectors for this system are given by

$$\lambda_1 = -3, \quad \lambda_2 = 2, \quad w_1 = \begin{bmatrix} 1 \\ -4 \end{bmatrix}, \quad w_2 = \begin{bmatrix} 1 \\ 1 \end{bmatrix},$$

respectively. By Theorem 1.2, the general solution is therefore given by

$$x(t) = c_1 e^{-3t} \begin{bmatrix} 1 \\ -4 \end{bmatrix} + c_2 e^{2t} \begin{bmatrix} 1 \\ 1 \end{bmatrix},$$

and a fundamental matrix solution is given by

$$\Psi(t) = \begin{bmatrix} e^{-3t} & e^{2t} \\ -4e^{-3t} & e^{2t} \end{bmatrix}.$$

By Proposition 1.1, the principal matrix solution is

$$\Phi(t) = \frac{1}{5}\begin{bmatrix} e^{-3t} + 4e^{2t} & -e^{-3t} + e^{2t} \\ -4e^{-3t} + 4e^{2t} & 4e^{-3t} + e^{2t} \end{bmatrix}.$$

To visualize the flow of this system, consider the parametrically defined circle

$$x_0(\theta) = \begin{bmatrix} 0.15 \\ -0.6 \end{bmatrix} + 0.1\begin{bmatrix} \cos\theta \\ \sin\theta \end{bmatrix}.$$

The center of this circle lies on the stable subspace. The flow of the perimeter of this region is determined by $x(t;\theta) = \varphi(t; x_0(\theta)) = \Phi(t)x_0(\theta)$. The following MATLAB program animates the evolution of this region through one unit of time.

examp2.m

```
clear
u = linspace(0,2*pi);
x = 0.15 + 0.1*cos(u);
y = -0.6 + 0.1*sin(u);
t=linspace(0,1);
x1 = linspace(-1,1);
y1 =-4*x1;
y2 = x1;
for(i=1:100);
    PHI(1,1) = exp(-3*t(i))/5+4*exp(2*t(i))/5;
    PHI(1,2) = -exp(-3*t(i))/5+exp(2*t(i))/5;
    PHI(2,1) = -4*exp(-3*t(i))/5+4*exp(2*t(i))/5;
    PHI(2,2) = 4*exp(-3*t(i))/5+exp(2*t(i))/5];
    R = PHI*[x;y];
    plot(R(1,:),R(2,:))
    hold on
    plot(x1,y1,'r--')
    plot(x1,y2,'r--')
    axis([-1 1 -1 1])
    hold off
    A(i) = getframe;
end
t = linspace(0,1,6);
hold on
```

```
plot(x1,y1,'r--')
plot(x1,y2,'r--')
for(i=1:6)
    PHI(1,1) = exp(-3*t(i))/5+4*exp(2*t(i))/5;
    PHI(1,2) = -exp(-3*t(i))/5+exp(2*t(i))/5;
    PHI(2,1) = -4*exp(-3*t(i))/5+4*exp(2*t(i))/5;
    PHI(2,2) = 4*exp(-3*t(i))/5+exp(2*t(i))/5];
    R = PHI*[x;y];
    plot(R(1,:),R(2,:))
    axis([-1 1 -1 1])
end
```

The output of the program examp2.m is plotted in Figure 1.6, which shows the stable and unstable subspaces (dashed lines) and six temporally equidistant snapshots of the flow of the initial circle. The two points on the initial circle that lie on the stable subspace remain on the stable subspace for all time. Points to either side of the stable subspace are stretched outward in the direction of the unstable subspace, causing the circle to become more and more compressed in that direction with time. Asymptotically, as $t \to \infty$, points on the initial circle will degenerate into a line that coincides with the unstable subspace. □

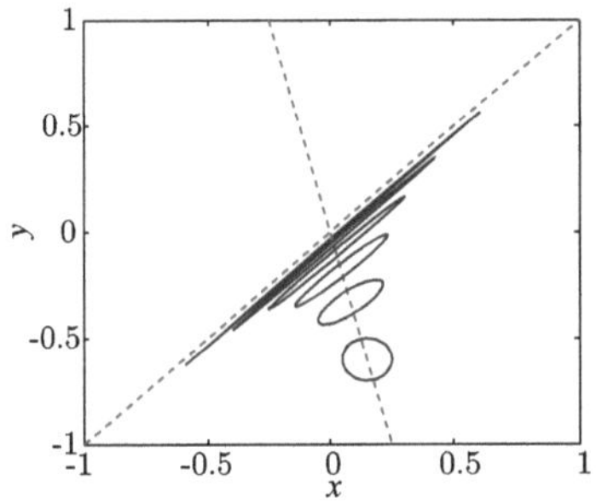

Figure 1.6: Flow near a saddle.

Exercise 1.22. Use the program provided in examp2.m to reproduce Figure 1.6. Investigate the evolution of other initial regions. What happens if a circle is initially situated such that it is centered on the unstable manifold?

Exercise 1.23. Find the general solution to the linear system

$$\dot{x} = \begin{bmatrix} -1 & 2 & 0 \\ -2 & -1 & 0 \\ 0 & 0 & -2 \end{bmatrix} x.$$

Determine the principal matrix solution. *Extra:* Write a program that produces a 3-D graph of the flow of an initial sphere centered at $(1, 1, 1)$ with radius 0.1.

1.6 Nonhomogeneous and Nonautonomous Systems

So far we have restricted our attention to *homogeneous*, *autonomous* linear systems. Homogeneous means that the system is of the form $\dot{x} = Ax$, as opposed to a nonhomogeneous system that would take the form $\dot{x} = Ax + b$. Autonomous means that both A and b are independent of time. In this paragraph, we broaden our perspective of linear systems to include nonautonomous, nonhomogeneous linear systems. For additional details, see [59].

Nonautonomous Nonhomogeneous Systems

Consider the class of nonautonomous, nonhomogeneous systems of the form

$$\dot{x} = Ax + b, \tag{1.22}$$

where $A : \mathbb{R} \to \mathbb{R}^{n\times n}$ and $b : \mathbb{R} \to \mathbb{R}^n$, i. e., A is a matrix-valued function of time and b is a vector-valued function of time. Given a nonhomogeneous system (1.22), we define its *associated homogeneous system* as

$$\dot{x}_h = Ax_h, \tag{1.23}$$

where $A = A(t)$ is the same matrix-valued function of time as appears in the nonhomogeneous system. Solutions to (1.23) are referred to as *associated homogeneous solutions*. They are denoted as $x_h(t)$ to remind us that they are *not* solutions of our original system (1.22).

Theorem 1.13. *If $\Psi(t)$ is any fundamental matrix solution of the associated homogeneous system* (1.23)*, then the solution of nonhomogeneous system* (1.22)*, with initial condition $x(0) = x_0$, is*

$$x(t) = \Psi(t)\Psi^{-1}(0)x_0 + \int_0^t \Psi(t)\Psi^{-1}(\tau)b(\tau)\,d\tau. \tag{1.24}$$

Proof. Clearly, the proposed solution (1.24) satisfies the initial condition $x(0) = x_0$. To show that it is also a solution to differential equation (1.22), let us differentiate. We have

$$\begin{aligned}\dot{x}(t) &= \dot{\Psi}(t)\Psi^{-1}(0)x_0 + \Psi(t)\Psi^{-1}(t)b(t) + \int_0^t \dot{\Psi}(t)\Psi^{-1}(\tau)b(\tau)\,d\tau \\ &= A(t)\Psi(t)\Psi^{-1}(0)x_0 + b(t) + A(t)\int_0^t \Psi(t)\Psi^{-1}(\tau)b(\tau)\,d\tau\end{aligned}$$

$$= A(t)\left[\Psi(t)\Psi^{-1}(0)x_0 + \int_0^t \Psi(t)\Psi^{-1}(\tau)b(\tau)\,d\tau\right] + b(t)$$
$$= A(t)x(t) + b(t).$$

Hence, (1.24) is the solution of (1.22) with initial condition $x(0) = x_0$. □

Note that if one uses the *principal* matrix solution $\Phi(t)$, (1.24) will appear slightly cleaner due to the fact that $\Phi(0) = I$. However, it is often more practical to use a general fundamental matrix solution $\Psi(t)$ instead of first forming the potentially more complicated principal matrix solution using (1.21).

Example 1.12. Consider the system

$$\dot{x} = \begin{bmatrix} -2t & \cos t \\ 0 & -2t \end{bmatrix} x + \begin{bmatrix} 2te^{-t^2} \\ e^{-t^2} \end{bmatrix}.$$

In Example 1.16 we will show that the *principal* matrix solution of the associated homogeneous equation is given by

$$\Phi(t) = e^{-t^2}\begin{bmatrix} 1 & \sin t \\ 0 & 1 \end{bmatrix}.$$

Note that $\Phi(0) = I$. The inverse of $\Phi(t)$ is given by

$$\Phi^{-1}(t) = e^{t^2}\begin{bmatrix} 1 & -\sin t \\ 0 & 1 \end{bmatrix}.$$

Using (1.24), we obtain

$$x(t) = \Phi(t)x_0 + \Phi(t)\int_0^t \Phi^{-1}(\tau)b(\tau)\,d\tau$$
$$= e^{-t^2}\begin{bmatrix} 1 & \sin t \\ 0 & 1 \end{bmatrix} x_0 + e^{-t^2}\begin{bmatrix} 1 & \sin t \\ 0 & 1 \end{bmatrix}\int_0^t \begin{bmatrix} 2\tau - \sin\tau \\ 1 \end{bmatrix} d\tau$$
$$= e^{-t^2}\begin{bmatrix} 1 & \sin t \\ 0 & 1 \end{bmatrix} x_0 + e^{-t^2}\begin{bmatrix} 1 & \sin t \\ 0 & 1 \end{bmatrix} \cdot \begin{bmatrix} t^2 + \cos t - 1 \\ t \end{bmatrix}$$
$$= e^{-t^2}\begin{bmatrix} 1 & \sin t \\ 0 & 1 \end{bmatrix} x_0 + e^{-t^2}\begin{bmatrix} t^2 + \cos t - 1 + t\sin t \\ t \end{bmatrix}.$$ □

Exercise 1.24. Determine the general solution to the nonhomogeneous linear system

$$\dot{x} = \begin{bmatrix} 1 & 1 \\ 4 & -2 \end{bmatrix} x + \begin{bmatrix} 15 \\ 30 \end{bmatrix}.$$

Example 1.13. The motion of a particle in a constant gravitational field, with constant, downward gravitational acceleration g, is governed by the second-order equations of motion

$$\begin{bmatrix} \ddot{x} \\ \ddot{y} \\ \ddot{z} \end{bmatrix} = \begin{bmatrix} 0 \\ 0 \\ -g \end{bmatrix}.$$

We can use a common trick when studying mechanical systems (which are second-order systems in nature) and transform this into a first-order system by introducing the additional state variables $u = \dot{x}$, $v = \dot{y}$, and $w = \dot{z}$. The preceding second-order system in $\mathbb{R}^3$ may therefore be rewritten as the following first-order, linear, nonhomogeneous system in $\mathbb{R}^6$:

$$\frac{d}{dt}\begin{bmatrix} x \\ y \\ z \\ u \\ v \\ w \end{bmatrix} = \begin{bmatrix} 0 & 0 & 0 & 1 & 0 & 0 \\ 0 & 0 & 0 & 0 & 1 & 0 \\ 0 & 0 & 0 & 0 & 0 & 1 \\ 0 & 0 & 0 & 0 & 0 & 0 \\ 0 & 0 & 0 & 0 & 0 & 0 \\ 0 & 0 & 0 & 0 & 0 & 0 \end{bmatrix}\begin{bmatrix} x \\ y \\ z \\ u \\ v \\ w \end{bmatrix} + \begin{bmatrix} 0 \\ 0 \\ 0 \\ 0 \\ 0 \\ -g \end{bmatrix}. \tag{1.25}$$

This equation is of the form $\dot{x} = Ax + b$, where $x = (x, y, z, u, v, w)^T$. In addition, the first three rows of this matrix equation simply define the additional state variables u, v, and w. This example is continued in Exercise 1.25. □

Exercise 1.25. Show that

$$e^{At} = \begin{bmatrix} 1 & 0 & 0 & t & 0 & 0 \\ 0 & 1 & 0 & 0 & t & 0 \\ 0 & 0 & 1 & 0 & 0 & t \\ 0 & 0 & 0 & 1 & 0 & 0 \\ 0 & 0 & 0 & 0 & 1 & 0 \\ 0 & 0 & 0 & 0 & 0 & 1 \end{bmatrix},$$

where A is the coefficient matrix defined in (1.25). Note that $\Phi(t) = e^{At}$ is the principal matrix solution of the associated homogeneous equation. Now use Theorem 1.13 to show that the general solution to (1.25) is given by

$$x(t) = \begin{bmatrix} x_0 + u_0 t \\ y_0 + v_0 t \\ z_0 + w_0 t - \frac{1}{2}gt^2 \\ u_0 \\ v_0 \\ w_0 - g \end{bmatrix},$$

where $x_0 = (x_0, y_0, z_0, u_0, v_0, w_0)$ are the initial conditions.

Theorem 1.13 relies on one's ability to determine a fundamental matrix solution of the associated homogeneous equation (1.23). However, we do not yet know how to determine the solutions of general *nonautonomous* homogeneous equations. In what follows, we will describe a method that achieves this.

Nonautonomous Homogeneous Systems

Previously, we saw that solving general, linear, nonhomogeneous equations can be done with the aid of a single integral equation as long as one can determine a fundamental matrix solution of the associated homogeneous equation. If the equation is nonautonomous, however, we do not as of yet have a way to solve for a fundamental matrix solution. In this section, we will discuss the Peano–Baker method, which yields a series solution for the principal matrix solution of any homogeneous system, including ones that are nonautonomous. Depending on the time-dependent nature of the coefficient matrix, however, this method is not always a practical one.

We first recall that the principal matrix solution, also known as the *state-transition matrix*, satisfies the following initial value problem:

$$\dot{\Phi}(t) = A(t)\Phi(t),$$
$$\Phi(0) = I.$$

The trick is to reexpress the above initial value problem as a matrix Volterra integral equation by writing

$$\Phi(t) = I + \int_0^t A(\tau)\Phi(\tau)\, d\tau. \tag{1.26}$$

Let us now rewrite this equation as

$$\Phi(\tau_1) = I + \int_0^{\tau_1} A(\tau_2)\Phi(\tau_2)\, d\tau_2$$

and then substitute this new expression into the integral on the right-hand side of (1.26), thereby obtaining

$$\Phi(t) = I + \int_0^t A(\tau_1)\, d\tau_1 + \int_0^t A(\tau_1)\int_0^{\tau_1} A(\tau_2)\Phi(\tau_2)\, d\tau_2 d\tau_1. \tag{1.27}$$

Let us again rewrite (1.26) to obtain

$$\Phi(\tau_2) = I + \int_0^{\tau_2} A(\tau_3)\Phi(\tau_3)\, d\tau_3.$$

Substituting this new expression into the integral on the right-hand side of (1.27) then yields the expression

$$\Phi(t) = I + \int_0^t A(\tau_1)\, d\tau_1$$
$$+ \int_0^t A(\tau_1) \int_0^{\tau_1} A(\tau_2)\, d\tau_2 d\tau_1$$
$$+ \int_0^t A(\tau_1) \int_0^{\tau_1} A(\tau_2) \int_0^{\tau_2} A(\tau_3)\Phi(\tau_3)\, d\tau_3 d\tau_2 d\tau_1.$$

This motivates the following definition, introduced by Peano [233] and further developed by Baker [17].

Definition 1.19. The *Peano–Baker series* of the principal matrix solution of the nonautonomous, homogeneous linear system (1.23) is defined by the series

$$\Phi(t) = I + \sum_{n=1}^{\infty} \int_0^t A(\tau_1) \int_0^{\tau_1} A(\tau_2) \cdots \int_0^{\tau_{n-1}} A(\tau_n) d\tau_n \cdots d\tau_1$$
$$= I + \int_0^t A(\tau_1)\, d\tau_1$$
$$+ \int_0^t A(\tau_1) \int_0^{\tau_1} A(\tau_2)\, d\tau_2 d\tau_1$$
$$+ \int_0^t A(\tau_1) \int_0^{\tau_1} A(\tau_2) \int_0^{\tau_2} A(\tau_3)\, d\tau_3 d\tau_2 d\tau_1 + \cdots. \qquad (1.28)$$

The series defined in Definition 1.19 converges uniformly and absolutely; see [152] for details.

Example 1.14. For *autonomous* systems, in which $A(t) = A$ is a constant, the Peano–Baker series for the state-transition matrix converges to the matrix exponential of At,

$$\Phi(t) = I + At + \frac{At^2}{2!} + \cdots = e^{At},$$

agreeing with Proposition 1.2. □

Exercise 1.26. Show that

$$\int_0^t \left[\int_0^{\tau_1} A(\tau_2)\, d\tau_2 \right] A(\tau_1) d\tau_1 = \frac{1}{2} \left[\int_0^t A(\tau)\, d\tau \right]^2.$$

Next, show that

$$\int_0^t \left[\int_0^{\tau_1} \left[\int_0^{\tau_2} A(\tau_3)d\tau_3\right] A(\tau_2)\, d\tau_2\right] A(\tau_1)\, d\tau_1 = \frac{1}{3!}\left[\int_0^t A(\tau)\, d\tau\right]^3.$$

Example 1.15. If the matrices $A(t)$ and $\int_0^t A(\tau)d\tau$ commute, then one may generalize the result of Exercise 1.26 using mathematical induction to show that (1.28) simplifies to

$$\begin{aligned}\Phi(t) &= I + \sum_{n=1}^{\infty} \int_0^t A(\tau_1) \int_0^{\tau_1} A(\tau_2) \cdots \int_0^{\tau_{n-1}} A(\tau_n)d\tau_n \cdots d\tau_1 \\ &= I + \sum_{n=1}^{\infty} \int_0^t \left[\int_0^{\tau_1} \cdots \left[\int_0^{\tau_{n-1}} A(\tau_n)d\tau_n\right] \cdots A(\tau_2)\right] A(\tau_1)d\tau_1 \\ &= \sum_{n=0}^{\infty} \frac{1}{n!}\left[\int_0^t A(\tau)\, d\tau\right]^n = \exp\left(\int_0^t A(\tau)d\tau\right).\end{aligned}$$

If $A(t)$ is a constant, this again produces the result of Example 1.14. □

Example 1.16. Consider the nonautonomous system

$$\dot{x} = \begin{bmatrix} -2t & \cos t \\ 0 & -2t \end{bmatrix} x.$$

It is straightforward to check that coefficient matrix $A(t)$ commutes with the integral

$$\int_0^t A(\tau)\, d\tau = \begin{bmatrix} -t^2 & \sin t \\ 0 & -t^2 \end{bmatrix}.$$

By Example 1.15 and Exercise 1.14, it follows that the principal matrix solution is given by

$$\Phi(t) = \exp\left(\int_0^t A(\tau)\, d\tau\right) = e^{-t^2}\begin{bmatrix} 1 & \sin t \\ 0 & 1 \end{bmatrix}.$$

Thus, the solution with initial condition $x(0) = x_0$ is

$$x(t) = e^{-t^2}\begin{bmatrix} 1 & \sin t \\ 0 & 1 \end{bmatrix} x_0.$$ □

Exercise 1.27. Determine the solution to the nonautonomous, nonhomogeneous linear system

$$\dot{x} = \begin{bmatrix} \sin t & 0 \\ 0 & \cos t \end{bmatrix} x - \begin{bmatrix} \sin t \\ \cos t \end{bmatrix}.$$

Exercise 1.28. Compute the principal matrix solution for the system

$$\dot{x} = \frac{1}{2t^2 - 2t + 1} \begin{bmatrix} 2-4t & -2 \\ 2 & 2-4t \end{bmatrix} x.$$

You should obtain

$$\Phi(t) = \frac{1}{(2t^2 - 2t + 1)^2} \begin{bmatrix} 2t - 2t^2 & 1-2t \\ 2t-1 & 2t-2t^2 \end{bmatrix}. \tag{1.29}$$

1.7 Application: Linear Control Theory

A control system is essentially a dynamical system with *controls*, i. e., parameters that may be varied at will. In control theory, one does more than passively determine the evolution of a system; one seeks to actively inject one's influence over the outcome. In general, the study of control theory is divided among several questions:

1. **Controllability.** Given two boundary conditions $x_1, x_2 \in \mathbb{R}^n$ in the state space, does there exist an admissible control that steers the system from x_1 to x_2 in finite time?
2. **Motion planning.** If so, can a simple expression for this control be found? (This is called an *open-loop control* if it can be predetermined and does not require updated knowledge of the state to self-adjust. On the other hand, a *closed-loop* control is one that requires state feedback.)
3. **Stability.** Given a state $x_0 \in \mathbb{R}^n$, does there exist a closed-loop feedback control that renders the state x_0 stable?
4. **Trajectory tracking.** Given a nominal trajectory $y : [0, T] \to \mathbb{R}^n$, does there exist a closed-loop feedback control that renders the trajectory y stable? (We discuss the stability of trajectories in Chapter 2.)
5. **Optimal control.** Given a controllable system, how does one determine optimal trajectories while steering the system state between two boundary points, for example, in a minimal amount of time or with a minimal amount of fuel?

In this paragraph, we discuss the controllability of linear systems. For additional references on control theory, see, for example, [24], [37], [225], or [272].

A linear control system is governed by a linear system of ordinary differential equations in the form

$$\dot{x} = Ax + Bu, \tag{1.30}$$

where $x \in \mathbb{R}^n$ is the *state* of the system and $u \in \Omega \subset \mathbb{R}^m$ is the *control*. (Alternatively, the various components of the vector u are referred to as the *controls*.) A and B are $n \times n$ and $n \times m$ matrices, respectively. The set $\Omega \subset \mathbb{R}^n$ is called the *control set*, and a control

$u \in \Omega$ is referred to as an *admissible* control, whereas a control in the complement of Ω in $\mathbb{R}^m$ is referred to as an *inadmissible* control.

In the study of system (1.30), one computes the $n \times (nm)$ *controllability matrix* R, defined by

$$R = \begin{bmatrix} B & AB & \cdots & A^{n-1}B \end{bmatrix}.$$

The following theorem on the controllability of linear systems was presented by Kalman [155, 156]; see also [157]. We will follow the proof given in [243] and [281].

Theorem 1.14. *The linear, time-invariant system* (1.30) *is controllable if and only if the classical* controllability rank condition *holds, i. e., the condition that* $\operatorname{rank}(R) = n$.

Proof. We first prove that controllability implies the controllability rank condition. Proceeding by contradiction, let us assume that the system is controllable but that the controllability rank condition fails. Since the system is controllable, the entire state space $\mathbb{R}^n$ is accessible from the origin: for a given $x_f \in \mathbb{R}^n$, there exists a control $u : [0, t_f] \to \mathbb{R}^m$ such that the boundary value problem (1.30), with boundary conditions $x(0) = 0$ and $x(t_f) = x_f$, is satisfied. The solution path through the state space is therefore determined by (1.24)

$$x(t) = e^{At} \int_0^t e^{-A\tau} Bu(\tau)\, d\tau.$$

The Cayley–Hamilton theorem, which is attributed to [46] and [121–123], states that a matrix satisfies its own characteristic equation; therefore, A^k is a linear combination of the matrices $I, A, \ldots, A^{n-1}$, for $k \geq n$. Therefore, due to the definition of the matrix exponential (1.9), it follows that $x(t) \in \operatorname{im}(R)$, a contradiction.

Next, let us assume that the controllability rank condition holds and prove controllability. For a fixed $t_f > 0$, we begin by considering the symmetric matrix

$$M = \int_0^{t_f} e^{-As} BB^T E^{-A^T s}\, ds.$$

Let us first show that M is nonsingular. Consider some $v \in \mathbb{R}^n$ such that $Mv = 0$. It follows that $v^T Mv = 0$, and hence

$$\int_0^{t_f} \psi(s)\psi(s)^T\, ds = 0,$$

where $\psi(s) = v^T e^{-As} B$ is a $1 \times m$ matrix. Since $\int_0^{t_f} \|\psi(s)\|^2 ds = 0$, we obtain $\psi(s) \equiv 0$. Evaluating the first $(n-1)$ derivatives of $\psi(s)$ at $s = 0$ yields

$$\psi(0) = v^T B = 0,$$

$$\psi'(0) = -v^T AB = 0,$$
$$\vdots$$
$$\psi^{n-1}(0) = \pm v^T A^{n-1} B = 0,$$

which implies that $v \in \operatorname{im}(R)^\perp = \{0\}$; M is therefore nonsingular.

Finally, we consider a control of the form

$$u(s) = B^T e^{-A^T s} w,$$

where $w \in \mathbb{R}^n$ is free to be selected later. The terminal point x_f of the solution to control problem (1.30) is therefore given by (1.24) as $x_f = e^{At_f}(x_0 + Mw)$. Since M is nonsingular, this relation can be inverted to obtain

$$w = M^{-1}\left(e^{-At_f} x_f - x_0\right).$$

The system is therefore controllable. □

Example 1.17. Let us consider the dynamical system of Example 1.10 with a control in the x-direction. We have

$$\dot{x} = \begin{bmatrix} -1 & -4 \\ 1 & -1 \end{bmatrix} x + \begin{bmatrix} 1 \\ 0 \end{bmatrix} u.$$

Recall that in the absence of any control, the origin is a stable spiral. The controllability matrix $R = [B \;\; AB]$ is given by

$$R = \begin{bmatrix} 1 & -1 \\ 0 & 1 \end{bmatrix}$$

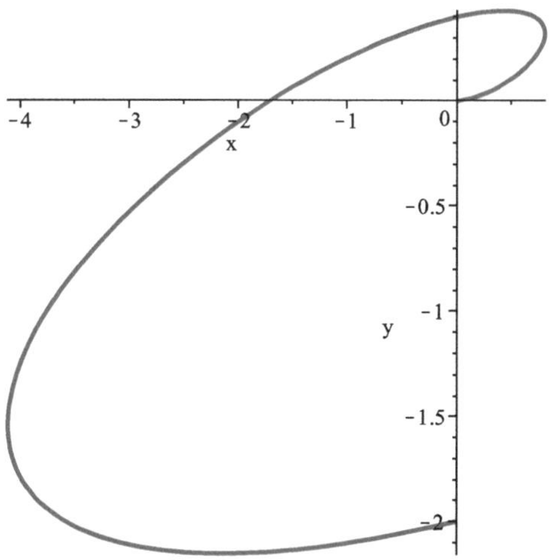

Figure 1.7: Example 1.17: steering the system from (0, 0) to (0, −2).

and clearly has full rank. The system is therefore controllable. This example demonstrates that one may escape a stable spiral given only a control in a single direction. □

Exercise 1.29. Show that the control that steers the system from $x_0 = (0,0)$ to $x_f = (0,-2)$ during the time span $t \in [0,\pi]$ is given by

$$u(t) = \frac{8e^{\pi+t}}{e^{2\pi}-1}\left(\cos(2t) + 3\sin(2t)\right).$$

You may use technology (*Maple* is especially useful for symbolic matrix computation). Plot the controlled trajectory.

The resultant trajectory from Exercise 1.29 is shown in Figure 1.7.

2 Linearization of Trajectories

In this chapter, we will begin our study of nonlinear dynamical systems, devoting our attention initially to the topic of *linearization* of nonlinear systems. In particular, we will discuss notions of stability for individual trajectories as well as for fixed points, linearization of trajectories and fixed points, Lyapunov functions, and Lyapunov exponents. Finally, we will explore the link between mechanics and dynamical systems and, as an example, discuss some of the mathematical aspects of the classical planar restricted three-body problem.

2.1 Introduction and Numerical Simulation

An autonomous, nonlinear system in $\mathbb{R}^n$ is defined by an autonomous vector field, yielding the differential equation

$$\dot{x} = f(x), \tag{2.1}$$

where $x \in \mathbb{R}^n$ and $f : \mathbb{R}^n \to \mathbb{R}^n$. A given differential equation can then be coupled with an initial condition

$$x(0) = x_0. \tag{2.2}$$

The *solution*, or *flow*, of (2.1) with initial condition (2.2) is a smooth function $\varphi : \mathbb{R} \times \mathbb{R}^n \to \mathbb{R}^n$ that satisfies

$$\frac{d\varphi(t; x_0)}{dt} = f(\varphi(t; x_0)), \quad \text{for all } t \in \mathbb{R}, \tag{2.3}$$

$$\varphi(0; x_0) = x_0, \quad \text{for all } x_0 \in \mathbb{R}^n. \tag{2.4}$$

During the course of our study of nonlinear systems, special trajectories called *equilibrium solutions* or *fixed points*, which remain constant in time, will be of particular importance.

Definition 2.1. A point $x^* \in \mathbb{R}^n$ is called a *fixed point* of system (2.1) if $f(x^*) = 0$.

An immediate corollary to the preceding definition is the following.

Corollary 2.1. *A point $x^* \in \mathbb{R}^n$ is a fixed point of* (2.1) *if and only if $\varphi(t; x^*) = x^*$ for all $t \in \mathbb{R}$.*

Proof. The solution flow $\varphi(t; x^*)$ is constant if and only if

$$\frac{d\varphi(t; x^*)}{dt} = 0,$$

which occurs if and only if $f(\varphi(t; x^*)) = f(x^*) = 0$. □

https://doi.org/10.1515/9783110597806-002

Numerical Simulation

The most straightforward and simple way to simulate a nonlinear differential equation is using MATLAB's built-in ode45 program, which is a variable time step Runge–Kutta method that approximates solutions to the initial value problem

$$\dot{x} = f(t, x),$$
$$x(0) = x_0.$$

This can be accomplished by writing your own m-file to execute the ode45 program.

We will assume basic familiarity with MATLAB software. As an example of what MATLAB's ode45 program can do, let us consider the Lorenz system

$$\dot{x} = \sigma(y - x), \tag{2.5}$$
$$\dot{y} = x(\rho - z) - y, \tag{2.6}$$
$$\dot{z} = xy - \beta z. \tag{2.7}$$

This system was proposed by American mathematician and meteorologist Edward N. Lorenz while studying atmospheric models for weather prediction [191]. As it turned out, this seemingly benign system possesses an unexpectedly rich dynamic. This apparently simple and docile system is known to exhibit chaotic behavior for the choice of parameters $\sigma = 10, \beta = 8/3, \rho = 28$. The solutions are bounded, yet the asymptotic behavior is chaotic. Solution trajectories never settle down by tending toward a fixed point or periodic orbit but rather hop around in a chaotic fashion asymptotically approaching a set known as a *strange attractor*. See [110] or [274] for a detailed analysis of this system.

Consider now the following two MATLAB programs that simulate the solution to the Lorenz equations. Each program is to be saved in a folder that must be selected as MATLAB's directory in order to execute the program in the command prompt.

```
lorenzmain.m
clear
global sigma rho beta
sigma = 10;
beta = 8/3;
rho = 28;
IC = [-10.0 -10 25];
Tspan = [0 20];
OPTIONS = odeset('RelTol',1e-6,'AbsTol',1e-9);
[T, X] = ode45(@lorenzfun, Tspan, IC, OPTIONS);
x = X(:,1); y = X(:,2); z = X(:,3);
plot3(x,y,z,'linewidth',2)
xlabel('x'), ylabel('y'), zlabel('z')
```

```
lorenzfun.m
function F = lorenzfun(T,X)
global sigma rho beta
x=X(1); y=X(2); z=X(3);
F = [sigma*(y-x); x*(rho-z)-y; x*y-beta*z];
```

To execute the program, one types *lorenzmain* into the MATLAB command prompt and clicks *Enter*. Upon successful execution, the graph shown in Figure 2.1 appears, showing the trajectory corresponding to the initial conditions specified in the program `lorenzmain.m`. For more on `ode45`, type `help ode45` into MATLAB's command prompt.

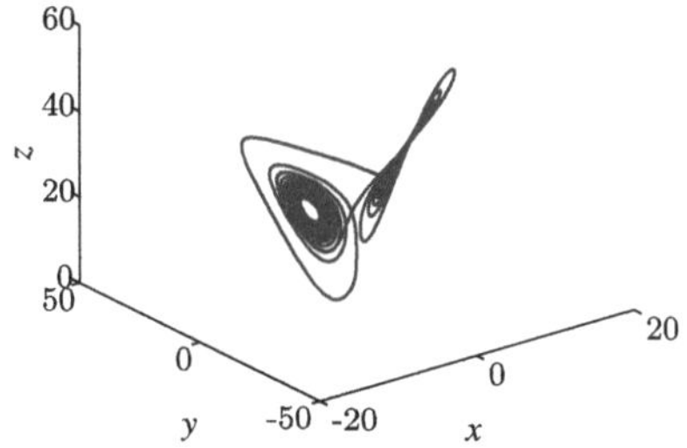

Figure 2.1: Chaotic motion in the Lorenz system.

Exercise 2.1. Use MATLAB to reproduce Figure 2.1. How does the figure change if you change the initial condition to $x_0 = 10.01$?

Exercise 2.2. Write a program that simulates the solution to the pendulum equations, $\dot{x} = y$, $\dot{y} = -\sin x$, with initial condition $(x_0, y_0) = (0, y_0)$, for $y_0 = 0.25, 0.5, 0.75, 1.00, 1.25, 1.5, 1.75, 1.99, 2.00, 2.01$. Use the `hold on` command to plot each solution concurrently on the same graph.

2.2 Linearization of Trajectories

In this section, we seek an answer to the following question: how can one describe the flow of the local neighborhood of a given nominal trajectory $\varphi(t; x_0)$ emanating from a given initial condition x_0? In other words, in the first order approximation, how do *neighboring* trajectories behave?

To begin, let us consider a given trajectory $\varphi(t; x_0)$ and a neighboring trajectory $\varphi(t; x_0 + \delta x_0)$, where δx_0 represents a small displacement to the initial condition. Since $\varphi(t; x_0 + \delta x_0)$ is by assumption a solution to differential equation (2.1), it satisfies

$$\frac{d}{dt}[\varphi(t; x_0 + \delta x_0)] = f(\varphi(t; x_0 + \delta x_0)). \tag{2.8}$$

Now we can use a first-order Taylor polynomial to expand the flow about the nominal trajectory, obtaining

$$\varphi(t;x_0+\delta x_0)=\varphi(t;x_0)+\frac{\partial\varphi(t;x_0)}{\partial x_0}\cdot\delta x_0+O(|\delta x_0|^2), \tag{2.9}$$

as $\delta x_0 \to 0$, for every fixed t. It is important to note that the $O(|\delta x_0|^2)$ term goes to zero as fast as $|\delta x_0|^2$, as $\delta x_0 \to 0$, for each fixed t; however, for a fixed initial variation δx_0, the $O(|\delta x_0|^2)$ term may, in general, grow with t, even possibly at an exponential rate. Hence, for any fixed t, a varied trajectory will approach the nominal trajectory plus its first-order correction as the initial variation δx_0 tends to zero; however, for a fixed initial variation δx_0, the first-order linearization may fail to accurately approximate the varied trajectory $\varphi(t;x_0+\delta x_0)$ within a finite amount of time. To be precise, the asymptotic approximation (2.9) literally means that the limit

$$\lim_{\delta x_0\to 0}\frac{|\varphi(t;x_0+\delta x_0)-\varphi(t;x_0)-\frac{\partial\varphi(t;x_0)}{\partial x_0}\cdot\delta x_0|}{|\delta x_0|^2}$$

equals a constant at any given fixed time t, though this constant may in fact depend on t.

The time-varying coefficient matrix of the linear term of the expansion (2.9) is important enough to warrant its own nomenclature, which we now interject before moving on.

Definition 2.2. Given a trajectory $\varphi(t;x_0)$ of system (2.1), its *state-transition matrix*, $\Phi:\mathbb{R}\to\mathrm{GL}(n;\mathbb{R})$, is a matrix-valued function of time

$$\Phi(t;x_0)=\frac{\partial\varphi(t;x_0)}{\partial x_0},$$

which describes the first-order variation of the state with respect to a differential change in the initial condition. Whenever the initial condition is understood, we denote it simply as $\Phi(t)$.

The state-transition matrix of a nominal trajectory is a matrix-valued function of time that maps variations in the initial condition to their variations at time t, i. e., to the first-order

$$\delta x(t)=\Phi(t)\cdot\delta x_0+O(|\delta x_0|^2),$$

where we have introduced the shorthand $\delta x(t)=\varphi(t;x_0+\delta x_0)-\varphi(t;x_0)$. Further, since $\varphi(0;x_0+\delta x_0)=x_0+\delta x_0$, (2.9) implies that

$$\Phi(0)=\frac{\partial\varphi}{\partial x_0}(0;x_0)=I_n, \tag{2.10}$$

the $n \times n$ identity matrix. Now, if we expand equation (2.8) to the first order, using expansion (2.9), we obtain

$$\frac{d\varphi(t;x_0)}{dt} + \frac{d}{dt}\frac{\partial\varphi(t;x_0)}{\partial x_0} \cdot \delta x_0 = f(\varphi(t;x_0)) + \frac{\partial f}{\partial x} \cdot \frac{\partial\varphi(t;x_0)}{\partial x_0} \cdot \delta x_0 + O(|\delta x_0|^2),$$

as $\delta x_0 \to 0$, for every fixed t. The first term on the left- and right-hand sides cancel each other out due to (2.3). Since δx_0 is an arbitrary, small, initial displacement vector, it follows that the state-transition matrix must satisfy the matrix differential equation

$$\dot{\Phi} = Df(x) \cdot \Phi \tag{2.11}$$

for all time, where coefficient matrix $Df(x)$ is the *Jacobian matrix* of vector field $f(x)$, given by

$$Df(x) = \frac{\partial f(x)}{\partial x}.$$

We therefore obtain the following theorem.

Theorem 2.1. *Given a trajectory $\varphi(t;x_0)$, its associated state-transition matrix, $\Phi : \mathbb{R} \to \mathrm{GL}(n;\mathbb{R})$, satisfies matrix differential equation (2.11) and initial condition (2.10).*

Now let $\xi : \mathbb{R} \to \mathbb{R}^n$ be defined by the relation

$$\xi(t) = \Phi(t) \cdot \delta x_0, \tag{2.12}$$

so that $x(t)$ can be thought of as a first-order approximation to the difference $\varphi(t;x_0 + \delta x_0) - \varphi(t;x_0)$. Note that by virtue of (2.10), we have $\xi(0) = \delta x_0$. The following theorem follows immediately from the discussion above.

Theorem 2.2. *The vector-valued function $\xi(t) = \Phi(t) \cdot \delta x_0$ satisfies the differential equation*

$$\dot{\xi} = Df(x) \cdot \xi \tag{2.13}$$

and initial condition $\xi(0) = \delta x_0$.

Equation (2.13) is referred to as the *linearization* of (2.1) (or the *linearized equations*). The state-transition matrix is simply the principal matrix solution of the linearized equations (2.13) (see Definition 1.18).

Example 2.1. Consider again the Lorenz equations (2.5)–(2.7). The Jacobian of the vector field that defines the flow is given by

$$Df(x) = \begin{bmatrix} -\sigma & \sigma & 0 \\ (\rho - z) & -1 & -x \\ y & x & -\beta \end{bmatrix}.$$

To compute the linearization, the nine differential equations (2.11) (one for each component of $\Phi(t)$) with initial conditions (2.10) must be integrated along with the equations for the nominal trajectory (2.5)–(2.7), yielding in total a coupled system of 12 equations. □

Exercise 2.3. Consider the system

$$\dot{x} = x^2 - y^2,$$
$$\dot{y} = 2xy.$$

(a) Plot the phase plane.
(b) Show that there is a solution of the form $x = \cos\theta$, $y = 1 + \sin\theta$, for some unknown $\theta(t)$, by substituting these into the differential equations and using the chain rule.
(c) Solve for $\theta(t)$ and use trigonometric identities to show that the trajectory

$$x(t) = \frac{1-2t}{2t^2 - 2t + 1},$$
$$y(t) = \frac{1}{2t^2 - 2t + 1}$$

constitutes a particular solution of the system.
(d) Compute the Jacobian matrix and conclude that the state-transition matrix for the solution obtained in part (c) is given by (1.29).
(e) Compute $\lim_{t\to\infty} \Phi(t)$ and state what this tells you about the fate of nearby trajectories.

2.3 Stability of Trajectories

We say that a given trajectory, emanating from a given initial condition, is *stable* if trajectories starting at neighboring initial conditions, in a certain sense, remain close to the original trajectory for all time. We formalize this notion with the following definition.

Definition 2.3. The trajectory $\varphi(t; x_0)$ is *Lyapunov stable* if, for each $\varepsilon > 0$, there exists a $\delta > 0$ such that $y_0 \in B_\delta(x_0) = \{x \in \mathbb{R}^n : |x - x_0| \leq \delta\}$ implies that

$$|\varphi(t; x_0) - \varphi(t; y_0)| < \varepsilon \quad \text{for } t > 0. \tag{2.14}$$

In essence, (2.14) states that the trajectory starting at a given $y_0 \in B_\delta(x_0)$ remains within a spatiotemporal tube of width $\varepsilon > 0$ that surrounds the forward half of the fiducial trajectory $\varphi(t; x_0)$. Hence, for every ε-tube about the fiducial trajectory, there exists some δ-ball about x_0 such that all trajectories emanating from that δ-ball remain within the ε-tube for all $t \geq 0$. See Figure 2.2 for a conceptual sketch of this idea.

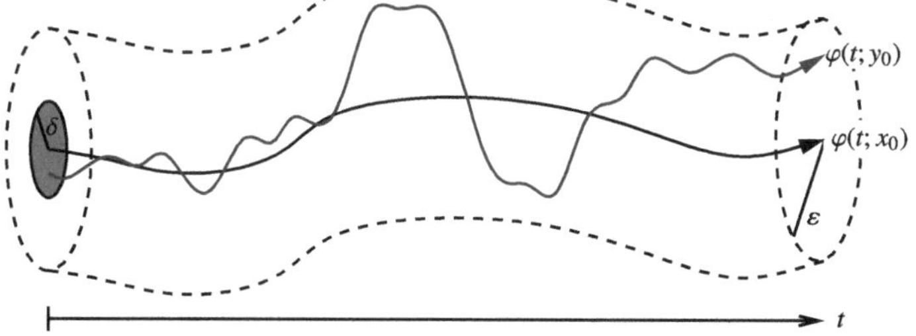

Figure 2.2: Lyapunov stability of a trajectory. All trajectories starting within the δ-ball around x_0 (shaded region on the left) remain within the ε-tube of the fiducial trajectory for all positive time.

A stronger notion of stability is given as follows.

Definition 2.4. A trajectory $\varphi(t; x_0)$ is said to be *asymptotically stable* if it is Lyapunov stable and if there exists $\delta > 0$ such that $y_0 \in B_\delta(x_0)$ implies

$$\lim_{t\to\infty} |\varphi(t; x_0) - \varphi(t; y_0)| = 0.$$

Essentially, Lyapunov stable trajectories are such that initially nearby trajectories remain nearby for all time but might not approach the fiducial trajectory, whereas asymptotically stable trajectories are such that initially nearby trajectories remain nearby for all time *and* approach the fiducial trajectory asymptotically as $t \to \infty$.

2.4 Lyapunov Exponents

We next discuss certain quantities called *Lyapunov exponents*, which provide a quantitative measure of a trajectory's stability. These quantities are further related to the linearization of individual trajectories and often serve as a measure of sensitivity in initial conditions, one of the indicators of chaos in dynamical systems. Simply put, a Lyapunov exponent in a particular direction is a measure of the exponential divergence between a nominal trajectory and one with an infinitesimal perturbation in the initial conditions in the prescribed direction. Some important references on the topic include [20, 21], [188], [192], and [229].

Earlier in this chapter we saw how the linearization of a system

$$\dot{x} = f(x) \tag{2.15}$$

about a single trajectory $x(t) = \varphi(t; x_0)$ emanating from the initial condition x_0 could be described by the linearized system

$$\dot{\xi} = Df(x(t)) \cdot \xi, \tag{2.16}$$

where $\xi(t) = \varphi(t; x_0 + \delta x_0) - \varphi(t; x_0) + O(|\delta x_0|^2)$ and where $x_0 + \delta x_0 \in B_\varepsilon(x_0)$. The state-transition matrix $\Phi(t)$ is the principal matrix solution of (2.16), which, by definition, is the fundamental matrix solution that satisfies the initial condition $\Phi(0) = I$, so that $\xi(t) = \Phi(t)\xi_0$. Let $u \neq 0$, $u \in \mathbb{R}^n$. Then the *finite-time coefficient of expansion in the direction of u along the trajectory through* x_0 is defined as

$$e_t(x_0, u) = \frac{\|\Phi(t) \cdot u\|}{\|u\|}.$$

Definition 2.5. The *Lyapunov characteristic element* (or *Lyapunov exponent*) in the direction of $u \neq 0$ along the trajectory through x_0 is defined by

$$\chi(x_0, u) = \limsup_{t\to\infty} \frac{1}{t} \log e_t(x_0, u). \tag{2.17}$$

Furthermore, we define $\chi(x_0, 0) = -\infty$.

Several remarks are in order following this definition. First, the Lyapunov exponents are *asymptotic quantities*, i. e., they depend on the limiting behavior of the trajectory. It is assumed that the trajectories also exist for $t \geq 0$. Since the state-transition matrix $\Phi(t)$ is associated with a particular trajectory, different trajectories, in general, have different Lyapunov exponents. The Lyapunov exponents are not, generally, continuous functions of the trajectories. When the initial condition x_0 is understood, we will simply denote the Lyapunov exponent of a given orbit as $\chi(u) = \chi(x_0, u)$.

Lemma 2.1. *For any $u, v \in \mathbb{R}^n$ and $c \in \mathbb{R}$, $c \neq 0$,*

$$\chi(u + v) \leq \max\{\chi(u), \chi(v)\}, \tag{2.18}$$

$$\chi(cu) = \chi(u). \tag{2.19}$$

Proof. The second statement, (2.19), follows directly from the definition of $e_t(x_0, u)$. Without loss of generality, we may therefore assume that $\|u\| = \|v\| = 1$ and that $u \neq \pm v$. The finite-time coefficient of expansion in the direction of $u + v$ is given by

$$\begin{aligned} e_t(u + v) &= \frac{\|\Phi(t) \cdot (u + v)\|}{\|u + v\|} \leq \frac{\|\Phi(t) \cdot u\| + \|\Phi(t) \cdot v\|}{\|u + v\|} \\ &\leq \frac{2\max\{\|\Phi(t) \cdot u\|, \|\Phi(t) \cdot v\|\}}{\|u + v\|}. \end{aligned}$$

Hence,

$$\begin{aligned} \chi(u + v) &\leq \limsup_{t\to\infty} \frac{1}{t} \log\left(\frac{2\max\{\|\Phi(t) \cdot u\|, \|\Phi(t) \cdot v\|\}}{\|u + v\|}\right) \\ &= \limsup_{t\to\infty} \left[\frac{1}{t} \log\left(\frac{2}{\|u + v\|}\right) + \frac{1}{t} \log\left(\max\{\|\Phi(t) \cdot u\|, \|\Phi(t) \cdot v\|\}\right)\right] \\ &= \max\{\chi(u), \chi(v)\}, \end{aligned}$$

which proves the claim. □

Proposition 2.1. *For any $r \in \mathbb{R}$, the set*

$$L(r) = \{u \in \mathbb{R}^n : \chi(u) \leq r\}$$

is a vector subspace of $\mathbb{R}^n$.

Proof. First, $0 \in L(r)$ for any $r \in \mathbb{R}$ since $\chi(0) = -\infty$ by definition. Second, suppose $u, v \in L(r)$. Then $\chi(u) \leq r$ and $\chi(v) \leq r$. But

$$\chi(u + v) \leq \max\{\chi(u), \chi(v)\} \leq r$$

by Lemma 2.1, so $u + v \in L(r)$. Finally, for $c \in \mathbb{R}$, $c \neq 0$, we have $\chi(cu) = \chi(u) \leq r$, so $cu \in L(r)$ whenever $u \in L(r)$. The result follows. □

Proposition 2.2. *The set of numbers*

$$\{\chi(u)\}_{u \in \mathbb{R}^n, u \neq 0}$$

takes at most n distinct values, which we denote $v_1 > \cdots > v_s$, where $1 \leq s \leq n$.

Exercise 2.4. Use Lemma 2.1 and Proposition 2.1 to prove Proposition 2.2.

Associated with the s distinct Lyapunov exponents, we have $s+1$ nested subspaces.

Proposition 2.3. *Let $L_i = \{u \in \mathbb{R}^n : \chi(u) \leq v_i\}$, for $i = 1, \ldots, s$. Define $L_{s+1} = \{0\}$. Then*

$$L_{s+1} \subset L_s \subset \cdots \subset L_1 = \mathbb{R}^n,$$

with $L_{i+1} \neq L_i$. Moreover, $\chi(u) = v_i$ if and only if $u \in L_i \backslash L_{i+1}$ for $i = 1, \ldots, s$.

Exercise 2.5. Prove Proposition 2.3.

A natural question is whether the limit in (2.17) itself exists, rather than the limit supremum. Before discussing this, we first establish a few definitions.

Definition 2.6. A basis $f_1, \ldots, f_n$ is called a normal basis for $\mathbb{R}^n$ if

$$\sum_{i=1}^{n} \chi(f_i) \leq \sum_{i=1}^{n} \chi(g_i),$$

for all other bases $\{g_1, \ldots, g_n\}$ for $\mathbb{R}^n$.

We now state a lemma due to [112], which states that the Euclidean volume of any parallelepiped may be no greater than the product of the length of its sides. More precisely, we have the following lemma.

Lemma 2.2 (Hadamard's Inequality). *Let A be an invertible matrix, comprised of column vectors $v_1, \ldots, v_n$. Then*

$$|\det A| \leq \prod_{i=1}^{n} \|v_i\|.$$

Exercise 2.6. Prove that, for any basis $g_1, \dots, g_n$ of $\mathbb{R}^n$, Hadamard's inequality implies the inequality

$$\limsup_{t\to\infty} \frac{1}{t} \ln|\det \Phi(t)| \le \sum_{i=1}^{n} \chi(g_i). \tag{2.20}$$

Definition 2.7. A state-transition matrix Φ is called *regular* if the limit

$$\lim_{t\to\infty} \frac{1}{t} \ln|\det \Phi(t)|$$

exists and is finite, and if there exists a basis $f_1, \dots, f_n$ for $\mathbb{R}^n$ such that

$$\lim_{t\to\infty} \frac{1}{t} \ln|\det \Phi(t)| = \sum_{i=1}^{n} \chi(f_i). \tag{2.21}$$

Clearly, due to relation (2.20), any basis satisfying relation (2.21) is a normal basis. We can now state the following theorem, due to [192], which states whether the limit in (2.17) exists, i. e., whether the limit supremum in the definition of a Lyapunov characteristic exponent may be replaced by a limit.

Theorem 2.3 (Lyapunov). *If state-transition matrix $\Phi(t)$ is regular, then the exact Lyapunov characteristic element in direction u,*

$$\chi(u) = \lim_{t\to\infty} \frac{1}{t} \log e_t(u),$$

exists and is finite for each nonzero $u \in \mathbb{R}^n$.

Proof. By Proposition 2.2, the Lyapunov characteristic elements take at most n distinct values. It therefore suffices to prove the result for these values. Let $\Phi(t)$ be the state-transition matrix for a trajectory $x(t) = \varphi(t; x_0)$ and define the matrix $\Psi(t) = \Phi(t)^T\Phi(t)$. Let $\lambda_1(t), \dots, \lambda_n(t)$ and $\xi_1(t), \dots, \xi_n(t)$ be the real eigenvalues and an associated set of orthonormal eigenvectors of the matrix $\Psi(t)$, respectively (note that $\Psi(t)$ is a symmetric matrix). Moreover, suppose that the eigenvalues are ordered in decreasing order, so that $\lambda_1(t) \ge \cdots \ge \lambda_n(t)$. Now let us compute the finite-time coefficient of expansion, at time t, in the direction of $\xi_i(t)$, for some $i \in \{1, \dots, n\}$. We obtain

$$e_t(x_0, \xi_i(t)) = \|\Phi(t)\xi_i(t)\| = \sqrt{\xi_i^T(t)\Phi^T(t)\Phi(t)\xi_i(t)} = \sqrt{\lambda_i(t)}.$$

We might then define the *finite-time Lyapunov exponents* as

$$\chi_i(t) = \frac{1}{2t} \ln \lambda_i(t), \tag{2.22}$$

where the numbers $\lambda_i(t)$ are the eigenvalues of the matrix $\Psi(t) = \Phi^T(t)\Phi(t)$. These are simply the natural logarithms of the eigenvalues of the matrix $\Psi(t)^{1/2t}$. Since the state-transition matrix is assumed to be regular, the limit

$$\Lambda = \lim_{t\to\infty} \left(\Phi^T(t)\Phi(t)\right)^{\frac{1}{2t}}$$

exists. Moreover, the natural logarithms of the eigenvalues of matrix Λ are the Lyapunov exponents for the flow, which proves the theorem. □

Note that the Lyapunov exponents may be approximated by computing the finite-time Lyapunov exponents

$$v_i = \lim_{t\to\infty} \chi_i(t) \tag{2.23}$$

over a sufficiently long period of time.

The question when a state-transition matrix is regular was taken up in [229]. The theorem below contains the main substance of the Oseledec noncommutative ergodic theorem.

Theorem 2.4 (Oseledec). *Let $M \subset \mathbb{R}^n$ be a compact, invariant set. Then, for almost every $x \in M$, the state-transition matrix associated with the trajectory through x is regular.*

Despite these nice results, various subtleties arise during the computation of Lyapunov exponents. The first issue is that one cannot literally take the limit as $t \to \infty$ during a numerical computation. The best one might hope for is to run a simulation long enough to obtain a fairly decent approximation to the limit. The second issue is that, due to the sensitivity of the initial conditions, the state-transition matrix grows in norm exponentially along any particular trajectory in a chaotic system. Therefore, in practice, one typically integrates a given trajectory iteratively over small intervals of time and reorthogonalizes the state-transition matrix using a Gram–Schmidt process at the end of each iteration. We will not discuss the details of this but rather refer the interested reader to the literature; see, for instance, [21], [69], and [188]. Work has also been done on utilizing properties of orthogonal matrices and QR factorization for the computation of Lyapunov exponents without rescaling and reorthogonalization; see, for instance, [248].

Exercise 2.7. Let $\Psi = \Phi^T\Phi$, where Φ is the state-transition matrix for a solution of the flow $\dot{x} = f$. Show that Ψ satisfies the matrix differential equation

$$\frac{d\Psi}{dt} = Df \cdot \Psi + \Psi \cdot Df^T,$$

where $Df = Df(x)$ is the matrix Jacobian of vector field f.

Exercise 2.8. Show that the state-transition matrix of the Lorenz equations (2.5)–(2.7) satisfies the matrix initial value problem

$$\dot{\Phi} = \begin{bmatrix} -\sigma & \sigma & 0 \\ (\rho - z) & -1 & -x \\ y & x & -\beta \end{bmatrix} \Phi, \quad \Phi(0) = I.$$

Modify the m-files provided in §2.1 so the resulting program simultaneously solves for the solution trajectory and state-transition matrix on the interval $t \in [0, 100]$. Use

$\sigma = 10$, $\rho = 8/3$, and $\beta = 28$. Let $\lambda_1(t), \lambda_2(t)$, and $\lambda_3(t)$ be the eigenvalues of $\Phi^T(t)\Phi(t)$. Estimate the limits (2.23) by plotting all three finite-time Lyapunov exponents (2.22) concurrently in the same graph, on the domain $0 \leq t \leq 100$.

2.5 Linearization and Stability of Fixed Points

Since fixed points in dynamical systems are themselves essentially trajectories, the technology we built in §2.2 immediately applies to the topic of understanding the linearized flow about a fixed point, with a few simplifications.

Suppose that x^* is a fixed point of system (2.1), i. e., the trajectory emanating from x^* is the constant trajectory $\varphi(t; x^*) = x^*$. Then the Jacobian of vector field f, when evaluated along the nominal trajectory, is a constant matrix

$$Df(\varphi(t; x^*)) = Df(x^*).$$

Thus, the state-transition matrix is the principal matrix solution of the autonomous linearized equations

$$\dot{\xi} = Df(x^*) \cdot \xi. \tag{2.24}$$

These are literally the same equations that appear in (2.13), except in this case, the Jacobian matrix $Df(x^*)$ is constant due to the fact that the nominal trajectory is a fixed point of the system. We therefore immediately obtain the principal matrix solution

$$\Phi(t) = e^{Df(x^*)t}.$$

Hence the solution set of the linearized equations,

$$\xi(t) = e^{Df(x^*)t}\xi_0,$$

is regarded as the *linearized flow* about the fixed point x^*. An immediate question to ask is whether the linearized flow, in some appropriate way, resembles the nonlinear flow in some neighborhood of the fixed point x^*. Before we answer this question, we must first broach the concept of topological equivalence of flows.

The Hartman–Grobman Theorem

In this section, we examine necessary and sufficient conditions for the linearized flow about a fixed point to resemble the picture of the nonlinear flow in some neighborhood about the fixed point. These conditions are the content of the Hartman–Grobman theorem, which was proved independently by Grobman [108] and Hartman [125]. Before stating the theorem, we must first define hyperbolic fixed points in nonlinear systems and local topological equivalence of flows.

Definition 2.8. A fixed point x^* of nonlinear system (2.1) is called a *hyperbolic fixed point* if the origin is a hyperbolic fixed point of the linearized equations (2.24).

In other words, a fixed point x^* of nonlinear system (2.1) is a hyperbolic fixed point if none of the eigenvalues of the Jacobian matrix $Df(x^*)$ has a zero real part, i. e., if none of the eigenvalues lies on the purely imaginary axis in the complex plane. In two dimensions, a linear flow with nonzero eigenvalues is hyperbolic if and only if the origin is a center.

The appropriate way to compare the nonlinear flow about a fixed point with its corresponding linearized flow about the origin is to use the notion of topological equivalence.

Definition 2.9. The autonomous systems

$$\dot{x} = f(x), \tag{2.25}$$

$$\dot{y} = g(y), \tag{2.26}$$

where $x, y \in \mathbb{R}^n$ are *(locally) topologically equivalent* about the fixed points x^* and y^* of systems (2.25) and (2.26), respectively, if there exist neighborhoods $U \ni x^*$ and $V \ni y^*$ and a homeomorphism $H : U \to V$, with the property that $H(x^*) = y^*$, which maps trajectories of system (2.25), restricted to neighborhood U, onto trajectories of system (2.26), restricted to neighborhood V, while preserving their orientation with respect to time.

Example 2.2. Determine which of the linear flows depicted in Figure 2.3 are topologically equivalent. (*The answer can be found at the end of this section.*) □

Our next theorem provides a sufficient condition for when a nonlinear flow in a neighborhood of a fixed point is topologically equivalent to its linearization.

Theorem 2.5 (Hartman–Grobman). *The flow in a neighborhood of a hyperbolic fixed point of nonlinear system* (2.1) *is topologically equivalent to the flow of its associated linearization.*

Proof. See, for example, [7], [126] or [235]. □

Though we omit the proof, there is a certain intuition regarding the foregoing result that we can discuss. Since the various types of linear flows form an exhaustive museum of topological equivalence classes about a fixed point, it suffices to ask how each might transform under a higher-order perturbation of its eigenvalues. We will further restrict our attention to 2-D flows, though the intuition we form generalizes straightaway to higher dimensions.

Let us consider the case of two nonzero, real, distinct eigenvalues, as depicted in Figure 2.4. The eigenvalues can be either of the same sign or of opposite signs. Since

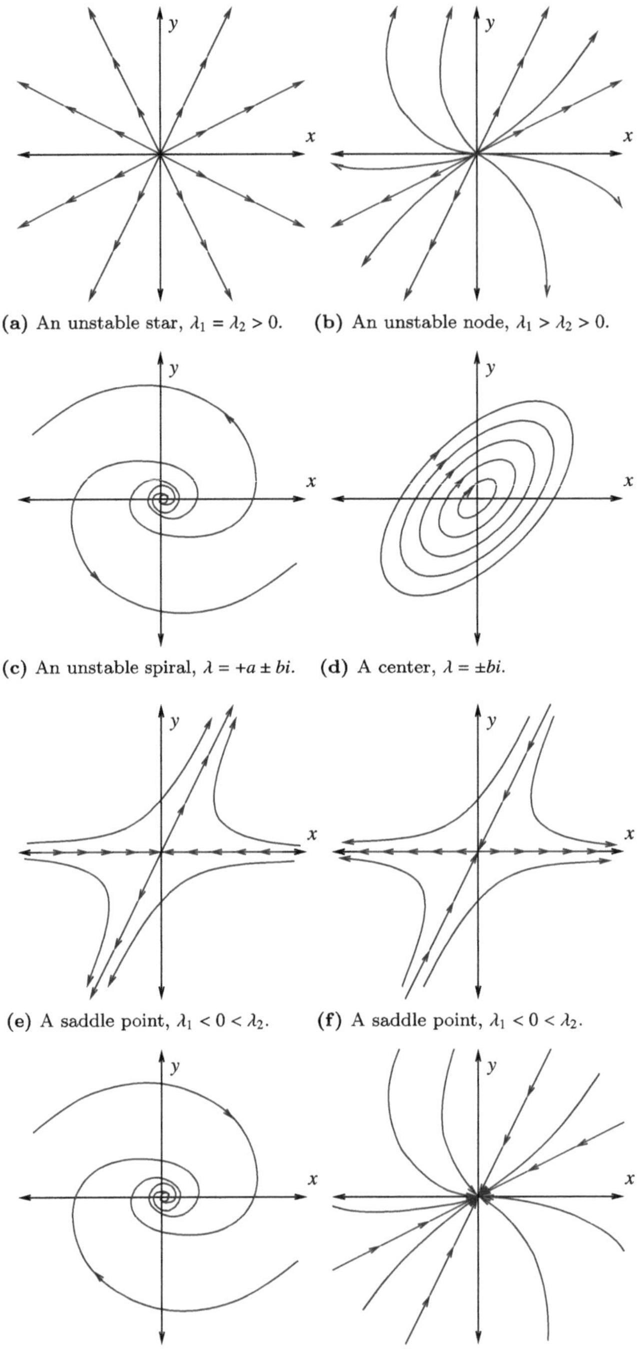

(a) An unstable star, $\lambda_1 = \lambda_2 > 0$. (b) An unstable node, $\lambda_1 > \lambda_2 > 0$.

(c) An unstable spiral, $\lambda = +a \pm bi$. (d) A center, $\lambda = \pm bi$.

(e) A saddle point, $\lambda_1 < 0 < \lambda_2$. (f) A saddle point, $\lambda_1 < 0 < \lambda_2$.

(g) A stable spiral, $\lambda = -a \pm bi$. (h) A stable node, $\lambda_1 < \lambda_2 < 0$.

Figure 2.3: Topological behavior of various linear systems.

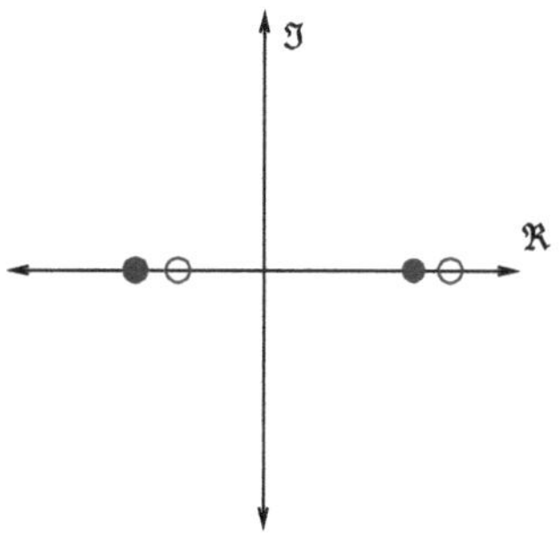

Figure 2.4: Two real eigenvalues in a complex plane shown before (solid dots) and after (dashed outline) a possible higher-order perturbation.

complex eigenvalues must occur in complex conjugate pairs, there is no way for distinct eigenvalues, separated by a finite distance on a real line, to become complex under a higher-order perturbation. Moreover, any higher-order perturbation will not be sufficient to cause either eigenvalue to change sign. Hence, a stable node (two distinct, negative eigenvalues) will remain a stable node; a saddle (two eigenvalues of opposite signs) will remain a saddle; and an unstable node (two distinct, positive eigenvalues) will remain an unstable node.

If the eigenvalues are real, nonzero, and repeated, then three possible outcomes can occur: under a higher-order perturbation, they can remain repeated, real, nonzero eigenvalues (stable or unstable star); they can become distinct, real, nonzero eigenvalues (stable or unstable node); or they can become complex conjugate pairs (stable or unstable spiral); see Figure 2.5. Either way, the sign of the real part will remain the same, meaning the stability of the fixed point will not be affected by the perturbation. Since unstable spirals, unstable nodes, and unstable stars are topologically equivalent, the perturbation will not affect the topological equivalence class of the flow.

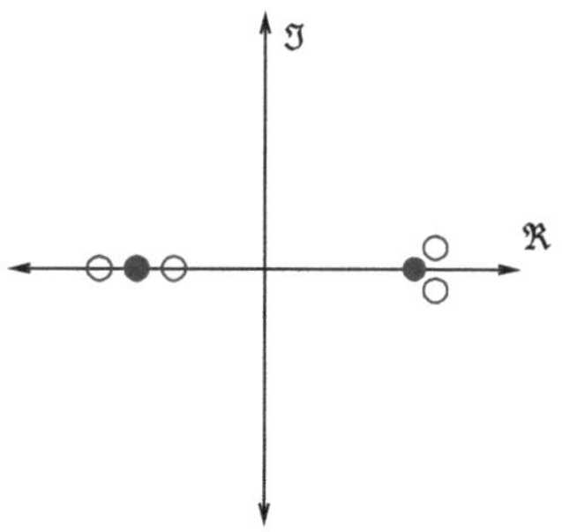

Figure 2.5: Two repeated, real eigenvalues in the complex plane shown before (solid dots) and after (dashed outline) a higher-order perturbation. Two separate cases are shown in the picture, one in the left-half plane and one in the right-half plane. Each solid dot is a *repeated*, real eigenvalue.

It is easy to imagine that if the eigenvalues are complex conjugate pairs with nonzero real parts (stable or unstable spiral, depending on the sign of the real part), they will remain complex conjugate pairs under a higher-order perturbation while retaining the sign of the real part. Hence stable spirals will remain stable spirals and unstable spirals will remain unstable spirals.

The last case to consider is that of nonhyperbolic fixed points, i. e., fixed points with an eigenvalue with zero real part. In this case, the unperturbed system is a center. Under the higher-order perturbation, the eigenvalues might remain purely imaginary, complex conjugate pairs, in which case the center will remain a center. However, the perturbation could push the eigenvalues off of the purely imaginary axis, either to the left or to the right. Since they must remain complex conjugate pairs, they must both move in the same direction (Figure 2.6). Hence, a *linear center*—a fixed point whose linearization is a center—might possibly be, in actuality, a stable or unstable spiral. Since stable spirals, unstable spirals, and centers represent distinct topological equivalence classes, the perturbed (actual) flow is not necessarily topologically equivalent to the unperturbed (linearized) flow.

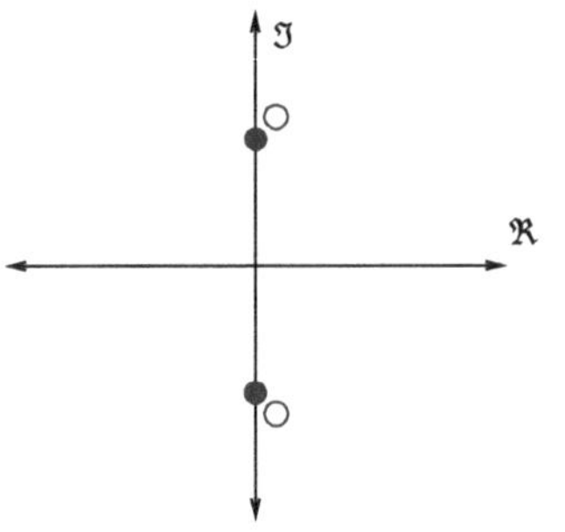

Figure 2.6: Two purely imaginary, complex conjugate pair eigenvalues in the complex plane shown before (solid dots) and after (dashed outline) a possible higher-order perturbation.

In the foregoing discussion, we cheated a bit, as nonlinear systems do not have eigenvalues except for the eigenvalues of their linearization. So asking what happens to the eigenvalues under a nonlinear perturbation is a bit of an ill-posed question. However, what we are actually doing is presupposing that by *wiggling* the eigenvalues of the linearization by a small amount, the altered linear system is of the topological class of the nonlinear flow, that is, we presupposed that the local topological class of the nonlinear picture was that of a linear picture whose eigenvalues were related to the eigenvalues of the linearized picture by only a small perturbation. And this, as it turns out, is a legitimate point of view.

Definition 2.10. If the flow of nonlinear system (2.1) in a neighborhood of a fixed point x^* is locally topologically equivalent to the flow of a center, then the fixed point x^* is referred to as a *nonlinear center*.

Stating that a fixed point x^* is a *linear center* implies that the linearized flow is that of a center. As we saw above, this could mean that the nonlinear flow is locally topologically equivalent to a center, a stable spiral, or an unstable spiral. If the *nonlinear* flow happens to be locally topologically equivalent to a center, then we call x^* a *nonlinear center*. Otherwise, we would state that its linearization is a center but that the nonlinear flow is locally topologically equivalent to a stable or unstable spiral. In the next section, we will learn of one particular method that is often employed in determining the topological equivalence class of the *nonlinear* flow about a linear center.

Exercise 2.9. Consider the nonlinear system

$$\dot{x} = y,$$
$$\dot{y} = -y + x - x^3.$$

Find the three fixed points and compute the linearization about each fixed point. Classify each fixed point in terms of its stability and type. Use these linearizations to draw a sketch of the phase plane.

Answer to Example 2.2: The equivalence classes are given by $A \sim B \sim C$, D, $E \sim F$, and $G \sim H$.

Lyapunov Functions and Stability

In this section, we present a technique that may be used to determine the stability of nonhyperbolic fixed points, the case for which linearization yields no information. The idea is to use a *Lyapunov function*, introduced by Lyapunov [192], which is a positive definite function whose material derivative along solution trajectories is monotonic or constant. For recent references, see, for example, [137], [154], [274], or [294].

Definition 2.11. Let U be an open neighborhood of an isolated fixed point x^* of the system $\dot{x} = f(x)$. A *Lyapunov function*, $V : U \to \mathbb{R}$, is a C^1 function that satisfies
(i) $V(x^*) = 0$;
(ii) $V(x) > 0$ for all $x \in U \backslash \{x^*\}$; and
(iii) $\nabla V(x) \cdot f(x) \leq 0$ for all $x \in U \backslash \{x^*\}$.

If the inequality in (iii) is a strict inequality, we refer to V as a *strict Lyapunov function*. If the inequality in (iii) is an equality, then function V is referred to as an *integral of motion*.

If $x(t)$ is a solution trajectory of system (2.1) that passes through open set U, then

$$\frac{DV}{Dt} = \nabla V(x(t)) \cdot f(x(t))$$

represents the time derivative of the function $V(x(t))$, i. e., it represents the rate of change of the values of V as viewed from the particle traveling along the given trajectory.

Theorem 2.6. *Suppose x^* is a fixed point of the nonlinear system $\dot{x} = f(x)$. If there exists a Lyapunov function $V : U \to \mathbb{R}$ in a neighborhood U containing x^*, then the fixed point x^* is Lyapunov stable. Moreover, if V is a strict Lyapunov function, then x^* is asymptotically stable. If V is an integral of motion, then x^* is a nonlinear center.*

Proof. Let $\varepsilon > 0$ be sufficiently small so that the closed ball

$$\overline{B}_\varepsilon(x^*) = \{x \in \mathbb{R}^n : |x - x^*| \le \varepsilon\}$$

is contained within the open neighborhood U. Since the boundary $\partial\overline{B}_\varepsilon(x^*)$ is compact, we can define the number

$$m = \min_{x \in \partial\overline{B}_\varepsilon(x^*)} V(x).$$

Finally, let U' be the set

$$U' = \{x \in \overline{B}_\varepsilon(x^*) : V(x) \le m\};$$

see Figure 2.7. Consider an arbitrary $x_0 \in U'$ and the trajectory $x(t) = \varphi(t; x_0)$ for $t \ge 0$. Since the material derivative of V along the trajectory is

$$\frac{DV}{Dt} = \nabla V \cdot f(x) \le 0,$$

we have $V(x(t)) \le V(x(0)) \le m$ for all $t \ge 0$. Hence $x(t) \in U'$ for all $t \ge 0$. We conclude that x^* is Lyapunov stable.

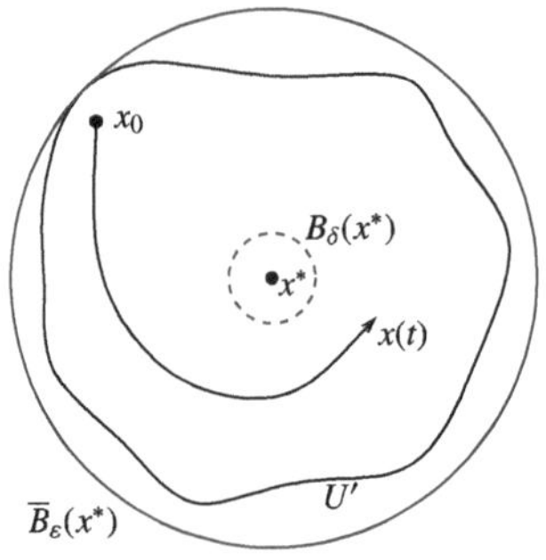

Figure 2.7: Proof of Lyapunov's theorem; the figure shows the closed ball $\overline{B}_\varepsilon(x^*) \subset U$ and its interior.

Next, suppose that V is a *strict* Lyapunov function. We aim to show that x^* is *asymptotically stable*. We will proceed by contradiction. Suppose there exists an initial condition $x_0 \in U'$ such that $x(t) = \varphi(t; x_0)$ does *not* approach x^* asymptotically as $t \to \infty$. Then there exists a $\delta > 0$ such that

$$\delta < \min_{x \in \partial U'} |x - x^*|$$

and such that the trajectory $x(t)$, for $t \ge 0$, never enters the open ball

$$B_\delta(x^*) = \{x \in \mathbb{R}^n : |x - x^*| < \delta\},$$

as depicted in Figure 2.7. Since U' is a closed set, the set $W = U' \backslash B_\delta(x^*)$ is also closed and, since it is bounded, compact. We may therefore define the constants

$$K = -\max_{x \in W} \nabla V \cdot f(x),$$
$$K' = \min_{x \in W} V(x).$$

Since $\nabla V \cdot f(x) < 0$ and $V(x) > 0$ for all $x \in W$, we conclude that $K, K' > 0$. The material derivative satisfies the inequality

$$\frac{DV}{Dt} \leq -K < 0$$

for all $t \geq 0$. From this we see

$$V(x(t)) \leq V(x_0) - Kt,$$

and hence for

$$t > \frac{V(x_0) - K'}{K}$$

it follows that $V(x(t)) < K'$, implying that the trajectory has entered the region $B_\delta(x^*)$, a contradiction. Therefore, $x(t)$ must approach x^* asymptotically as $t \to \infty$ for all initial conditions $x_0 \in U'$, allowing us to conclude that x^* is asymptotically stable.

Finally, suppose that the Lyapunov function V is an integral of motion for the system. Solution trajectories are therefore confined to the level sets of V. Since $V \in C^1$ and since $V > 0$ for all $x \in U \backslash \{x^*\}$ and $V(x^*) = 0$, it follows that the function V may be described locally by its second-order Taylor polynomial as

$$V(x) = \frac{1}{2}(x - x_0)^T V_{xx}(x - x_0) + O(|x - x_0|^3).$$

(Note that the Hessian matrix V_{xx} is positive definite.) Hence, in a neighborhood of x^*, those level sets are, to the second order, merely ellipsoids centered at x^*. Since trajectories are confined to these level sets, x^* must be a nonlinear center. □

Example 2.3. Consider the nonlinear system

$$\dot{x} = xy^4 - y,$$
$$\dot{y} = -x^2 y + x,$$

which has equilibrium points at $(-1, 1)$, $(0, 0)$, and $(1, 1)$. The Jacobian matrix for this system is

$$Df(x, y) = \begin{bmatrix} y^4 & 4xy^3 - 1 \\ -2xy + 1 & -x^2 \end{bmatrix}.$$

Evaluating the Jacobian at each fixed point yields

$$Df(0,0) = \begin{bmatrix} 0 & -1 \\ 1 & 0 \end{bmatrix} \quad \text{and} \quad Df(-1,-1) = Df(1,1) = \begin{bmatrix} 1 & 3 \\ -1 & -1 \end{bmatrix}.$$

The eigenvalues of the linearization at the origin are $\lambda = \pm i$, and the eigenvalues at the fixed points $(-1,-1)$ and $(1,1)$ are $\lambda = \pm\sqrt{2}i$. For all three cases, the linearization is that of a center. Thus all three fixed points of this system are *nonhyperbolic* fixed points, so we cannot determine the topological type based on the linearization. We can, however, use Theorem 2.6 to determine the stability of the origin. Consider the function

$$V(x,y) = \frac{1}{2}(x^2 + y^2),$$

whose graph and contour plot are plotted in Figure 2.8(a) and Figure 2.8(b), respectively. As it turns out, $V(x,y)$ is a Lyapunov function for the origin in the open neighborhood $B_1(0) = \{x \in \mathbb{R}^2 : |x| < 1\}$. The level sets of this function are concentric circles centered at the origin. We compute the material derivative of V along the solution flow as

$$\frac{DV}{Dt} = \nabla V \cdot f = x^2y^4 - x^2y^2 = x^2y^2(y^2 - 1).$$

We conclude that $DV/Dt \leq 0$ along the portion of any trajectory that lies within the infinite horizontal strip $|y| \leq 1$. The largest level contour of V that lies entirely within this strip is the boundary of the open ball $U = B_1(0)$. Therefore, V is a Lyapunov function for the origin in the open neighborhood U, and thus we conclude that the origin is Lyapunov stable.

One can, however, do a little bit better. Notice that, but for the x- and y-axes, V would be a strict Lyapunov function. The only way for the origin to fail to be asymptotically stable is if solution trajectories manage to get "caught up" on these axes. However, for points on the y-axis we have $\dot{x} = -y$, $\dot{y} = 0$, and for points on the x-axis we have

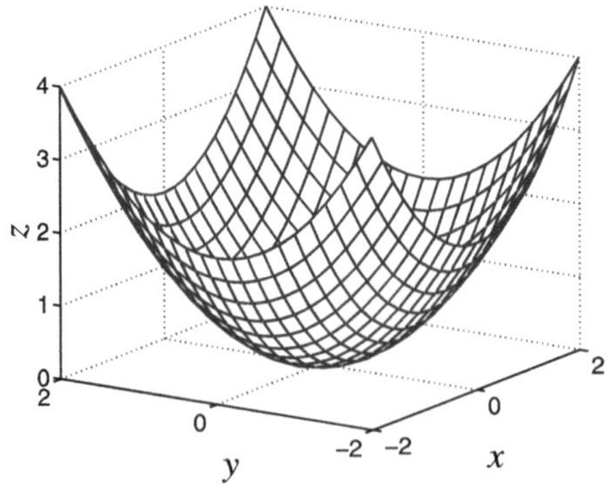

(a) Graph of $V(x,y)$.

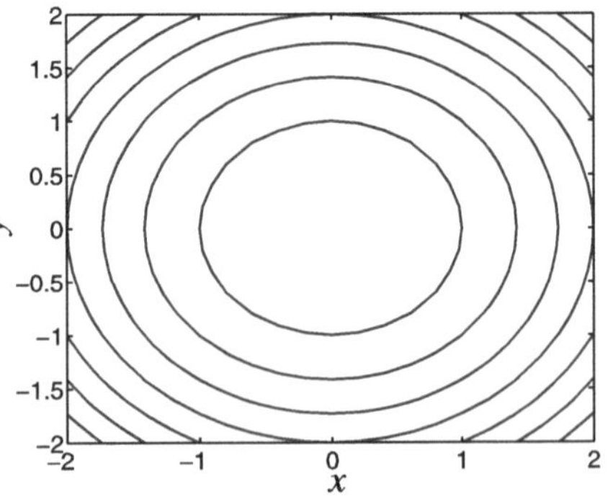

(b) Contour plot of $V(x,y)$.

Figure 2.8: The function $V = \frac{1}{2}(x^2 + y^2)$ plotted as a graph (a) and as a contour plot (b).

$\dot{x} = 0, \dot{y} = x$. Thus trajectories flow past these axes unabated. The vanishing of the material derivative of V is accounted for by the fact that the trajectory lies tangent to the contours of V during the instants at which the trajectories pass the x- and y-axes. We might therefore reasonably conjecture that the origin is indeed asymptotically stable, which is confirmed by numerical simulation. Figure 2.9 shows the phase portrait of our system and a sample trajectory in the phase plane. The plot was created with the aid of the free software `pplane8.m`.[1] The plotted trajectory is not shown in its entirety; it spirals very tightly away from the fixed point at $(-1, -1)$ and then turns and spirals into the fixed point at the origin. Such a trajectory is an example of a *heteroclinic orbit*. (We will discuss these in §3.1.) □

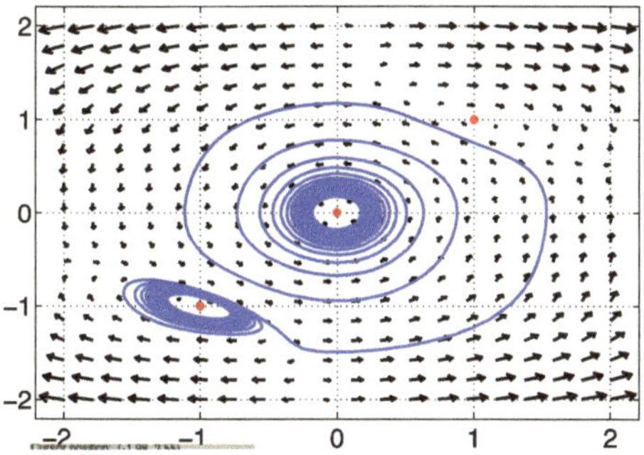

Figure 2.9: Phase portrait: portion of a heteroclinic orbit from $(-1, -1)$ to $(0, 0)$.

Exercise 2.10. Show that the same Lyapunov function used in Example 2.3 may be used to show that the origin of the nonlinear system

$$\dot{x} = xy^4,$$
$$\dot{y} = -x^2 y$$

is Lyapunov stable. Plot the phase portrait and several sample trajectories using `pplane8.m` and conclude that the origin is Lyapunov stable but *not* asymptotically stable. How is this different from Example 2.3?

Exercise 2.11. Show that if $\rho < 1$, the function

$$V(x, y, z) = \frac{1}{2}\left(\frac{x^2}{\sigma} + y^2 + z^2\right)$$

is a strict Lyapunov function for the fixed point at the origin of the Lorenz system (2.5)–(2.7) and conclude that the origin is asymptotically stable. *Hint*: complete the square to show that

$$\frac{DV}{Dt} = -\left(x - \frac{\rho + 1}{2}y\right)^2 - \left[1 - \left(\frac{\rho + 1}{2}\right)^2\right]y^2 - \beta z^2.$$

1 Available at http://math.rice.edu/~dfield/

Exercise 2.12. Consider the system

$$\dot{x} = y,$$
$$\dot{y} = -\sin x.$$

Compute the linearization about the fixed points $z = (0,0), (\pm\pi, 0)$. Show that the function

$$V = \frac{y^2}{2} + (1 - \cos x)$$

is a Lyapunov function on some open neighborhood of $(0,0)$. What type of fixed point is in the linearization of $(0,0)$? What type of *nonlinear* fixed point is in the nonlinear equations? Draw a sketch of the phase portrait.

2.6 Dynamical Systems in Mechanics

In this paragraph, we consider the broader context of *where* systems of first-order ordinary differential equations come from. It is not surprising that many such systems arise from the discipline of physics. All the laws of physics are fundamentally differential in nature, i. e., they relate infinitesimal, *local* changes of one variable to others. (Un)fortunately, they commonly arise as systems of *second*-order ordinary differential equations, not first-order ones. To wit, *Newton's second law of motion* relates the total force F acting on a particle with its mass m and position x by the relation

$$F = \frac{d(m\dot{x})}{dt}, \tag{2.27}$$

that is, the net force is equal to the time rate of change of the particle's momentum $m\dot{x}$. It may invariably be cast in the form

$$\ddot{x} = f(x, \dot{x}, t). \tag{2.28}$$

The functional dependence on $\dot{x}$ on the right-hand side of this equation could arise due to a temporal variation in the particle's mass (e. g., a leaking cart or a rocket ship) or due to velocity-dependent forces. Velocity-dependent forces may arise due to friction, air resistance, magnetism, or inertial forces, such as the Coriolis force, that arise when the coordinate system is noninertial. Despite the ubiquity of second-order systems in nature, they can always be recast in an equivalent set of first-order differential equations. To do this, one introduces a secondary variable, u. The pair $(x, u) \in \mathbb{R}^{2n}$ describes the position in "phase space," or coordinate-velocity space. The *first*-order system associated with (2.28) is then given by

$$\frac{d}{dt}\begin{bmatrix} x \\ u \end{bmatrix} = \begin{bmatrix} u \\ f(x, u, t) \end{bmatrix}.$$

Alas! This system (in the common case of time-*independent* forces) is precisely equivalent to (2.1). Hence, all of the theory on systems of first-order equations follows through directly to analyses regarding the phase space dynamics of many common mechanical or physical systems.

Example 2.4 (The Pendulum). A hallmark example of a mechanical system is the simple pendulum, as depicted in Figure 2.10. The pendulum has mass m and is attached to a fixed point by a massless rod of length l. The pendulum swings under the influence of a uniform gravitational field with acceleration g. The force balance is shown in Figure 2.10 and is comprised of a downward gravitational force, $F_g = mg$, and a force, T, due to tension in the rod. The downward gravitational force, F_g, may be resolved into two separate components: one, $(F_g)_n$, in the direction of the massless rod and the other, $(F_g)_t$, tangential to the massless rod. We may describe the position of the rod by an angle θ, so that $x = (l \sin\theta, -l\cos\theta)$ for a coordinate system centered at the fixed point. Since the massless rod is a fixed length, there can be no motion of the pendulum bob in the direction of the rod, so that $T + (F_g)_n = 0$.

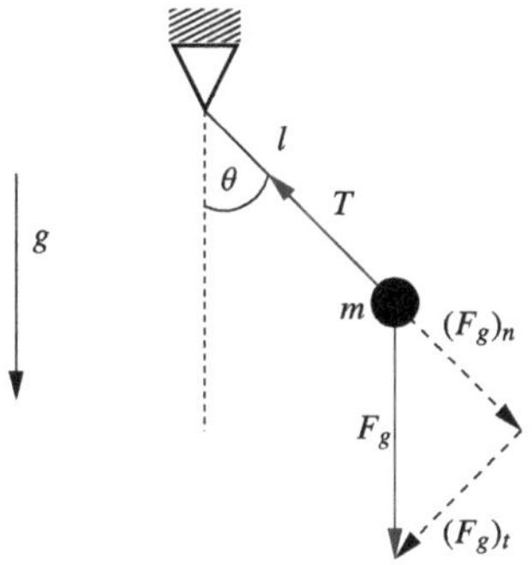

Figure 2.10: Diagram of a simple pendulum.

Exercise 2.13. Defining unit vectors $\hat{n} = (\sin\theta, -\cos\theta)$ and $\hat{t} = (-\cos\theta, -\sin\theta)$, respectively, shows that

$$\begin{aligned} F_g &= mg\cos\theta\hat{n} + mg\sin\theta\hat{t}, \\ \ddot{x} &= -l\dot{\theta}^2\hat{n} - l\ddot{\theta}\hat{t}. \end{aligned}$$

Use this result to conclude that $\theta(t)$ satisfies the second-order differential equation

$$\ddot{\theta} + \frac{g}{l}\sin\theta = 0. \tag{2.29}$$

Express the tension in the rod as a function of θ and $\dot{\theta}$.

Equation (2.29) is known as the *pendulum equation*. By defining a new variable, $y = \dot{\theta}$, this second-order equation is precisely equivalent to the first-order system studied in Exercise 2.12. We will return to this example in §3.1 and again in Chapter 7. □

Conserved quantities, such as energy and momentum, play a key role in the analysis of many mechanical systems. Such quantities for dynamical systems in general are referred to as *integrals of motion*.

Definition 2.12. A function $J : \mathbb{R} \times \mathbb{R}^n \to \mathbb{R}$ is called an *integral of motion*, or *first integral*, of the first-order system (2.1) if it is a conserved quantity along individual trajectories, i. e., if $J(t, \varphi(t; x_0)) = J(0, x_0)$ for all $x_0 \in \mathbb{R}^n$ and all $t \in \mathbb{R}$.

An immediate consequence of the foregoing definition is that a function $J : \mathbb{R} \times \mathbb{R}^n \to \mathbb{R}$ is an integral of motion if and only if its material derivative along the flow vanishes, i. e.,

$$\frac{DJ}{Dt} = \nabla J \cdot \dot{x} + \frac{\partial J}{\partial t} = \nabla J \cdot f(x) + \frac{\partial J}{\partial t} = 0.$$

We will see various examples of integrals of motion during our discussion of applications in the final two sections of this chapter.

2.7 Application: Elementary Astrodynamics

The first application of the concepts discussed in this chapter is focused on the motion of two gravitationally attracting bodies in space. Let us suppose that the first body, P_1, has mass m_1 and position vector $\mathbf{r}_1$ and that the second body, P_2, has mass m_2 and position vector $\mathbf{r}_2$. We further define the *relative position vector* $\mathbf{r}_{12} = \mathbf{r}_2 - \mathbf{r}_1$ between the two bodies as depicted in Figure 2.11; by convention, $\mathbf{r}_{21} = -\mathbf{r}_{12}$.

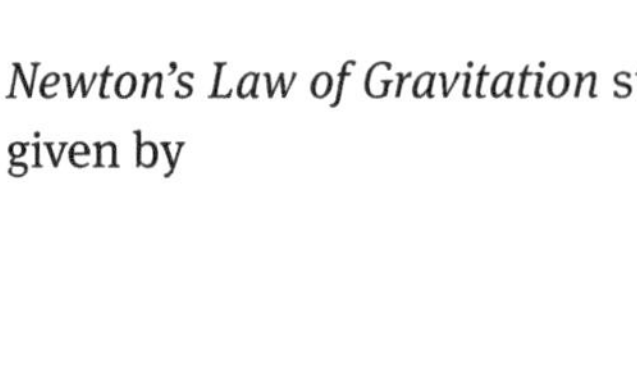

Figure 2.11: Two-body problem.

Newton's Law of Gravitation states that the force $\mathbf{F}_{12}$ that body P_1 exerts on body P_2 is given by

$$\mathbf{F}_{12} = \frac{-Gm_1m_2}{|\mathbf{r}_{12}|^3}\mathbf{r}_{12},$$

where $G \approx 6.672 \times 10^{-11}\,\mathrm{m}^3\,\mathrm{kg}^{-1}\,\mathrm{s}^{-2}$ is *Newton's gravitational constant*. Similarly, the force that body P_2 exerts on body P_1 is an equal and opposite force $\mathbf{F}_{21} = -\mathbf{F}_{12}$. Given each body's initial position and velocity, determining $\mathbf{r}_1(t)$ and $\mathbf{r}_2(t)$ is known as the *two-body problem* in astrodynamics.

Exercise 2.14. Prove the vector identity

$$\frac{\partial}{\partial \mathbf{r}}\left(\frac{1}{|\mathbf{r}|}\right) = -\frac{\mathbf{r}}{|\mathbf{r}|^3},$$

where $\mathbf{r} = (x, y, z)$ and where $\partial/\partial\mathbf{r}$ represents the gradient operator.

Exercise 2.15. Defining a *mutual force potential*

$$U_{12} = U_{21} = \frac{Gm_1m_2}{|\mathbf{r}_{12}|},$$

show that the equations of motion for the two-body problem may be rewritten as the following coupled first-order nonlinear system in $\mathbb{R}^{12}$:

$$\dot{\mathbf{r}}_1 = \mathbf{v}_1, \qquad m_1\dot{\mathbf{v}}_1 = \frac{\partial U_{21}}{\partial \mathbf{r}_1}, \tag{2.30}$$

$$\dot{\mathbf{r}}_2 = \mathbf{v}_2, \qquad m_2\dot{\mathbf{v}}_2 = \frac{\partial U_{12}}{\partial \mathbf{r}_2}. \tag{2.31}$$

You may use the result of Exercise 2.14.

The nonlinear dynamical system (2.30)–(2.31) on $\mathbb{R}^{12}$ can be solved by determining a set of 12 algebraically independent integrals of motion. We begin by defining the center of mass and the total linear momentum by the relations

$$\mathbf{R} = \frac{m_1\mathbf{r}_1 + m_2\mathbf{r}_2}{m_1 + m_2} \quad \text{and} \quad \mathbf{P} = m_1\dot{\mathbf{r}}_1 + m_2\dot{\mathbf{r}}_2,$$

respectively. A straightforward calculation shows

$$\dot{\mathbf{P}} = m_1\ddot{\mathbf{r}}_1 + m_2\ddot{\mathbf{r}}_2 = \mathbf{F}_{21} + \mathbf{F}_{12} = 0;$$

hence the vector-valued function $\mathbf{P}(t)$ constitutes three integrals of motion. Moreover, by integrating

$$\int_{\mathbf{R}_0}^{\mathbf{R}(t)} d\mathbf{R} = \int_{t_0}^{t} \frac{\mathbf{P}}{m_1 + m_2}\, dt,$$

one may solve for the center of mass as a function of time, i. e.,

$$\mathbf{R}(t) = \mathbf{R}_0 + \frac{\mathbf{P}}{m_1 + m_2}(t - t_0).$$

The center of mass $\mathbf{R}_0$ at the epoch t_0 constitutes three additional integrals of motion. Our originally 12-D system is therefore reduced to the problem of determining the *relative* motion of the bodies about the center of mass. By defining the relative position vector $\mathbf{r} = \mathbf{r}_{12} = \mathbf{r}_2 - \mathbf{r}_1$ and the *reduced force potential*

$$U = \frac{\mu}{|\mathbf{r}|},$$

where $\mu = G(m_1 + m_2)$ is the combined *gravitational parameter* of the two bodies, we may consider the following reduced system on $\mathbf{R}^6$:

$$\dot{\mathbf{r}} = \mathbf{v}, \quad \dot{\mathbf{v}} = \frac{\partial U}{\partial \mathbf{r}}. \tag{2.32}$$

Exercise 2.16. Show that the total mechanical energy and total angular momentum,

$$E = \frac{1}{2}\mathbf{v} \cdot \mathbf{v} - U(\mathbf{r}) \quad \text{and} \quad \mathbf{H} = \mathbf{r} \times \mathbf{v},$$

are integrals of motion for the reduced system (2.32).

Conservation of angular momentum implies that the relative position vector $\mathbf{r}(t)$ is constrained to lie on the fixed plane in space, known as the *orbital plane*, that is perpendicular to vector $\mathbf{H}$.

Exercise 2.17. This exercise is concerned with a quantity known as the *Laplace–Runge–Lenz vector* (also known as the *eccentricity vector*).

(a) Show that the Laplace–Runge–Lenz vector

$$\mathbf{B} = \mathbf{v} \times \mathbf{H} - \frac{\mu \mathbf{r}}{|\mathbf{r}|}$$

constitutes an integral of motion for the reduced system (2.32). However, as you will subsequently demonstrate, this vector only adds *one* new algebraically independent integral of motion to the system—not three.

(b) Show that vector $\mathbf{B}$ must lie in the orbital plane.

(c) Show that the magnitude of $\mathbf{B}$ may be expressed in terms of the previous integrals of motion, in particular, $B^2 = \mu^2 + 2H^2E$.

The new piece of information that eccentricity vector $\mathbf{B}$ adds to the system is its *orientation* within the orbital plane; that is, it defines a preferred, invariant direction, in the orbital plane, that is somehow characteristic of the ensuing motion.

Exercise 2.18. Defining angle f to be the angle between vectors $\mathbf{r}$ and $\mathbf{B}$ and computing the dot product $\mathbf{r} \cdot \mathbf{B}$, show that

$$r = \frac{H^2/\mu}{1 + e\cos f},$$

where $e = B/\mu$ is the *eccentricity* of the orbit. Angle f is referred to as the *true anomaly*.

From the previous exercise we see that the relative motion is that of a conic section: the orbit is circular, elliptic, parabolic, or hyperbolic depending on whether $e = 0$, $e \in (0, 1)$, $e = 1$, or $e \in (1, \infty)$, respectively.

A more general problem is the *many-body problem*: determine the resulting motion of a collection of N mutually gravitating bodies $P_1, \ldots, P_N$ with positions $\mathbf{r}_1, \ldots, \mathbf{r}_N$

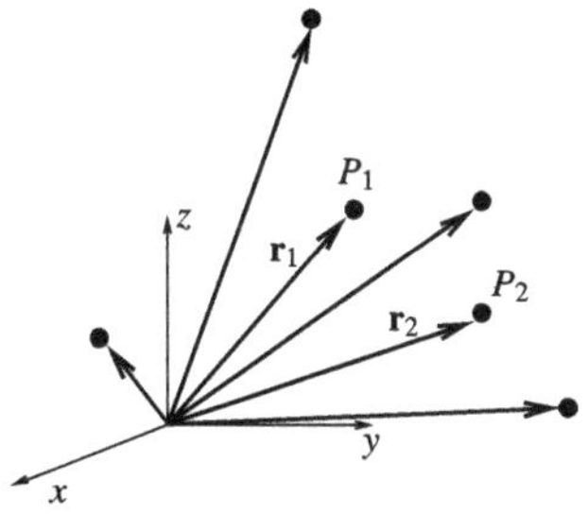

Figure 2.12: Many-body problem.

and masses $m_1, \dots, m_N$, as shown in Figure 2.12. For further study, see, for example, [8], [36], and [256]. For a treatise on astrodynamics problems in strongly perturbed environments (satellite motion about irregularly shaped asteroids, for example), see [261], and for a discussion on applications of Hamiltonian mechanics and perturbation theory to astrophysical systems, see [92].

2.8 Application: Planar Circular Restricted Three-Body Problem

Continuing our theme of astrodynamics, we now turn to a special case of a three-body problem: the *planar, circular, restricted three-body problem* (PCR3BP). This problem is further discussed in, for example, [11], [163], and [256]. This system is a simplification of the general three-body problem in astrodynamics, in which three point masses form a self-gravitating triad, which takes advantage of the following simplifying assumptions:

1. Two of the bodies, referred to as the *primaries*, have much larger masses than the third, so that the third body does not affect the orbit of the two primaries.
2. The two primaries are in a circular orbit about their center of mass.
3. Length is normalized so that the distance between the two primaries is one nondimensionalized unit of length.
4. Time is normalized so that the period of the circular orbit is 2π.
5. The equations are written with respect to a noninertial, rotating reference frame with respect to which the primaries always lie on the x-axis and such that the center of mass of the primaries is the origin.

These assumptions would be valid, for instance, for a spacecraft in the Earth–Moon system or for asteroids in the Sun–Jupiter system, etc.

Let m_1, m_2 be the masses of the two primaries, with $m_1 > m_2$. Next, define the mass parameter

$$\mu = \frac{m_2}{m_1 + m_2},$$

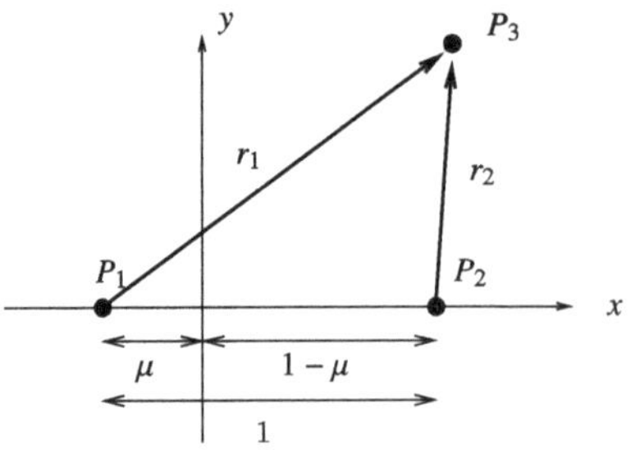

Figure 2.13: Particles in the planar, circular, restricted three-body problem.

so that particle P_1, with mass m_1, is located at $(-\mu, 0)$ and particle P_2, with mass m_2, is located at $(1-\mu, 0)$, as depicted in Figure 2.13. The mass of particle P_3, m_3, is taken to be small enough so as not to influence the motion of the two primaries; hence, its precise value is irrelevant when deriving the equations of motion for the third particle. Now let

$$r_1 = \sqrt{(x+\mu)^2 + y^2},$$
$$r_2 = \sqrt{(x-1+\mu)^2 + y^2}$$

be the distance from P_3 to P_1 and the distance from P_3 to P_2, respectively. Then define the function

$$U(x,y) = \frac{1}{2}(x^2 + y^2) + \frac{1-\mu}{r_1} + \frac{\mu}{r_2}. \tag{2.33}$$

This function is *not* the potential energy of the system, as the term $\frac{1}{2}(x^2 + y^2)$ arises to take into account the centripetal acceleration due to the rotation of the reference frame. However, we will see that this function behaves analogously to a potential energy and that we may nevertheless think of $U(x,y)$ as a potential-energy-like function. The equations of motion for this system are given as the following second-order system:

$$\ddot{x} - 2\dot{y} = \frac{\partial U}{\partial x}, \tag{2.34}$$
$$\ddot{y} + 2\dot{x} = \frac{\partial U}{\partial y}. \tag{2.35}$$

In both cases, the force (i. e., acceleration) contains a potential term and a term $(2\dot{y}, -2\dot{x})^T$ that depends on the velocity of the particle. This velocity-dependent term corresponds to the *Coriolis forces* acting on the particle, which arise due to the fact that we are expressing the equations of motion relative to a noninertial frame; see, for example, [102] and [106] for more details on the kinematics of motion expressed relative to noninertial frames.

As we discussed in §2.6, one may convert an n-dimensional, second-order mechanical system into a $2n$-dimensional first-order system by introducing a set of velocity variables. Defining $u = \dot{x}$ and $v = \dot{y}$, we may therefore rewrite the system (2.34)–(2.35) as the first-order system

$$\dot{x} = u, \tag{2.36}$$

$$\dot{y} = v, \tag{2.37}$$

$$\dot{u} = 2v + \frac{\partial U}{\partial x}, \tag{2.38}$$

$$\dot{v} = -2u + \frac{\partial U}{\partial y}. \tag{2.39}$$

Proposition 2.4. *The function*

$$J(x, y, u, v) = \frac{1}{2}(u^2 + v^2) - U(x, y), \tag{2.40}$$

known as the Jacobi integral, where $U(x, y)$ *is defined in* (2.33), *is an integral of motion for the system* (2.36)–(2.39).

Proof. Computing the material derivative of (2.40) along the vector field defined in (2.36)–(2.39), we obtain

$$\begin{aligned}\frac{DJ}{Dt} &= \frac{\partial J}{\partial x}\dot{x} + \frac{\partial J}{\partial y}\dot{y} + \frac{\partial J}{\partial u}\dot{u} + \frac{\partial J}{\partial v}\dot{v} \\ &= -u\frac{\partial U}{\partial x} - v\frac{\partial U}{\partial y} + 2vu + u\frac{\partial U}{\partial x} - 2uv - v\frac{\partial U}{\partial y} = 0.\end{aligned}$$ □

Since $U(x, y)$ is not the potential energy, the Jacobi integral does not represent the total energy. However, it is a conserved quantity that is similar to the total mechanical energy in the sense that it is given by the kinetic energy plus a function of position whose gradient contributes to the particle's acceleration.

The value the Jacobi integral takes along any particular orbit is referred to as the *Jacobi constant*, C. Since J is an integral of motion, the ensuing motion is restricted to the level contour $J(x, y, u, v) = C$ for the entire length of the trajectory. The Jacobi integral is of fundamental importance because, for any given value of the Jacobi constant, the condition that the kinetic energy of the particle must be nonnegative works in combination with the Jacobi integral in dividing the orbital x-y plane into admissible and forbidden regions. To see how this comes about, set Jacobi integral (2.40) equal to C and rearrange to form the inequality

$$U(x, y) + C = \frac{1}{2}(u^2 + v^2) \geq 0.$$

Thus, for a given value of Jacobi constant C, it is only physically possible to find particle P_3 in the region

$$\mathcal{A}_C = \{(x, y) \in \mathbb{R}^2 : U(x, y) + C \geq 0\}. \tag{2.41}$$

The boundary of this region, $\partial\mathcal{A}_C$, is referred to as the *zero-velocity surface* because, for the given value of the Jacobi constant C, this boundary is precisely the set of points for which the particle has zero velocity at the instant its path comes into contact with the surface. Take, for example, the gravitational field of the Earth. For a given value of energy, the zero-velocity surface is the surface at which a projectile with this amount of total mechanical energy ceases its upward ascent, pauses at rest for an instant, and begins to fall back toward Earth. Moreover, it is impossible to go beyond that surface with only the set amount of total energy. Similarly, in the restricted three-body problem, the zero-velocity surface, for a given value of the Jacobi constant, divides the plane into *admissible* and *forbidden* regions.

The forbidden regions for various values of the Jacobi constant C are plotted in Figure 2.14. For these plots, the mass parameter was taken to be $\mu = 0.15$, and the position of the two primaries (at $(-\mu, 0)$ and $(1-\mu, 0)$) is denoted by a small black x. For concreteness, let us suppose that we are in the context of the Earth–Moon system, so that we will think of the first primary P_1 as the Earth and the second primary

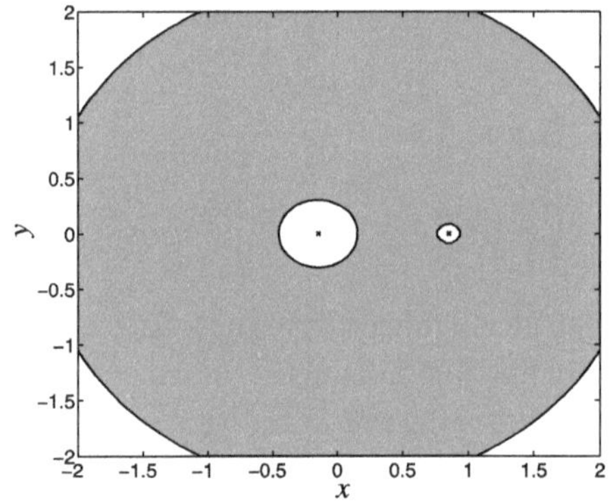

(a) $C = -3$.

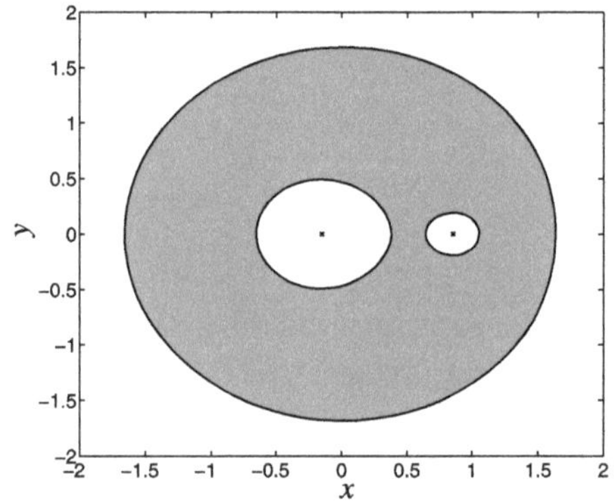

(b) $C = -2$.

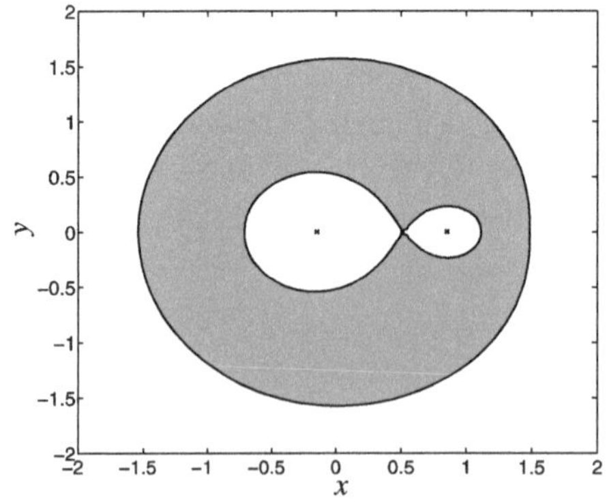

(c) $C = -1.86$.

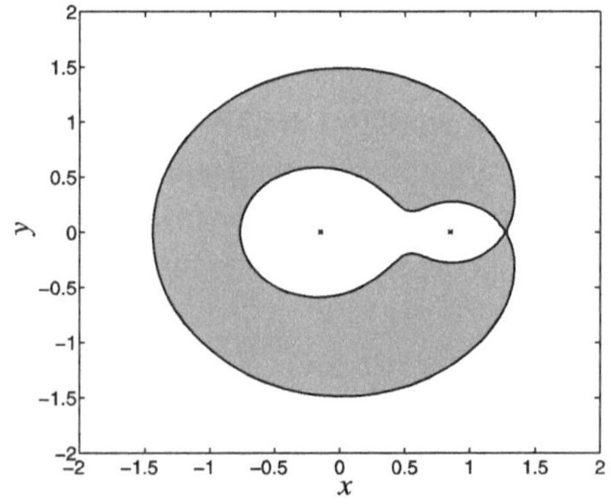

(d) $C = -1.763$.

Figure 2.14: Forbidden region for various values of Jacobi constant C.

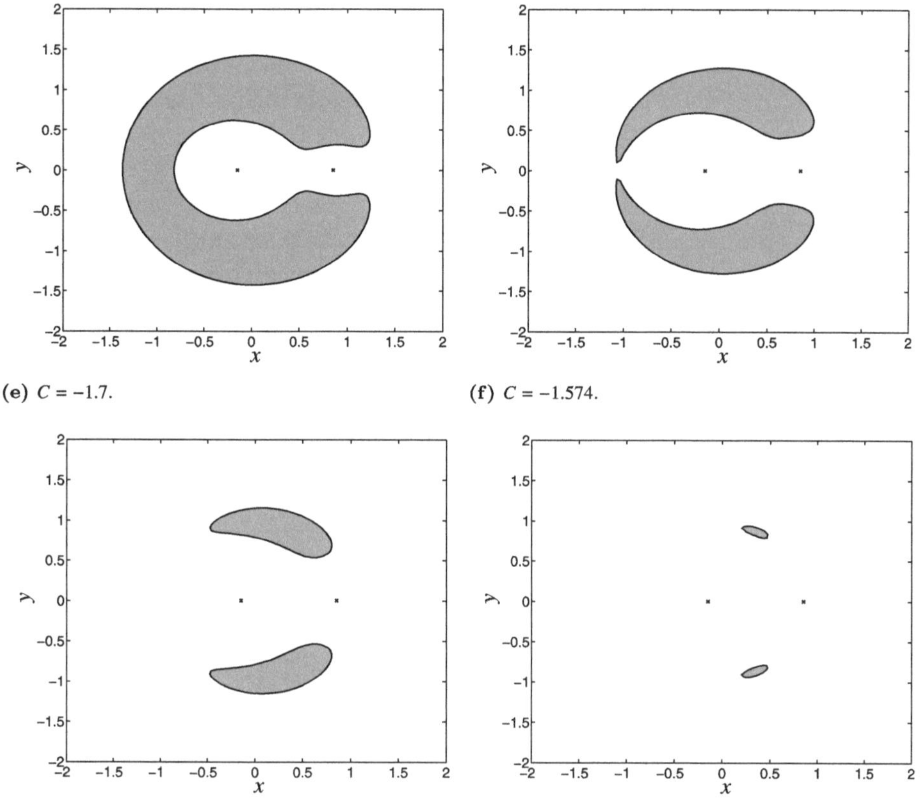

(e) $C = -1.7$.

(f) $C = -1.574$.

(g) $C = -1.5$.

(h) $C = -1.44$.

Figure 2.14: (continued).

P_2 as the Moon. For large, negative values of C, such as $C = -3$ or $C = -2$, the third particle does not have enough energy to escape from the vicinity of either primary, as shown in Figures 2.14(a) and 2.14(b). Hence, there is a circular-appearing "pocket" around each primary in which motion of the third particle is allowed. Interestingly, the pocket outside the Earth, for a fixed value of C, is larger than the pocket outside of the Moon. This seeming contradiction with intuition is again due to the fact that the Jacobi integral is *not* the total energy of the particle but the total energy plus the $\frac{1}{2}(x^2 + y^2)$ term, which one may identify as the "potential" for the centripetal acceleration. Hence, for the same value of C, the actual mechanical energy of the particle near the Earth is greater than the energy of its counterpart in the vicinity of the Moon, which accounts for the larger accessible space about the environment of the Earth. In addition, for these large, negative values of C, motion is allowed outside a very large bubble that contains both the Earth and the Moon. So it is also allowable that a comet, for example, with this value of the Jacobi constant, passes by the Earth–Moon system as long as it does not enter the forbidden region.

Slowly increasing the Jacobi constant, we eventually arrive at the value $C = -1.86$, at which the two previously separate admissible regions outside the Earth and Moon coalesce into one (Figure 2.14(c)). Let us call this point, at which one sees a cusp in the forbidden region, L_1. Point L_1 is plotted in Figure 2.15. We will see that the coalescence of the Earth's and Moon's admissible regions will forever brand the dynamics with its mark because for every value of the Jacobi constant $C \geq -1.86$, there is a relative equilibrium point precisely at the location of this coalescence, L_1. Point L_1 is referred to as a *libration point* or *Lagrange point*. Point L_1 is a *relative equilibrium point* since, if you place a particle at this point with zero velocity relative to our rotating frame, its configuration relative to the primaries will remain constant with time. In other words, as the two primaries revolve around their center of mass (as viewed in an inertial frame), the third particle situated at L_1 will remain collinear with the primaries for all time, thereby maintaining its *relative* configuration.

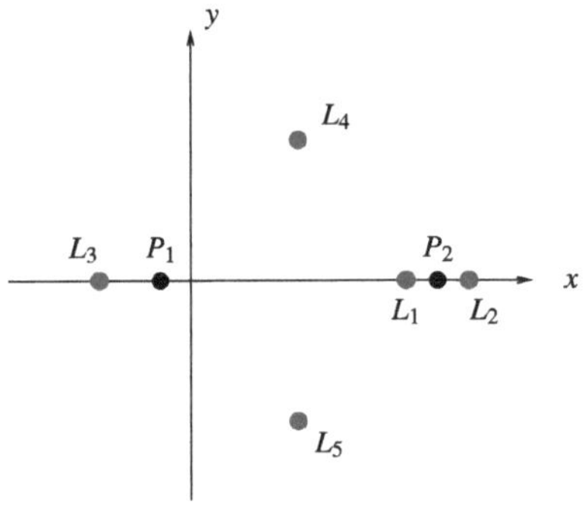

Figure 2.15: Libration points in the restricted three-body problem.

By further increasing the Jacobi constant, one encounters similar changes of topology in the forbidden region near the values $C = -1.763$, $C = -1.574$, and $C = -1.44$, which leave behind the libration points L_2, L_3, L_4, and L_5; see Figures 2.14(d), 2.14(f), and 2.14(h), respectively, and Figures 2.14(e) and 2.14(g) for intermediary values of C. Each of these five libration points are relative equilibrium points of the restricted three-body problem; their positions are plotted relative to the two primaries in Figure 2.15. To show that these five locations are actually equilibrium points, simply note that each one represents a critical point of the function $f(x,y) = U(x,y) + C$, which defines the boundary of the admissible region, (2.41), and that the equilibrium points for the system (2.36)–(2.39) are precisely the points at which $\nabla U = 0$, $u = 0$, and $v = 0$.

Location of Libration Points

As was mentioned previously, the libration points of the restricted three-body problem correspond to critical points of the potential function $U(x,y)$ and to the relative equilibrium points for the system. Differentiating $U(x,y)$, (2.33), with respect to x and y, we obtain

$$\frac{\partial U}{\partial x} = x - \frac{(1-\mu)(x+\mu)}{[(x+\mu)^2+y^2]^{3/2}} - \frac{\mu(x-1+\mu)}{[(x-1+\mu)^2+y^2]^{3/2}},$$
$$\frac{\partial U}{\partial y} = y - \frac{(1-\mu)y}{[(x+\mu)^2+y^2]^{3/2}} - \frac{\mu y}{[(x-1+\mu)^2+y^2]^{3/2}}.$$

The point $(x,0)$ on the x-axis is an equilibrium point if and only if x is a root of the function

$$f(x) = x - \frac{(1-\mu)(x+\mu)}{|x+\mu|^3} - \frac{\mu(x-1+\mu)}{|x-1+\mu|^3} = 0. \tag{2.42}$$

It can be shown that this function has a distinct root on each of three intervals, $(-\infty,-\mu)$, $(-\mu,1-\mu)$, and $(1-\mu,\infty)$. These regions are separated by the two primaries. It may further be shown that, for small values of μ, the positions of these three equilibrium points, L_1, L_2, and L_3, may be approximated by

$$\begin{aligned} x_1 &= 1 - \left(\frac{\mu}{3}\right)^{1/3} + O(\mu^{2/3}), \\ x_2 &= 1 + \left(\frac{\mu}{3}\right)^{1/3} + O(\mu^{2/3}), \\ x_3 &= -1 - \left(\frac{\sqrt{2}-1}{3}\right)\mu + O(\mu^2). \end{aligned} \tag{2.43}$$

When additional precision is required, or when the value of μ is not close to zero, one may use a computer to accurately estimate the roots of (2.42).

As was indicated previously, there are a total of five relative equilibrium points in the restricted three-body problem. To determine their location, first note that if $r_1 = 1$ and $r_2 = 1$, then the partials of $U(x,y)$ simplify to

$$\frac{\partial U}{\partial x} = x - (1-\mu)(x+\mu) - \mu(x+\mu-1) = 0,$$
$$\frac{\partial U}{\partial y} = 1 - (1-\mu) - \mu = 0.$$

A given point with $y \neq 0$ is therefore an equilibrium point if and only if $r_1 = r_2 = 1$. We conclude that both libration points L_4 and L_5 form an equilateral triangle with P_1 and P_2; that is,

$$x_{4,5} = \frac{1}{2} - \mu, \quad y_{4,5} = \pm\frac{\sqrt{3}}{2}.$$

Stability of Equilibria

In this section, we determine the stability of each libration point by considering the linearization of the vector field (2.36)–(2.39). The Jacobian for this system is easily computed as

$$\frac{\partial f}{\partial x} = \begin{bmatrix} 0 & 0 & 1 & 0 \\ 0 & 0 & 0 & 1 \\ U_{xx} & U_{xy} & 0 & 2 \\ U_{yx} & U_{yy} & -2 & 0 \end{bmatrix}. \tag{2.44}$$

The characteristic equation is therefore

$$\det\left(\frac{\partial f}{\partial x} - \lambda I\right) = \lambda^4 + \lambda^2(4 - U_{xx} - U_{yy}) + 4\lambda U_{xy} + (U_{xx}U_{yy} - U_{xy}^2) = 0.$$

Moreover, one can easily compute that, for points on the x-axis, $U_{xy}(x, 0) = 0$. Therefore, the eigenvalues for libration points L_1, L_2, and L_3 satisfy

$$\lambda^4 + \lambda^2(4 - U_{xx} - U_{yy}) + U_{xx}U_{yy} = 0.$$

Applying the quadratic formula, we obtain, for L_1, L_2, and L_3, the equation

$$\lambda^2 = -\frac{1}{2}(4 - U_{xx} - U_{yy}) \pm \frac{1}{2}\sqrt{(4 - U_{xx} - U_{yy})^2 - 4U_{xx}U_{yy}}. \tag{2.45}$$

To determine the stability of the linearization about L_1, L_2, and L_3, we calculate

$$U_{xx} = 1 + \frac{2(1-\mu)}{|x+\mu|^3} + \frac{2\mu}{|x-1+\mu|^3} > 0,$$
$$U_{yy} = 1 - \frac{(1-\mu)}{|x+\mu|^3} - \frac{\mu}{|x-1+\mu|^3}.$$

Observing that the value of x at each of the points L_1, L_2, and L_3 must be a root of equation (2.42), we may rewrite U_{yy} as

$$U_{yy} = \frac{\mu(1-\mu)}{x}\left[\frac{1}{|x+\mu|^3} - \frac{1}{|x-1+\mu|^3}\right].$$

Thus, $U_{yy} < 0$ at points L_1 and L_2 and $U_{yy} > 0$ at point L_3. We conclude from (2.45) that the linearizations at both L_1 and L_2 have two real eigenvalues of opposite sign and two purely imaginary, complex conjugate eigenvalues. Also, the linearization at L_3 has four purely imaginary, complex conjugate eigenvalues. Therefore, each of the linearized systems at L_1 and L_2 has a 2-D saddle subspace as well as a 2-D center subspace, whereas the linearization at L_3 is a 4-D center.

Exercise 2.19. Show that at L_4 and L_5,

$$U_{xx} = \frac{3}{4}, \quad U_{xy} = \pm\frac{3\sqrt{3}}{4}(1-2\mu), \quad U_{yy} = \frac{9}{4},$$

and show that, moreover, the characteristic polynomial for the linearization at these points simplifies to

$$\lambda^4 + \lambda^2 \pm \lambda 3\sqrt{3}(1-2\mu) + \frac{27}{4}\mu(1-\mu) = 0.$$

It can be further shown that the eigenvalues of the linearization at L_4 and L_5 are purely imaginary and distinct only when

$$27\mu(1-\mu) < 1.$$

This inequality holds for the Earth–Moon and Sun–Jupiter systems. Since equilibrium points L_4 and L_5 are nonhyperbolic, their stability cannot be determined using linearization alone. However, it can be shown that these equilibrium points are indeed Lyapunov stable; see, for example, [11]. Interestingly, when Lagrange discovered the existence of the five libration points in the restricted three-body problem, he thought of them as mere mathematical curiosities. However, as it turns out, in the Sun–Jupiter system, small pockets of asteroids have been discovered at the L_4 and L_5 points, called the "Greeks" and the "Trojans," respectively. These asteroids always maintain their relative positions with respect to the Sun and Jupiter. Moreover, the L_1 and L_2 points in the Sun–Earth system have been destinations for unmanned satellite missions. The center subspace of the linearization of these points can be used to park a satellite in a periodic orbit about either L_1 or L_2 that maintains its relative position to the Sun and Earth throughout the year. In particular, since the L_2 point is always in the shadow of the Earth, it has been utilized by telescopes to take panoramic photos of the surrounding universe without the interference of sunlight. For more detailed analysis on the geometry near the libration points and how one might utilize known geometric structures of the phase space for space mission planning, see [163].

Exercise 2.20. What is the minimum speed a spacecraft requires at launch to reach the Moon? Take the Earth's radius to be 0.016 times the distance between the centers of the Earth and the Moon, one dimensionless velocity unit to be about 1.05 km/s, and $\mu = 0.012$. Is it more economical to launch from the point on the Earth's surface between the Earth's center and the Moon or from this point's antipodal point?

3 Invariant Manifolds

In this chapter, we discuss the nonlinear analog of the stable, unstable, and center subspaces that we encountered during our discussion of linear systems. In particular, we will show that tangent to the stable, unstable, and center subspaces of the linearization of a fixed point of a nonlinear system there exist invariant stable, unstable, and center manifolds, respectively.

3.1 Asymptotic Behavior of Trajectories

We begin by discussing several fundamental notions regarding the limiting behavior of trajectories.

Definition 3.1. A point $x^* \in \mathbb{R}^n$ is called an *ω-limit point* of the trajectory $y : \mathbb{R} \to \mathbb{R}^n$ if there exists an unbounded, increasing sequence $(t_i)_{i=1}^{\infty}$ such that

$$\lim_{i\to\infty} y(t_i) = x^*.$$

Furthermore, the *ω-limit set*, $\omega(y)$, of trajectory y is the set of all ω-limit points of that trajectory.

Definition 3.2. A point $x^* \in \mathbb{R}^n$ is called an *α-limit point* of the trajectory $y : \mathbb{R} \to \mathbb{R}^n$ if there exists an unbounded, decreasing sequence $(t_i)_{i=1}^{\infty}$ such that

$$\lim_{i\to\infty} y(t_i) = x^*.$$

Furthermore, the *α-limit set*, $\alpha(y)$, of trajectory y is the set of all α-limit points of that trajectory.

Remark. When the flow $\varphi : \mathbb{R} \times \mathbb{R}^n \to \mathbb{R}^n$ of a system is understood, we will sometimes use the shorthand $\alpha(x_0)$ for $\alpha(\varphi(t; x_0))$ and the shorthand $\omega(x_0)$ for $\omega(\varphi(t; x_0))$, i. e., $\alpha(x_0)$ and $\omega(x_0)$ represent the α- and ω-limit sets of the trajectory passing through point x_0, respectively.

Example 3.1. Consider the planar trajectory $y : \mathbb{R} \to \mathbb{R}^2$ defined by

$$y(t) = \left(\frac{1}{2} + \frac{1}{\pi}\arctan t, \sin t\right),$$

as shown in Figure 3.1. The α- and ω-limit sets are $\alpha(y) = \{0\} \times [-1, 1]$ and $\omega(y) = \{1\} \times [-1, 1]$, respectively. □

Example 3.2. Consider the planar trajectory $y : \mathbb{R} \to \mathbb{R}^2$ defined by $y(t) = (\sin t, \cos t)$. The α- and ω-limit sets are $\alpha(y) = y$ and $\omega(y) = y$, respectively. Both the α- and ω-limit sets coincide with the orbit itself. □

https://doi.org/10.1515/9783110597806-003

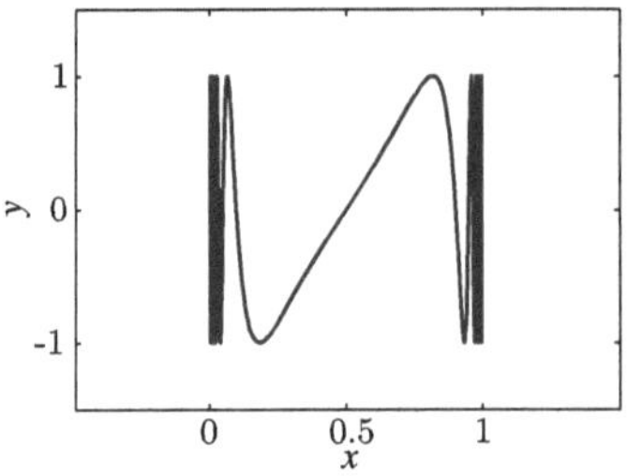

Figure 3.1: Trajectory of Example 3.1.

Definition 3.3. Let $\gamma : \mathbb{R} \to \mathbb{R}^n$ be a solution to nonlinear system (2.1) with initial condition $\gamma(0) = x_0$. Then γ is a *periodic orbit* if the set $\mathcal{S} = \{t > 0 : \gamma(t) = x_0\}$ is nonempty. Moreover, if φ is a periodic orbit, then we call the number $T = \inf \mathcal{S}$ the *period* of the orbit.

Remark. According to Definition 3.3, fixed points of system (2.1) are periodic orbits with period $T = 0$. We will call fixed points *trivial* periodic orbits. If γ is a nontrivial periodic orbit, then $\inf \mathcal{S} = \min \mathcal{S}$.

Definition 3.4. A *limit cycle* of system (2.1) is a nontrivial periodic orbit that, when considered as a set, is also an α- or ω-limit set of at least one other orbit of the system besides itself.

We will discuss limit cycles further in Chapter 5.

Example 3.3. Let γ be a periodic orbit with period T. Then $\alpha(\gamma) = \gamma$ and $\omega(\gamma) = \gamma$. To see this, fix $x \in \gamma$. Then there exists a number $\tau \in \mathbb{R}$ such that $x = \gamma(\tau)$. Since γ is a periodic orbit, it follows that $\gamma(\tau + iT) = x$ for all $i \in \mathbb{N}$. Let $(t_i)_{i=1}^{\infty} = (\tau + iT)_{i=1}^{\infty}$ be an increasing, unbounded sequence satisfying the conditions of Definition 3.1. Since $\gamma(t_i) = x$ for all $i \in \mathbb{N}$, it follows that $\lim_{i\to\infty} \gamma(t_i) = x$, and therefore x is an ω-limit point of γ. Since $x \in \gamma$ was arbitrary, every point on γ is an ω-limit point, and therefore $\omega(\gamma) = \gamma$. A similar argument shows $\alpha(\gamma) = \gamma$. □

Example 3.4. Consider the 2-D *affine linear* system depicted in Figure 3.2, centered about the fixed point x^*. Trajectories coinciding with the stable and unstable subspaces are shown. Suppose the eigenvectors are $v_1 = (1, 1)^T$ and $v_2 = (1, -1)^T$, with corresponding eigenvalues $\lambda_1 = -1$ and $\lambda_2 = 1$, respectively. Then the general solution is given by

$$x(t) - x^* = c_1 v_1 e^{-t} + c_2 v_2 e^{t}$$

for arbitrary constants c_1 and c_2, which represent a trajectory's initial conditions with respect to the basis $\{v_1, v_2\}$. The equilibrium solution $x(t) = x^*$ has α- and ω-limit sets $\alpha(x^*) = \omega(x^*) = x^*$. Moreover, the solutions

$$x_1(t) = x^* + v_1 e^{-t},$$

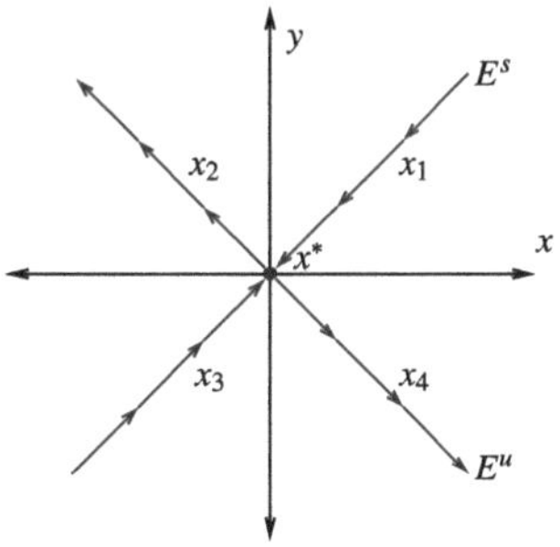

Figure 3.2: Invariant subspaces of a saddle point.

$$x_2(t) = x^* - v_2 e^t,$$
$$x_3(t) = x^* - v_1 e^{-t},$$
$$x_4(t) = x^* + v_2 e^t$$

have nonempty α- and ω-limit sets $\omega(x_1) = \omega(x_3) = x^*$ and $\alpha(x_2) = \alpha(x_4) = x^*$. Intuitively, this means that trajectories x_1 and x_3 approach the fixed point x^* as $t \to \infty$, while trajectories x_2 and x_4 emanate from the fixed point x^* as $t \to -\infty$. Moreover, $\alpha(x_1) = \alpha(x_3) = \omega(x_2) = \omega(x_4) = \emptyset$ since there exists no point in $\mathbb{R}^2$ that, for example, x_1 approaches at discrete points in time as $t \to -\infty$, and similarly for x_3, as well as for x_2 and x_4 as $t \to \infty$. □

In the above example, we saw how the invariant subspaces of a fixed point in a linear system had as a fixed point either an α- or an ω-limit set of the given subspace. In particular, if E^s and E^u are, respectively, the stable and unstable subspaces of a fixed point x^* in a linear system, then we have $\omega(E^s) = x^*$ and $\alpha(E^u) = x^*$, whereas $\alpha(E^s) = \omega(E^u) = \emptyset$. A natural question one may now pose is as follows. For a given nonlinear system, are there invariant sets or individual trajectories that have both an α-limit and an ω-limit set consisting only of fixed points of the system? The answer to this question is yes, as we will see in the following discussion.

Definition 3.5. If $x^* \in \mathbb{R}^n$ is a fixed point of nonlinear system (2.1) and if $y : \mathbb{R} \to \mathbb{R}^n$ is a solution curve of the aforementioned system, with the properties that $\alpha(y) = \omega(y) = x^*$, then trajectory y is referred to as a *homoclinic orbit.*

Definition 3.6. If $x_1^*, x_2^* \in \mathbb{R}^n$ are distinct fixed points of nonlinear system (2.1) and if $y : \mathbb{R} \to \mathbb{R}^n$ is a solution curve of the aforementioned system, with the properties that $\alpha(y) = x_1^*$ and $\omega(y) = x_2^*$, then y is referred to as a *heteroclinic orbit.*

Example 3.5. Examine the phase portrait of the 2-D, nonlinear system depicted in Figure 3.3. Three separate orbits, x_1, x_2, and x_3, are shown. The nonempty α- and ω-limit sets of these orbits are $\omega(x_1) = x^*$, $\alpha(x_2) = \omega(x_2) = x^*$, and $\alpha(x_3) = x^*$, respectively. The orbit x_2 is a *homoclinic orbit*, sometimes referred to as a *homoclinic loop*, as it both emanates from and culminates at the single fixed point x^*.

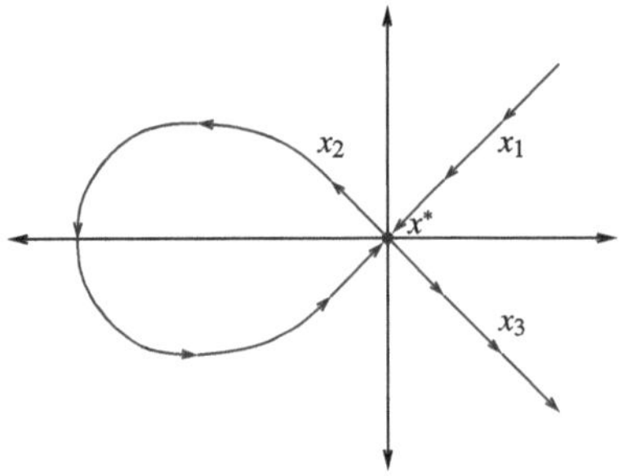

Figure 3.3: Example of a homoclinic loop.

Note that the phase portrait shown in Figure 3.3 must necessarily contain an additional fixed point located within the region bounded by the homoclinic loop $x_2 \cup x^*$. This statement follows directly from the Poincaré–Bendixson theorem, which we will discuss in Chapter 5. □

Example 3.6. Consider the *pendulum equations*, introduced in Example 2.4:

$$\dot{x} = y,$$
$$\dot{y} = -\sin x.$$

The phase portrait for this system is plotted in Figure 3.4. This system exhibits fixed points at $(n\pi, 0)$ for all $n \in \mathbb{Z}$. These fixed points are nonlinear centers for even values of n and saddle points for odd values of n. Orbit x_1 emanates from $\alpha(x_1) = (-\pi, 0)$ and culminates at $\omega(x_1) = (\pi, 0)$. The orbit is therefore a *heteroclinic orbit* that connects these two fixed points. Similarly, orbit x_2 emanates from $\alpha(x_2) = (\pi, 0)$ and culminates at $\omega(x_2) = (-\pi, 0)$. Therefore, orbit x_2 is also a heteroclinic orbit.

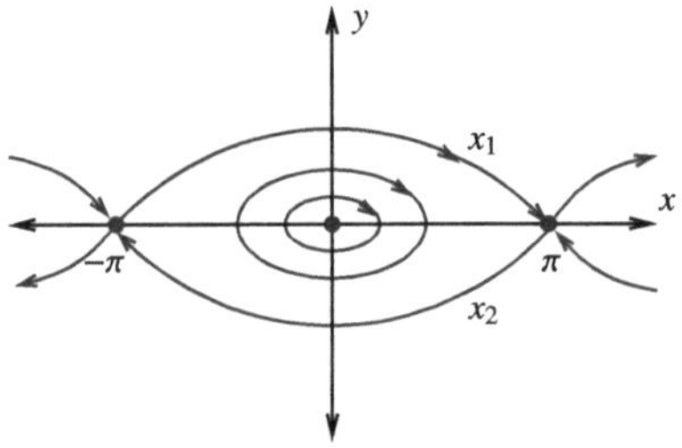

Figure 3.4: Phase portrait for a pendulum.

Interestingly, one may think of the dynamics of the pendulum as instead living on the topology of the cylinder $S^1 \times \mathbb{R}$. To construct this topology, we identify the lines $x = -\pi$ and $x = \pi$ and then consider x to be a cyclic variable. The pendulum dynamics in this topology is depicted in Figure 3.5. The lines $x = -\pi$ and $x = \pi$ now coincide and are represented by the vertical dashed line. Orbits x_1 and x_2, as seen from this topology, now each represent a *homoclinic orbit* of the fixed point $(\pi, 0)$. □

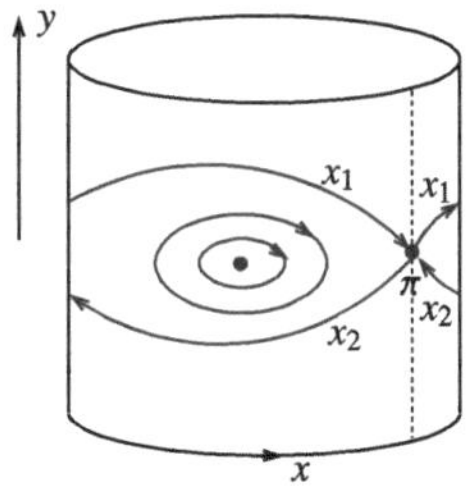

Figure 3.5: Phase portrait for a pendulum in cylinder topology $S^1 \times \mathbb{R}$.

These are examples of invariant sets that emanate from or culminate at fixed points. They are also examples of stable and unstable manifolds of these fixed points, a topic we will introduce in Section 3.2.

3.2 Invariant Manifolds in $\mathbb{R}^n$

Definition 3.7. A set $\mathcal{S} \subset \mathbb{R}^n$ is said to be *invariant* under the flow of the nonlinear system $\dot{x} = f(x)$ if, for all $x_0 \in \mathcal{S}$, the property $\varphi(t; x_0) \in \mathcal{S}$ holds for all $t \in \mathbb{R}$. Moreover, $\mathcal{S}$ is called *positively invariant* if the property $\varphi(t; x_0) \in \mathcal{S}$ holds for all $t \in [0, \infty)$ and *negatively invariant* if the same property holds for all $t \in (-\infty, 0]$.

Example 3.7. Fixed points are invariant sets. Suppose that x^* is a fixed point of system (2.1). Then the set $\mathcal{S} = \{x^*\}$ is invariant since $\varphi(t; x^*) = x^* \in \mathcal{S}$ for all $t \in \mathbb{R}$. □

Example 3.8. Periodic orbits are invariant sets. Suppose that $\gamma : \mathbb{R} \to \mathbb{R}^n$ is a periodic orbit of system (2.1). Then, clearly, $x_0 \in \gamma$ implies that $\varphi(t; x_0) \in \gamma$ for all $t \in \mathbb{R}$. In fact, for the same reason, every individual solution curve γ of system (2.1) is an invariant set. □

Example 3.9. *Trapping regions* are positively invariant sets. A *trapping region*, R, is a closed, connected set such that vector field $f(x)$ points "inward" for all points on the boundary of R. □

As we will see in Chapter 5, trapping regions will be useful in proving the existence of periodic orbits in certain nonlinear planar flows.

Example 3.10. Consider the 2-D linear system

$$\dot{x}_1 = \frac{-x_1}{2}, \quad \dot{x}_2 = -2x_2.$$

The region $\mathcal{S} = \{(x_1, x_2) \in \mathbb{R}^2 : \frac{x_1^2}{4} + x_2^2 \leq 1\}$ is positively invariant. To see this, consider the Lyapunov function $V(x) = \frac{x_1^2}{4} + x_2^2$. The material derivative of V along the solution flow is given by

$$\frac{DV}{Dt} = \nabla V \cdot \dot{x} = \nabla V \cdot f(x) = -\nabla V \cdot \nabla V = -|\nabla V|^2 \leq 0$$

since $f(x) = (-\frac{x_1}{2}, -2x_2) = -\nabla V$. Hence vector field f points "inward" along each contour of V. Since the boundary of $\mathcal{S}$ is a level contour of V and since $\mathcal{S}$ is closed and connected, it constitutes a *trapping region* and, thus, a positively invariant set. □

We will now introduce the notion of a manifold. What follows should be regarded as a quasidefinition, or a working definition, as it will be usurped later in Chapter 6 with a more mathematically rigorous one. However, both definitions are equivalent, and the more classical definition presented here is better suited for our present purposes.

Definition 3.8. The closed, connected set $M \subset \mathbb{R}^n$ is a *k-dimensional, C^r-differential manifold* if, for every point $p \in M$, there exists an open neighborhood $U \subset M$, containing p, such that M, when restricted to U, may be *locally* represented as the graph of a C^r function $h : \mathbb{R}^k \to \mathbb{R}^{n-k}$ relative to a certain basis $(e_1, \ldots, e_n)$.

In other words, for every point $p \in M$ there exists a neighborhood $U \subset M$ containing p, a basis $(e_1, e_2, \ldots, e_n)$ of $\mathbb{R}^n$ with coordinates (x, y), and a C^r function h : $\mathbb{R}^k \to \mathbb{R}^{n-k}$ with the property that $(x, y) \in U \subset M$ implies that $y = h(x)$. (Note: $x \in \mathbb{R}^k$ and $y \in \mathbb{R}^{n-k}$.) The numbers $(x_1, \ldots, x_k)$ are called *local coordinates* for M in neighborhood U. Thus manifolds are simply sets that may be *locally* described as the graph of a function but not necessarily *globally*. Typically, we will take a given manifold to be a C^∞-differential manifold and refer to it simply as a k-dimensional differential manifold.

Given a function $h : \mathbb{R}^k \to \mathbb{R}^{n-k}$ whose graph locally coincides with a differential manifold M, we say that the related function $\psi : \mathbb{R}^k \to \mathbb{R}^n$, defined by $\psi(x) = (x, h(x))$, is a *local parameterization* of manifold M.

Again, we will present a more modern definition of a manifold in Chapter 6. One problem with our present definition is that a k-dimensional manifold relies on the existence of an ambient Euclidean space. Such a definition will suffice in our present context; however, we will soon replace it with an *intrinsic* one that does not rely on the crutch of an ambient space.

Example 3.11. Any smooth curve $\gamma : \mathbb{R} \to \mathbb{R}^n$ is a 1-D differential manifold. Let $t_0 \in \mathbb{R}$ be arbitrary, and define $e_1 = \gamma'(t_0)/\|\gamma'(t_0)\|$. Let $(e_2, \ldots, e_n)$ be an orthonormal basis for the orthogonal complement of the subspace span$\{e_1\}$. Let $(x_1(t), y_2(t), \ldots, y_n(t))$ be the components of $\gamma(t)$ with respect to the basis $(e_1, \ldots, e_n)$. Since $x_1'(t_0) > 0$, by construction, there exists an $\varepsilon > 0$ such that the map $x_1 : \mathbb{R} \to \mathbb{R}$ is a diffeomorphism for $t \in (t_0 - \varepsilon, t_0 + \varepsilon)$ by the inverse function theorem. Therefore, γ may be parameterized as $\gamma(x_1) = (x_1, y_2(t(x_1)), \ldots, y_n(t(x_1)))$ for $x_1 \in (x_1(t - \varepsilon), x_1(t + \varepsilon))$. Since point t_0 was arbitrary, γ is a 1-D manifold. □

Example 3.12. The graph of a smooth function $h : \mathbb{R}^k \to \mathbb{R}^{n-k}$ is a k-dimensional differential manifold. Since the surface may *globally* be represented as the graph $(x, h(x))$ in $\mathbb{R}^n$, it may also be represented locally by the same graph. It follows that graphs of smooth functions are themselves manifolds. For instance, the graph of the function $z = \sin(x + y)$ is a smooth, 2-D manifold. □

Example 3.13. Let $f : \mathbb{R}^n \to \mathbb{R}$ be a smooth function on $\mathbb{R}^n$. Then the level contours of f are smooth, $(n-1)$-dimensional manifolds, as they may be represented *locally* as the graph of a function of $(n-1)$ variables. □

Definition 3.9. Let $M \subset \mathbb{R}^n$ be a k-dimensional C^r-differential manifold, and, for $p \in M$, let $h : \mathbb{R}^k \to \mathbb{R}^{n-k}$ be a C^r function whose graph, relative to some basis $(e_1, \ldots, e_n)$, coincides with M in a neighborhood U about p, as in Definition 3.8. Define the function $\psi : \mathbb{R}^k \to \mathbb{R}^n$ by the relation $\psi(x) = (x, h(x))$. Then the *tangent space*, T_pM, to manifold M at p is the affine subspace passing through p and spanned by the vectors $\frac{\partial \psi}{\partial x_i}|_p$ for $i = 1, \ldots, k$.

Suppose that $x_0 \in \mathbb{R}^k$ is such that $\psi(x_0) = p \in M$. Now let

$$u_1 = \left(1, 0, \ldots, 0, \frac{\partial h_{k+1}}{\partial x_1}, \ldots, \frac{\partial h_n}{\partial x_1}\right)\bigg|_{x_0},$$

$$\vdots \quad \vdots$$

$$u_k = \left(0, \ldots, 0, 1, \frac{\partial h_{k+1}}{\partial x_k}, \ldots, \frac{\partial h_n}{\partial x_k}\right)\bigg|_{x_0}.$$

Then the *tangent space* to M at p is given by the affine subspace $T_pM = p + \text{span}\{u_1, \ldots, u_k\}$.

Example 3.14. The graph of the function $y = 4 - x_1^2 - x_2^2$ is a 2-D manifold in $\mathbb{R}^3$, globally parameterized by the function $\psi(x_1, x_2) = (x_1, x_2, 4 - x_1^2 - x_2^2)$. The basis tangent vectors at the point $p = (1, 1, 2) \in M$ are therefore given by

$$u_1 = \frac{\partial \psi}{\partial x_1} = (1, 0, -2x_1)|_{(1,1)} = (1, 0, -2),$$

$$u_2 = \frac{\partial \psi}{\partial x_2} = (0, 1, -2x_2)|_{(1,1)} = (0, 1, -2).$$

Therefore, the tangent space at p is given by

$$T_pM = p + \text{span}(u_1, u_2) = \{(1+\alpha, 1+\beta, 2-2\alpha-2\beta) \in \mathbb{R}^3 : \alpha, \beta \in \mathbb{R}\}.$$ □

Example 3.15. Consider the 2-D manifold $S^2 = \{x \in \mathbb{R}^3 : |x| = 1\}$. The "north pole" is defined as the point $N = (0, 0, 1)$. Find the tangent space to the north pole. This scenario is plotted in Figure 3.6. *Locally* we can represent manifold S^2 as the graph of the function $h(x, y) = \sqrt{1 - x^2 - y^2}$. Hence the manifold is *locally* parameterized by the function $\psi(x, y) = (x, y, \sqrt{1 - x^2 - y^2})$. A short calculation shows that the vectors $(1, 0, 0)$ and $(0, 1, 0)$ span the tangent space to the north pole. □

Exercise 3.1. Find an equation for the tangent space of the unit sphere S^2 at the point $(\frac{1}{\sqrt{3}}, \frac{1}{\sqrt{3}}, \frac{1}{\sqrt{3}})$.

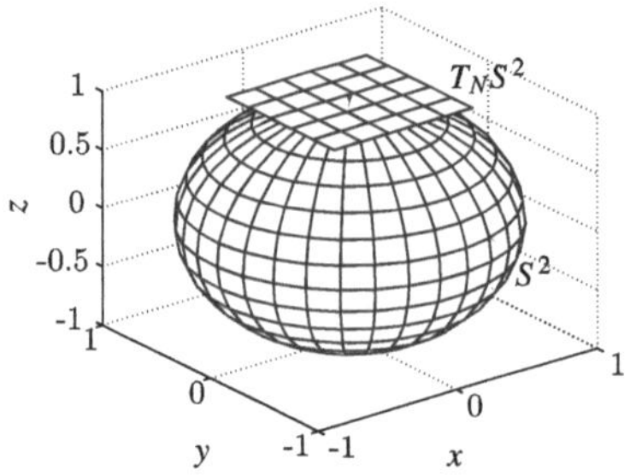

Figure 3.6: Manifold S^2 and tangent space T_NS^2 at the north pole.

3.3 Stable Manifold Theorem

In this section, we generalize the notions of stable and unstable subspaces of linear systems to their nonlinear analogs. Recall that in linear systems, these invariant subspaces are attached to the origin, the location of the system's only fixed point. For the linearization of a nonlinear system about a fixed point, the stable and unstable subspaces of the linearization are affine subspaces attached to the fixed point. We will discover invariant manifolds, attached to the fixed points of nonlinear systems, that lie tangent to the stable and unstable subspaces of the linearization. For the moment, we will content ourselves with considering only the case of *hyperbolic* fixed points, such that each eigenvalue of the linearization has a nonzero real part.

Let us begin by considering the system $\dot{x} = f(x)$ with fixed point x^*. The linearization about x^* is given by $\dot{z} = Df(x^*)z$, where $z = x - x^*$. We can reorder the eigenvalues of $Df(x^*)$ such that $\lambda_1, \ldots, \lambda_k$ have negative real parts and $\lambda_{k+1}, \ldots, \lambda_n$ have positive real parts. Let $w_1 = u_1 + iv_1, \ldots, w_n = u_n + iv_n$ be the corresponding eigenvectors. Then the stable and unstable subspaces are given by

$$E^s = \operatorname{span}\{u_1, v_1, \ldots, u_k, v_k\},$$
$$E^u = \operatorname{span}\{u_{k+1}, v_{k+1}, \ldots, u_n, v_n\},$$

respectively, so that $\mathbb{R}^n = E^s \oplus E^u$.

Theorem 3.1 (Stable Manifold Theorem). *Let x^* be a hyperbolic fixed point of the nonlinear system $\dot{x} = f(x)$ such that $Df(x^*)$ has k eigenvalues with negative real part, arranged as above. Then*

1. *there exists a k-dimensional invariant manifold, $W^s(x^*)$, such that $x^* \in W^s(x^*)$, $T_{x^*}W^s(x^*) = x^* + E^s$, and $\omega(x_0) = x^*$ for all $x_0 \in W^s(x^*)$; and*
2. *there exists an $(n-k)$-dimensional invariant manifold, $W^u(x^*)$, such that $x^* \in W^u(x^*)$, $T_{x^*}W^u(x^*) = x^* + E^u$, and $\alpha(x_0) = x^*$ for all $x_0 \in W^u(x^*)$.*

The manifolds $W^s(x^*)$ and $W^u(x^*)$ are referred to, respectively, as the stable and unstable manifolds of x^*.

The first requirement of the stable manifold is $x^* \in W^s(x^*)$, which states that the fixed point x^* must actually be a point on the manifold $W^s(x^*)$. Next, the condition $T_{x^*}W^s(x^*) = x^* + E^s$ states that the manifold $W^s(x^*)$ lies tangent to the (affine) stable subspace of the linearization at x^*. The final requirement, $\omega(x_0) = x^*$ for all $x_0 \in W^s(x^*)$, states that trajectories originating at all points on the stable manifold asymptotically approach the fixed point x^* as $t \to \infty$. A similar set of interpretations applies to each condition of the unstable manifold.

In going about proving the stable manifold theorem, it suffices to prove the existence of a *local* stable and unstable manifold. Let $U \subset \mathbb{R}^n$ be a sufficiently small open neighborhood of the fixed point x^*. We can then define the *local* stable and unstable manifolds by

$$\begin{aligned} W^s_{\text{loc}}(x^*) &= U \cap W^s(x^*), \\ W^u_{\text{loc}}(x^*) &= U \cap W^u(x^*). \end{aligned}$$

In other words, one proves that a *positively invariant* manifold $W^s_{\text{loc}}(x^*)$ exists in a neighborhood of x^* that lies tangent to the stable subspace and whose trajectories asymptotically approach the fixed point as $t \to \infty$, and, similarly, that a *negatively invariant* manifold $W^u_{\text{loc}}(x^*)$ exists in a neighborhood of x^* that lies tangent to the unstable subspace and whose trajectories asymptotically emanate from x^* as $t \to -\infty$. Once the existence of a *local* stable and unstable manifold is established, one may simply use the flow $\varphi : \mathbb{R} \times \mathbb{R}^n \to \mathbb{R}^n$ of the nonlinear system to define the *global* stable and unstable manifolds of x^* as follows:

$$W^s(x^*) = \bigcup_{\substack{t<0 \\ x_0 \in W^s_{\text{loc}}}} \varphi(t; x_0), \tag{3.1}$$

$$W^u(x^*) = \bigcup_{\substack{t>0 \\ x_0 \in W^u_{\text{loc}}}} \varphi(t; x_0). \tag{3.2}$$

Clearly, given a local stable and unstable manifold, these definitions yield global stable and unstable manifolds that satisfy the conditions of Theorem 3.1. It is therefore only compulsory that we prove the existence of such manifolds *locally*.

Example 3.16. Let us consider the fixed point x^* of a nonlinear system in $\mathbb{R}^3$. Suppose the linearization $Df(x^*)$ of the fixed point has two real, negative eigenvalues and one real, positive eigenvalue. Figure 3.7 is a depiction of how the *local* stable and unstable manifolds may appear. Notice that $W^s_{\text{loc}}(x^*)$ may be represented as a graph of a function of two variables and that, similarly, $W^u_{\text{loc}}(x^*)$ may be represented as a graph of a function of one variable. When we follow the reverse phase flow of the local stable manifold, extending the trajectories infinitely far back in negative time, the global stable manifold, $W^s(x^*)$, is revealed, as depicted in Figure 3.8. Similarly, following the trajectory segment defining the local unstable manifold forward, we reveal the appearance of the global unstable manifold, $W^u(x^*)$. The global unstable manifold may not

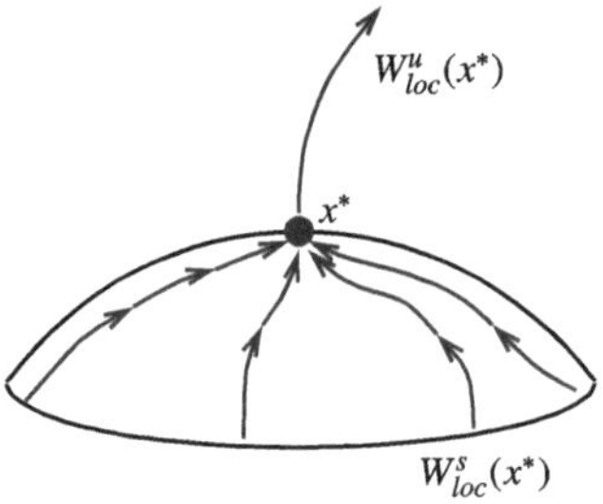

Figure 3.7: Local stable and unstable manifolds of a fixed point.

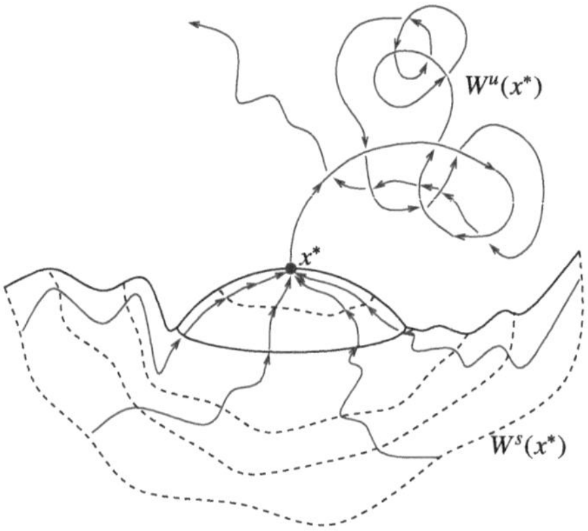

Figure 3.8: Global stable and unstable manifolds of a fixed point.

be thought of as a graph of any function $h : \mathbb{R} \to \mathbb{R}^2$. In addition, the unstable manifold never crosses its own path. Each "apparent intersection" actually takes place at different depths into the page. This follows from the fundamental feature of dynamical systems that solution paths never cross. Similarly, even though the depiction of the global stable manifold in the figure appears well behaved enough, as one continues its extension by following each of the infinite spectra of trajectories that comprise it far back into negative time, one finds that it may quite feasibly twist and stretch and fail the vertical line test from every imaginable direction, souring one's plan of viewing it *globally* as a graph of a function of two variables. □

Example 3.17. Consider again the phase portrait depicted in Figure 3.3. The stable and unstable manifolds are

$$W^s(x^*) = x_1 \cup x_2 \cup x^*,$$
$$W^u(x^*) = x_2 \cup x_3 \cup x^*.$$

The homoclinic loop coincides with the intersection of the stable and unstable manifolds. □

Exercise 3.2. Find the global stable and unstable manifolds of the origin for the system

$$\dot{x} = -x, \quad \dot{y} = 2y - 5x^3.$$

Example 3.18. Let us consider again the restricted circular three-body problem discussed in §2.8. The mass parameter for the Earth–Moon system is approximately $\mu = 0.012$. The location of the L_1 libration point may be approximated using equation (2.43). Using this value as an initial guess, the precise location of the first libration point may be computed numerically, for example, using the built-in MATLAB function `fzero.m`. The location of this libration point is approximately $x_1 = 0.8377$. One next evaluates the Jacobian matrix (2.44) and determines its eigenvalues and eigenvectors (for instance, using the built-in MATLAB function `eigs.m`). As we discussed earlier, the Jacobian at x_1 has two real eigenvalues of opposite signs and two purely imaginary eigenvalues. The eigenvector corresponding to the positive real eigenvalue is the vector $v_1 = (-0.2934, 0.1351, -0.8597, 0.3958)^T$. Taking a differential step off of equilibrium point L_1 toward direction v_1 should therefore place us directly on the unstable manifold of L_1. Using `ode45.m` to integrate the trajectory with initial condition $(x_0, y_0, u_0, v_0)^T = (x_1, 0, 0, 0)^T + 0.01v_1$, we obtain the plot shown in Figure 3.9. As it turns out, the unstable manifold of libration point L_1 makes several pedal-shaped orbits about the Earth and then returns to L_1, approaching L_1 in the direction tangent to its stable subspace. Hence, half of the unstable manifold actually coincides with half of the stable manifold. (Remember, both the stable and unstable manifolds have another half that is not plotted since the linearization of L_1 has a 2-D saddle subspace.) Since both the α- and ω-limit sets of this orbit consist solely of point L_1 itself, this orbit is a homoclinic orbit. □

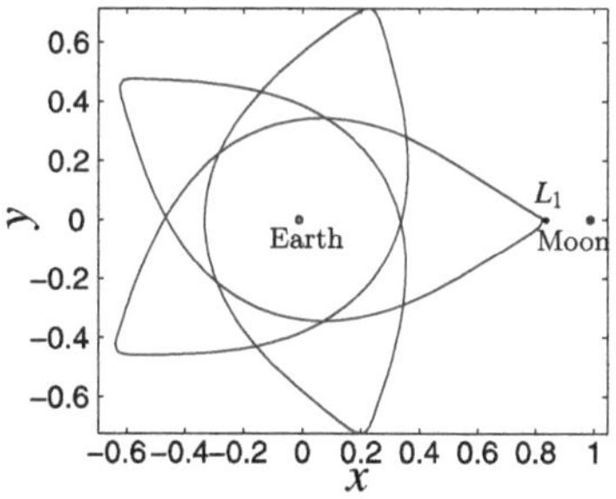

Figure 3.9: Homoclinic orbit of libration point L_1 in the Earth–Moon system.

Exercise 3.3. Use MATLAB to determine the remaining three eigenvectors of the L_1 libration point in the Earth–Moon system of Example 3.18.

Exercise 3.4. Use `ode45.m` to reproduce the plot of Figure 3.9. Then plot the trajectory with initial condition $(x_0, y_0, u_0, v_0)^T = (x_1, 0, 0, 0)^T - 0.01v_1$, thereby determining the Moon-half of the unstable manifold.

3.4 Contraction Mapping Theorem

In this section, we introduce an important theorem of applied mathematics. We will use this theorem in the next paragraph to prove the existence of local stable and unstable manifolds of a fixed point. The setting for the theorem is a complete, normed linear space, or *Banach* space.

Definition 3.10. A mapping $T : X \to X$, where X is a subset of a normed linear space, is called a *contraction mapping* if there exists a positive $a < 1$ such that

$$\|Tx - Ty\| \leq a\|x - y\| \quad \text{for all } x, y \in X.$$

Theorem 3.2 (Contraction Mapping Theorem). *If $T : X \to X$ is a contraction mapping of a closed subset X of a Banach space $\mathcal{B}$, then there exists exactly one $x \in X$ such that $Tx = x$. Moreover, for any $x_0 \in X$, the sequence $(x_n)_{n=0}^{\infty}$, defined recursively by $x_{n+1} = Tx_n$, converges to x.*

To prove this theorem, we must prove three separate claims:
1. The sequence (x_n) converges to some $\overline{x} \in X$.
2. $\overline{x}$ is a fixed point of T.
3. T has a unique fixed point.

Exercise 3.5. Let $x_0 \in X$ be arbitrary, X a subset of a Banach space, and $T : X \to X$ a contraction mapping. Set $x_n = T^n x_0$ and let $a \in (0, 1)$, as in Definition 3.10. Prove that if $m > n$, then

$$\|x_m - x_n\| \leq \frac{a^n}{1-a}\|x_1 - x_0\|.$$

Proof. 1. Our first step is to show that the previously defined sequence (x_n) converges to some $\overline{x} \in X$. By the result of Exercise 3.5, it follows that $\|x_m - x_n\| \to 0$ as $n \to \infty$. Therefore, (x_n) is a Cauchy sequence. Since $\mathcal{B}$ is complete, (x_n) converges to some $\overline{x} \in \mathcal{B}$. Since X is closed in $\mathcal{B}$, it follows that $\overline{x} \in X$.

2. Next we wish to show that $\overline{x}$ is indeed a fixed point of T, i. e., that $T\overline{x} = \overline{x}$. To do this, first notice that, for any n,

$$\begin{aligned}\|T\overline{x} - \overline{x}\| &\leq \|T\overline{x} - Tx_n\| + \|Tx_n - \overline{x}\| \\ &\leq a\|\overline{x} - x_n\| + \|\overline{x} - x_{n+1}\|.\end{aligned}$$

But since $(x_n) \to \overline{x}$, for every $\varepsilon > 0$ there exists $N \in \mathbb{N}$ such that $\|\overline{x} - x_n\| < \frac{\varepsilon}{1+a}$ for all $n \geq N$. Hence, if $n \geq N$, it follows that

$$\|T\overline{x} - \overline{x}\| < \frac{a\varepsilon}{a+1} + \frac{\varepsilon}{1+a} = \varepsilon.$$

Therefore, $\|T\overline{x} - \overline{x}\| < \varepsilon$ for all $\varepsilon > 0$. The only way this can be true is if $\|T\overline{x} - \overline{x}\| = 0$. By the properties of a norm, this implies that $T\overline{x} = \overline{x}$, and thus $\overline{x}$ is a fixed point of T.

3. Finally, we wish to show that T has a unique fixed point. Suppose that both $Tx = x$ and $Ty = y$ for some $x, y \in X$ such that $x \neq y$. Then

$$\|x - y\| = \|Tx - Ty\| \leq a\|x - y\|.$$

Since $x \neq y$ and since $0 < a < 1$, this is a contradiction. Thus the proposition that there exists more than one distinct fixed point must fail. This proves that any fixed point of T must be unique. □

Exercise 3.6. Show that a sufficient condition for a continuously differentiable function $g : \mathbb{R} \to \mathbb{R}$ to be a contraction mapping is

$$|g'(x)| \leq \alpha < 1.$$

Exercise 3.7 (Newton's Method). Let f be a real-valued, twice continuously differentiable function on $[a, b] \subset \mathbb{R}$ and let $\hat{x}$ be a simple zero of f on (a, b). Show that the function

$$g(x) = x - \frac{f(x)}{f'(x)}$$

is a contraction mapping on some neighborhood J of $\hat{x}$. Conclude that the sequence defined recursively by $x_{n+1} = g(x_n)$, for an initial $x_0 \in J$, converges to $\hat{x}$.

Application: Picard Iteration (Optional)

As an application to the contraction mapping theorem, we now discuss the existence and uniqueness of solutions to nonlinear ordinary differential equations. Let us consider the nonlinear initial value problem on $\mathbb{R}$, given by

$$\dot{x} = f(t, x), \tag{3.3}$$

$$x(t_0) = x_0. \tag{3.4}$$

The following theorem is due to [189] and [236].

Theorem 3.3 (Picard–Lindelöf). *Let f be continuous on the rectangle*

$$\mathcal{R} = \{(t, x) \in \mathbb{R}^2 : |t - t_0| \leq a, |x - x_0| \leq b\},$$

with $|f(t, x)| \leq c$ and suppose further that f is Lipschitz continuous on $\mathcal{R}$ with respect to its second argument, i. e., there exists a Lipschitz constant $k > 0$ such that, for all $(t, x), (t, y) \in \mathcal{R}$,

$$|f(t, x) - f(t, y)| \leq k|x - y|.$$

Then the initial value problem given by (3.3) *and* (3.4) *has a unique solution that exists on an interval* $[t_0 - \beta, t_0 + \beta]$, *where*

$$\beta < \min\left\{a, \frac{b}{c}, \frac{1}{k}\right\}.$$

Proof. Let $J = [t_0 - \beta, t_0 + \beta]$ and consider the Banach space $C^0(J)$ of continuous, real-valued functions on J, with the sup norm and the following closed subset:

$$X = \{x \in C^0(J) : |x(t) - x_0| \le c\beta\}.$$

Now define a mapping on X by the relation

$$Tx(t) = x_0 + \int_{t_0}^{t} f(\tau, x(\tau))\, d\tau. \tag{3.5}$$

If we can show that T is a contraction mapping on X, then this will prove the theorem.

First, let us show that $T : X \to X$. Note that

$$|Tx(t) - x_0| = \left|\int_{t_0}^{t} f(\tau, x(\tau)) d\tau\right| \le \int_{t_0}^{t} |f(\tau, x(\tau))| d\tau \le \int_{t_0}^{t} c\, d\tau = c|t - t_0|.$$

Hence, $|Tx(t) - x_0| \le c\beta$, which implies that $Tx \in X$.

Finally, let us prove that $T : X \to X$ is a contraction mapping. For any $x, y \in X$ we have

$$\begin{aligned}
|Tx(t) - Ty(t)| &= \left|\int_{t_0}^{t} [f(\tau, x(\tau)) - f(\tau, y(\tau))]\, d\tau\right| \\
&\le \int_{t_0}^{t} |f(\tau, x(\tau)) - f(\tau, y(\tau))|\, d\tau \\
&\le \int_{t_0}^{t} k|x(\tau) - y(\tau)| d\tau \\
&\le k|t - t_0| \sup_{t \in J} |x(\tau) - y(\tau)| \\
&\le k\beta \|x - y\|.
\end{aligned}$$

Therefore,

$$\|Tx - Ty\| \le k\beta \|x - y\|.$$

But $\beta k < 1$ by our hypothesis, so T is a contraction mapping. Therefore, T has a unique fixed point $x \in X$, satisfying $Tx = x$. Note that this proves the existence of a unique, continuous function x on J that satisfies the relation

$$x(t) = x_0 + \int_{t_0}^{t} f(\tau, x(\tau))\, d\tau.$$

Since f is continuous, we can differentiate this relation with respect to t in order to obtain

$$\dot{x}(t) = f(t, x(t)),$$

and hence x satisfies differential equation (3.3). It clearly also satisfies the initial condition (3.4). □

Exercise 3.8. The purpose of selecting $\beta < \frac{1}{k}$ in Theorem 3.3 is clear. Show that by also requiring $\beta < \min\{a, \frac{b}{c}\}$, the solution of the initial value problem given by (3.3) and (3.4), if it exists, on the interval $J = [t - \beta, t + \beta]$ must remain in the rectangle $\mathcal{R}$. *Hint*: use the fact that $|\dot{x}(t)| \le c$ and consider the two cases $a < b/c$ and $a > b/c$ separately.

In addition to the local existence and uniqueness of a solution to differential equation (3.3), the preceding theorem furthermore provides us a way of approximating the solution. Recall that since $T : X \to X$ is a contraction mapping, the sequence of functions $\{x_n\}$ defined recursively by $x_{n+1} = Tx_n$ must converge to the unique fixed point of T. Hence, the solution to the initial value problem may be approximated by taking, for instance, $x_0(t) = x_0$ and then using (3.5) to iteratively compute subsequently better approximations to the solution curve. This technique is known as *Picard iterations*.

3.5 Graph Transform Method

In this section, we will make use of the contraction mapping theorem to prove the stable manifold theorem using a technique originally due to [113] and more recently discussed in [294].

Let x^* be a hyperbolic fixed point with stable and unstable subspaces E^s and E^u of the linearization $Df(x^*)$. Suppose $\dim(E^s) = k$. Then a point in the neighborhood of x^* can be described by the local coordinates (x, y), where $x \in E^s$ and $y \in E^u$. Let $\pi_x : \mathbb{R}^n \to E^s$ and $\pi_y : \mathbb{R}^n \to E^u$ be the canonical projection operators, so that if $z = (x, y)$, then $\pi_x(z) = x$ and $\pi_y(z) = y$. Next define the set

$$S_\delta = \{h : \mathbb{R}^k \to \mathbb{R}^{n-k} : h(0) = 0,\ |h(x) - h(x')| \le \delta |x - x'| \text{ for all } |x| < \varepsilon\}.$$

This is the set of locally Lipschitz continuous functions attached to the origin with Lipschitz constant $\delta > 0$. The parameters δ and ε must be set so as to make our result work. The term S_δ represents a subset of $C^0(\mathbb{R}^k, \mathbb{R}^{n-k})$, the set of continuous functions from $\mathbb{R}^k$ to $\mathbb{R}^{n-k}$. The norm

$$\|h\| = \sup_{|x| \le \varepsilon} |h(x)|,$$

known as the *sup norm*, makes $C^0(\mathbb{R}^k, \mathbb{R}^{n-k})$ a Banach space (see, for example, [95]). It is possible to show that, given the metric topology induced by the sup norm, the set S_δ is a *closed* subset of $C^0(\mathbb{R}^k, \mathbb{R}^{n-k})$.

Definition 3.11. Let $h \in S_\delta$ and define $\psi : \mathbb{R}^k \to \mathbb{R}^n$ by the relation $\psi(x) = (x, h(x))$. Then the *graph transform* of function h is the function $\mathcal{G}h : B_\varepsilon(0) \subset \mathbb{R}^k \to \mathbb{R}^{n-k}$ defined by the relation

$$\mathcal{G}h = \pi_y \circ \varphi_{-t} \circ \psi \circ \pi_x \circ \varphi_t \circ \psi. \tag{3.6}$$

Our first aim after understanding the meaning of this definition will be to show that $\mathcal{G} : S_\delta \to S_\delta$. We will then show that $\mathcal{G}$ is a contraction mapping and conclude that it must have a unique fixed point, thereby proving the existence of the k-dimensional, local stable manifold of x^*.

Let us attempt to understand the *graph transform* diagrammatically. The graph transform takes a function h and returns a new function $\mathcal{G}h$. To understand this transformation, let us calculate the y-value corresponding to a particular input x. The following description is depicted graphically in Figure 3.10. For a given value of x, the first rung of (3.6) is the mapping ψ, which sends x to its corresponding point $(x, h(x))$ on the graph of h. Next, we apply the phase flow φ_t to that point for a fixed time $t > 0$, e. g., for $t = 1$. Next, the mapping $\psi \circ \pi_x$ literally drops the point $\varphi_t(x, h(x))$ *vertically* back onto the graph of h. The reverse phase flow φ_{-t} is then applied to this point. Upon flowing backward in time for t units, the final point is then projected onto the unstable subspace via the projection operator π_y. This final y-value is, by definition, the value of the function $\mathcal{G}h$ when evaluated at the original point x.

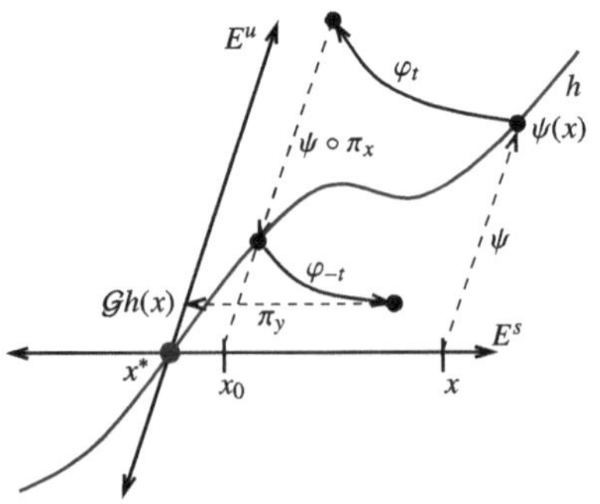

Figure 3.10: Graph transform.

Lemma 3.1. *Let $h \in S_\delta$. For ε and δ small enough, the function $\mathcal{G}h \in S_\delta$, i. e., $\mathcal{G} : S_\delta \to S_\delta$.*

Proof. By supposition, the graph of any $h \in S_\delta$ must lie within the box $\mathcal{B} = (-\varepsilon, \varepsilon) \times (-\delta\varepsilon, \delta\varepsilon)$. Let us choose ε and δ small enough so that the following estimates obtained using the local linearization are robust under higher-order perturbations.

Let $x_0, x_0' \in E^s$ such that $|x_0|, |x_0'| < \varepsilon$. Consider now the corresponding points on the graph $\psi(x_0) = (x_0, h(x_0))$ and $\psi(x_0') = (x_0', h(x_0'))$. Let $x_1 = \pi_x \circ \varphi_t(x_0, h(x_0))$ and $x_1' = \pi_x \circ \varphi_t(x_0', h(x_0'))$. Finally, define $y_1 = h(x_1)$ and $y_1' = h(x_1')$ and define $(x_2, y_2) = \varphi_{-t}(x_1, y_1)$ and $(x_2', y_2') = \varphi_{-t}(x_1', y_1')$, so that $y_2 = \mathcal{G}h(x_0)$ and $y_2' = \mathcal{G}h(x_0')$. The parameter $t > 0$ should be chosen small enough so that $\varphi_t(x_0, y_0), \varphi_t(x_0', y_0') \in \mathcal{B}$. We now claim that

$$|x_1 - x_1'| < |x_0 - x_0'|, \tag{3.7}$$

$$|y_2 - y_2'| < |y_1 - y_1'|. \tag{3.8}$$

To see this, note that

$$(x_0, y_0) = (c_1 w_1 + \cdots + c_k w_k, c_{k+1} w_{k+1} + \cdots + c_n w_n),$$
$$(x_0', y_0') = (c_1' w_1 + \cdots + c_k' w_k, c_{k+1}' w_{k+1} + \cdots + c_n' w_n),$$

where $w_1, \ldots, w_k$ constitute an eigenbasis for the stable subspace of the linearization and $w_{k+1}, \ldots, w_n$ constitute a basis for the unstable subspace of the linearization. Hence, to first order,

$$x_1 = \pi_x \circ \varphi_t(x_0, y_0) = c_1 w_1 e^{\lambda_1 t} + \cdots + c_k w_k e^{\lambda_k t},$$
$$x_1' = \pi_x \circ \varphi_t(x_0', y_0') = c_1' w_1 e^{\lambda_1 t} + \cdots + c_k' w_k e^{\lambda_k t},$$

where $\mathbb{R}\{\lambda_1\}, \ldots, \mathbb{R}\{\lambda_k\} < 0$. Now let

$$\alpha = - \min_{1 \le i \le k} \mathbb{R}\{\lambda_i\}.$$

It follows that

$$|x_1 - x_1'| < e^{-\alpha t} |x_0 - x_0'|.$$

Since $0 < e^{-\alpha t} < 1$, we conclude that the less stringent inequality (3.7) holds under the higher-order perturbations. Similarly,

$$(x_1, y_1) = (d_1 w_1 + \cdots + d_k w_k, d_{k+1} w_{k+1} + \cdots + d_n w_n),$$
$$(x_1', y_1') = (d_1' w_1 + \cdots + d_k' w_k, d_{k+1}' w_{k+1} + \cdots + d_n' w_n).$$

Therefore, to first order,

$$y_2 = \pi_y \circ \varphi_{-t}(x_1, y_1) = d_{k+1} w_{k+1} e^{-\lambda_{k+1} t} + \cdots + d_n w_n e^{-\lambda_n t},$$

$$y_2' = \pi_y \circ \varphi_{-t}(x_1', y_1') = d_{k+1}' w_{k+1} e^{-\lambda_{k+1} t} + \cdots + d_n' w_n e^{-\lambda_n t}.$$

Defining

$$\beta = \max_{k+1 \le i \le n} \Re\{\lambda_i\},$$

we conclude that

$$|y_2 - y_2'| < e^{-\beta t} |y_1 - y_1'|. \tag{3.9}$$

Since $e^{-\beta t} < 1$, we further conclude that inequality (3.8) is robust under the higher-order, nonlinear perturbations.

Now, to complete the proof of the lemma, notice that

$$|\mathcal{G}h(x_0) - \mathcal{G}h(x_0')| = |y_2 - y_2'| < |y_1 - y_1'|,$$

by (3.8). Since $y_1 = h(x_1)$, $y_1' = h(x_1')$, and $h \in S_\delta$, it follows that

$$|y_1 - y_1'| < \delta |x_1 - x_1'|.$$

Next,

$$|x_1 - x_1'| < |x_0 - x_0'|$$

by (3.7). Combining the preceding three inequalities yields the inequality

$$|\mathcal{G}h(x_0) - \mathcal{G}h(x_0')| < \delta |x_0 - x_0'|,$$

and thus $\mathcal{G}h$ is Lipschitz continuous with Lipschitz constant δ. Finally, since $x^* = (0, 0)$ is a fixed point, $\mathcal{G}h(0) = 0$ for any $h \in S_\delta$. Therefore, $\mathcal{G}h \in S_\delta$, which proves our lemma. □

Lemma 3.2. *The graph transform $\mathcal{G} : S_\delta \to S_\delta$ is a contraction mapping.*

Proof. Consider $h, h' \in S_\delta$. Let $y_0 = h(x_0)$ and $y_0' = h'(x_0)$; $(x_1, y_1) = \psi \circ \pi_x \circ \varphi_t(x_0, y_0)$ and $(x_1, y_1') = \psi \circ \pi_x \circ \varphi_t(x_0, y_0')$; and $(x_2, y_2) = \varphi_{-t}(x_1, y_1)$ and $(x_2, y_2') = \varphi_{-t}(x_1', y_1')$.

We will need an inequality slightly stronger than (3.8). From (3.9), there exists an $a \in (e^{-\beta|t|}, 1)$ such that

$$|y_2 - y_2'| < a|y_1 - y_1'|.$$

Thus

$$\begin{aligned} |\mathcal{G}h(x_0) - \mathcal{G}h'(x_0)| = |y_2 - y_2'| &< a|h(x_1) - h'(x_1)| \\ &< a \sup_{|x|<\varepsilon} |h(x) - h'(x)| = a\|h - h'\|. \end{aligned}$$

Therefore,

$$\|\mathcal{G}h - \mathcal{G}h'\| \le a\|h - h'\|,$$

and thus $\mathcal{G} : S_\delta \to S_\delta$ is a contraction mapping. □

It follows from Lemmas 3.1 and 3.2 and the contraction mapping theorem (Theorem 3.2) that $\mathcal{G} : S_\delta \to S_\delta$ has a unique fixed point, h, proving the existence of a local, stable manifold. A simple modification of the foregoing arguments can also be used to prove the existence of the local, unstable manifold. The global stable and unstable manifolds can then be defined using equations (3.1) and (3.2).

Example 3.19. To illustrate the preceding proof, we will now consider the system

$$\dot{x} = -x,$$
$$\dot{y} = y - x^2,$$

similar to one proposed in [294]. The linearization has eigenvalues $\lambda_1 = -1$ and $\lambda_2 = 1$ and associated eigenvectors $v_1 = (1,0)^T$ and $v_2 = (0,1)^T$, respectively, so that the stable subspace coincides with the x-axis. The solution to this system is readily identified as

$$x(t) = x_0 e^{-t}, \tag{3.10}$$
$$y(t) = y_0 e^t + \frac{1}{3}x_0^2 e^t (e^{-3t} - 1), \tag{3.11}$$

where x_0 and y_0 are the initial conditions. To make sense of (3.6), we decompose the terms one by one. Consider an arbitrary function $h \in S_\delta$. Then, using the flow given by (3.10) and (3.11) we obtain

$$\begin{aligned}
\psi(x) &= (x, h(x)),\\
\varphi_t \circ \psi(x) &= \left(xe^{-t}, h(x)e^t + \frac{1}{3}x^2 e^t (e^{-3t} - 1)\right),\\
\pi_x \circ \varphi_t \circ \psi(x) &= xe^{-t},\\
\psi \circ \pi_x \circ \varphi_t \circ \psi(x) &= (xe^{-t}, h(xe^{-t})),\\
\varphi_{-t} \circ \psi \circ \pi_x \circ \varphi_t \circ \psi(x) &= (x, h(xe^{-t})e^{-t} + \frac{1}{3}x^2(1 - e^{-3t})),\\
\pi_y \circ \varphi_{-t} \circ \psi \circ \pi_x \circ \varphi_t \circ \psi(x) &= h(xe^{-t})e^{-t} + \frac{1}{3}x^2(1 - e^{-3t}),
\end{aligned}$$

so that

$$\mathcal{G}h(x) = h(xe^{-t})e^{-t} + \frac{1}{3}x^2(1 - e^{-3t}). \tag{3.12}$$

To use the contraction mapping, we will first show that $\mathcal{G} : S_\delta \to S_\delta$. Consider any $x, x' \in [-\varepsilon, \varepsilon]$. Then

$$\begin{aligned}
|\mathcal{G}h(x) - \mathcal{G}h(x')| &= \Big| h(xe^{-t})e^{-t} - h(x'e^{-t})e^{-t}\\
&\quad + \frac{1}{3}x^2(1 - e^{-3t}) - \frac{1}{3}(x')^2(1 - e^{-3t}) \Big|\\
&\le e^{-t}|h(xe^{-t}) - h(x'e^{-t})| + \frac{1}{3}(1 - e^{-3t})|x^2 - (x')^2|
\end{aligned}$$

$$\le \left(\delta e^{-t} + \frac{1}{3}(1 - e^{-3t})|x + x'|\right)|x - x'|$$
$$\le \left(\delta e^{-t} + \frac{2\varepsilon}{3}(1 - e^{-3t})\right)|x - x'|.$$

Therefore, if we select an $\varepsilon > 0$ that satisfies the inequality

$$\varepsilon < \frac{3\delta}{2}\frac{(1 - e^{-t})}{(1 - e^{-3t})},$$

then we obtain the inequality

$$|\mathcal{G}h(x) - \mathcal{G}h(x')| \le \delta|x - x'|.$$

Hence, $\mathcal{G}h \in S_\delta$.

Finally, we show that $\mathcal{G} : S_\delta \to S_\delta$ is a contraction mapping. Consider any two $h_1, h_2 \in S_\delta$. Using (3.12) we obtain

$$|\mathcal{G}h_1(x) - \mathcal{G}h_2(x)| \le e^{-t}|h_1(xe^{-t}) - h_2(xe^{-t})|$$
$$\le e^{-t}\|h_1 - h_2\|.$$

It follows that

$$\|\mathcal{G}h_1 - \mathcal{G}h_2\| \le e^{-t}\|h_1 - h_2\|,$$

and hence $\mathcal{G}$ is a contraction mapping on S_δ, which has a unique fixed point by Theorem 3.2.

Starting with the initial guess $h_0(x) = 0$, we can now use (3.12) to form an iterative sequence of functions by the recursion relation $h_{n+1} = \mathcal{G}h_n$. By the contraction mapping theorem, this sequence of functions converges to the stable manifold as $n \to \infty$. The iterations may be efficaciously computed with the aid of MAPLE, and we have plotted the first five iterations, using $h_0(x) = 0$ and our initial guess, in Figure 3.11. □

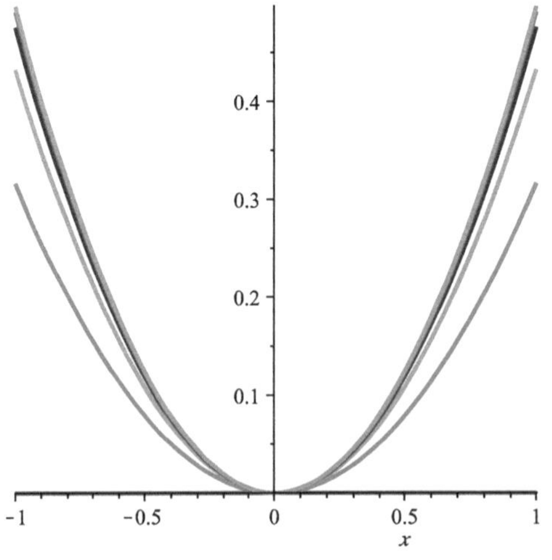

Figure 3.11: Graph of iterations $h_0, h_1, h_2, h_3, h_4, h_5$.

3.6 Center Manifold Theory

In our discussion of linear systems, we discussed the existence of invariant stable and unstable subspaces, spanned by the real and imaginary parts of the eigenvectors corresponding to the eigenvalues with negative and positive real parts, respectively. For nonhyperbolic linear systems, we then showed the additional existence of an invariant *center* subspace, spanned by the real and imaginary parts of the eigenvectors corresponding to purely imaginary eigenvalues. Whereas motion on the stable and unstable subspaces is hyperbolic, motion on the center subspace for a linear system is periodic. In this section, we generalize our discussion to nonlinear systems and discuss the existence of a *center manifold* in the vicinity of a nonhyperbolic fixed point x^* that lies tangent to the center subspace of the linearization at x^*. We further discuss ways to describe motion on this manifold; these motions, unlike their linear analogs, may be either periodic or hyperbolic depending on the effect of the nonlinear terms of the flow. For more details on center manifold theory, see [24], [42], and [54].

We begin by generalizing Theorem 3.1 to the case of nonhyperbolic fixed points. Consider the system $\dot{x} = f(x)$, with nonhyperbolic fixed point x^*. The linearization about x^* is given by $\dot{z} = Df(x^*)z$, where $z = x - x^*$. We can reorder the eigenvalues of $Df(x^*)$ so that $\lambda_1, \ldots, \lambda_k$ have a negative real part, $\lambda_{k+1}, \ldots, \lambda_{n-p}$ have a positive real part, and $\lambda_{n-p+1}, \ldots, \lambda_n$ are purely imaginary. Let $w_1 = u_1 + iv_1, \ldots, w_n = u_n + iv_n$ be the corresponding eigenvectors. Then the stable, unstable, and center subspaces of the linearization are given by

$$\begin{aligned}
E^s &= \operatorname{span}\{u_1, v_1, \ldots, u_k, v_k\}, \\
E^u &= \operatorname{span}\{u_{k+1}, v_{k+1}, \ldots, u_{n-p}, v_{n-p}\}, \\
E^c &= \operatorname{span}\{u_{n-p+1}, v_{n-p+1}, \ldots, u_n, v_n\},
\end{aligned}$$

respectively, so that $\mathbb{R}^n = E^s \oplus E^u \oplus E^c$.

Theorem 3.4 (Center Manifold Theorem: Part 1). *Let x^* be a nonhyperbolic fixed point of the nonlinear system $\dot{x} = f(x)$, such that $Df(x^*)$ has k eigenvalues with negative real part and p purely imaginary eigenvalues, arranged as above. Then:*

1. *there exists a k-dimensional invariant manifold, $W^s(x^*)$, such that $x^* \in W^s(x^*)$, $T_{x^*}W^s(x^*) = x^* + E^s$, and $\omega(x_0) = x^*$ for all $x_0 \in W^s(x^*)$;*
2. *there exists an $(n-p-k)$-dimensional invariant manifold, $W^u(x^*)$, such that $x^* \in W^u(x^*)$, $T_{x^*}W^u(x^*) = x^* + E^u$, and $\alpha(x_0) = x^*$ for all $x_0 \in W^u(x^*)$; and*
3. *there exists a p-dimensional invariant manifold, $W^c(x^*)$, such that $x^* \in W^c(x^*)$ and $T_{x^*}W^c(x^*) = x^* + E^c$.*

The manifolds $W^s(x^*)$, $W^u(x^*)$, and $W^c(x^*)$ are referred to as the stable, unstable, and center manifolds of x^*, respectively.

Example 3.20. As an illustrative example, let us suppose that for a given nonlinear system on $\mathbb{R}^3$, the linearization at the origin has eigenvalues $\lambda_1 = -1$, $\lambda_{2,3} = \pm i$ and associated eigenvectors $v_1 = (0,0,1)$ and $v_{2,3} = (1, \pm i, 0)$, respectively. Hence, we may conclude that there exists a 1-D stable manifold, tangent to the z-axis, that passes through the origin and that there exists a 2-D center manifold, tangent to the x-y plane, that also passes through the origin (Figure 3.12). It is not necessarily true, however, that the motion on the center manifold resembles that of a center; the trajectories *may* be periodic orbits constituting a nonlinear center, or they might be stable or unstable spirals that either approach or emanate from the origin as $t \to \pm\infty$, respectively. □

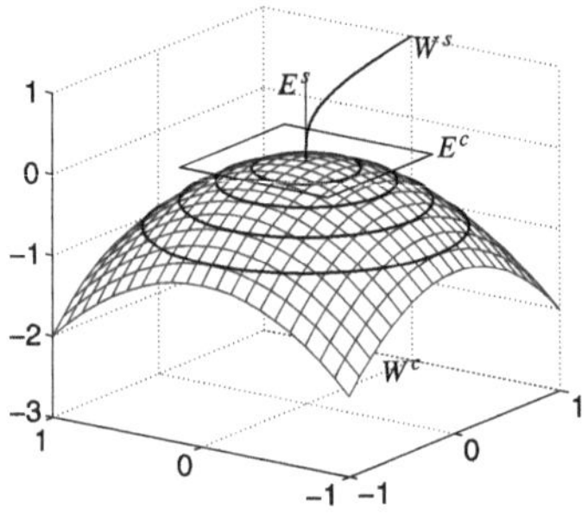

Figure 3.12: Stable and center manifolds of the origin.

To further describe the motions on the center manifold, we note that a given nonlinear system with a nonhyperbolic fixed point can always be described using coordinates $x \in E^c \cong \mathbb{R}^p$ and $y \in E^s \otimes E^u \cong \mathbb{R}^{n-p}$, so that the equations of motion may be written as follows:

$$\dot{x} = Ax + X(x, y), \tag{3.13}$$

$$\dot{y} = By + Y(x, y), \tag{3.14}$$

where X and Y are twice continuously differentiable functions of x and y, where $X(0,0) = DX(0,0) = Y(0,0) = DY(0,0) = 0$, and where the eigenvalues of matrices A and B have zero and nonzero real parts, respectively. Given such a setup, the motions of trajectories on the center manifold may be described by the equations presented in the following theorem.

Theorem 3.5 (Center Manifold Theorem: Part 2). *The flow on the local, p-dimensional center manifold $y = h(x)$ of the system given by* (3.13) *and* (3.14) *is governed by the equation*

$$\dot{x} = Ax + X(x, h(x)).$$

Proof. See [42] or [54]. □

Example 3.21. Consider the system

$$\dot{x}_1 = x_2 + x_2 y, \tag{3.15}$$

$$\dot{x}_2 = -x_1 - x_1 y, \tag{3.16}$$

$$\dot{y} = -y + x_1^2 + x_2^2. \tag{3.17}$$

The linearization for this system about the origin is clearly

$$\dot{z} = \begin{bmatrix} 0 & 1 & 0 \\ -1 & 0 & 0 \\ 0 & 0 & -1 \end{bmatrix} z,$$

where $z = (x_1, x_2, y)^T$. The eigenvalues and eigenvectors are clearly $\lambda_{1,2} = \pm i$, $\lambda_3 = -1$, $v_{1,2} = (1, \pm i, 0)$, and $v_3 = (0,0,1)$, so that the center and stable subspaces of the linearization are $E^c = \text{span}\{(1,0,0),(0,1,0)\}$ and $E^s = \text{span}\{(0,0,1)\}$, respectively. The origin of this system is therefore a nonhyperbolic equilibrium point, so that the system must possess a 2-D center manifold.

Exercise 3.9. Show that the graph of $y = h(x_1, x_2) = x_1^2 + x_2^2$ is an invariant manifold for the system given by (3.15)–(3.17).

Since the invariant manifold $y = x_1^2 + x_2^2$ lies tangent to the x_1-x_2 plane at $(0,0,0)$, which coincides with the center subspace of the linearization at the origin, it must be a center manifold of the origin. Therefore, according to Theorem 3.5, the motion on the center manifold is governed by the system

$$\dot{x}_1 = x_2 + x_2\left(x_1^2 + x_2^2\right),$$

$$\dot{x}_2 = -x_1 - x_1\left(x_1^2 + x_2^2\right).$$

As would be expected, the Lyapunov function $V(x_1, x_2) = x_1^2 + x_2^2$ turns out to be an integral of motion for the reduced system, since the material derivative of V along the flow vanishes identically. We have

$$\frac{1}{2}\frac{DV}{Dt} = x_1\dot{x}_1 + x_2\dot{x}_2 = x_1x_2 + x_2x_1^3 + 2x_1x_2^3 - x_1x_2 - x_2x_1^3 - x_1x_2^3 = 0.$$

Therefore, the motions on the center manifold $W^c(0)$, whose graph is $y = x_1^2 + x_2^2$, are those of a nonlinear center, i. e., the motion on the center manifold is a center even in the nonlinear picture.

By inspection of the system given by (3.15)–(3.17), the stable manifold coincides with the stable subspace of the linearization, i. e., $W^s(0) = E^s = \text{span}\{(0,0,1)\}$, so that the y-axis itself is an invariant manifold whose trajectories simply decay asymptotically toward the origin.

For a *general* trajectory, with initial condition on neither the y-axis nor the invariant center manifold $y = x_1^2 + x_2^2$, we might then expect the motion to decay *to the center manifold* and to asymptotically approach one of the periodic orbits of the center manifold as $t \to \infty$. That is, the ω-limit sets of generic orbits for this system consist solely of the periodic orbits on the center manifold. This is verified with the aid of MATLAB. We may run ode45 to numerically integrate trajectories for the system (3.15)–(3.17). The trajectories with initial conditions $(1, 0, 0)$ and $(1, 0, 2)$ are plotted in Figure 3.13.

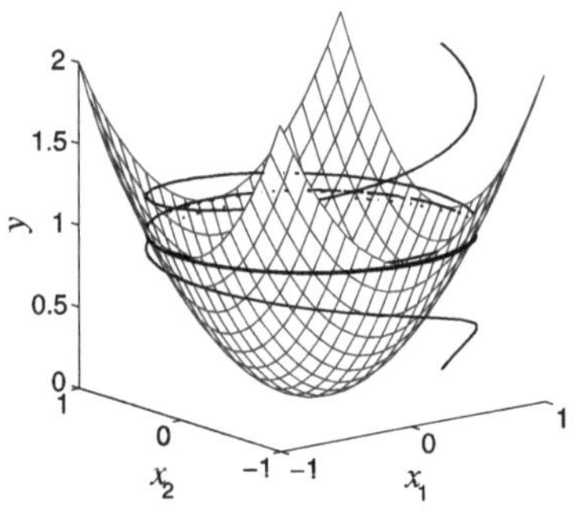

Figure 3.13: Sample trajectories and center manifold for Example 3.21.

Both trajectories remain on the invariant cylinder $Z_1 = \{(x_1, x_2, 0) \in \mathbb{R}^3 : x_1^2 + x_2^2 = 1\}$ and asymptotically approach the periodic orbit $Z_1 \cap W^c(0)$ as $t \to \infty$, as expected. We thus recognize the interesting fact that, for this particular system, each periodic orbit on the center manifold is also a *limit cycle* for the system (§3.1). □

Exercise 3.10. Show that the graph of the function $y = x_1^2 - x_2^2$ is an invariant manifold for the system

$$\begin{aligned} \dot{x}_1 &= x_2 + x_1 y, \\ \dot{x}_2 &= -x_1 + x_2 y, \\ \dot{y} &= -y + 2y^2 + x_1^2 - x_2^2 + 4x_1 x_2. \end{aligned}$$

Use Theorem 3.5 to write the reduced equations of motion that govern the dynamics on the center manifold. Use pplane8.m to plot the phase portrait on the x_1-x_2 plane for the reduced dynamics. Is the origin a nonlinear center or a stable or unstable spiral? Use ode45.m to reproduce the trajectories depicted in Figure 3.14. The initial conditions for the three trajectories are $(0.25, 0, 0.25^2)$, $(0.5, 0, 0.5^2)$, and $(0.75, 0, 0.75^2)$.

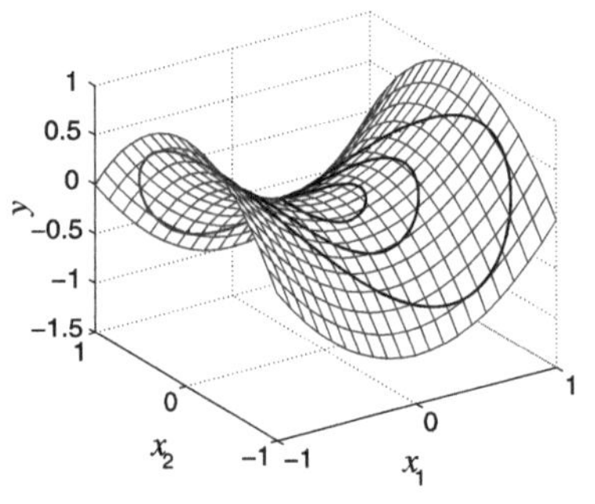

Figure 3.14: Sample trajectories on the center manifold of Exercise 3.10.

3.7 Application: Stability in Rigid-Body Dynamics

In this paragraph, we consider the dynamics of rigid-body motion. However, for now, we will restrict our attention to the simple problem of understanding the dynamics of a rigid body's angular velocity vector relative to a body-fixed frame. See [102], [105, 106], [174], and [196] for additional developments of rigid-body dynamics. We will discuss the derivation of these equations, along with the equations of motion one uses to determine the actual orientation of the rigid body relative to inertial space, later, in Chapter 9. For the present discussion, we will consider the resulting dynamics produced by *Euler's equations for rigid-body motion*:

$$I_x\dot{\Omega}_x + (I_z - I_y)\Omega_y\Omega_z = M_x, \tag{3.18}$$
$$I_y\dot{\Omega}_y - (I_z - I_x)\Omega_x\Omega_z = M_y, \tag{3.19}$$
$$I_z\dot{\Omega}_z + (I_y - I_x)\Omega_x\Omega_y = M_z. \tag{3.20}$$

In this context, I_x, I_y, and I_z are the three principal moments of inertia, $\Omega = (\Omega_x, \Omega_y, \Omega_z)$ is the angular velocity of the rigid body relative to the body-fixed principal axes, and $M = (M_x, M_y, M_z)$ is the net applied moment relative to the same body-fixed frame. The principal axes of a rigid body are literally just the orthonormal eigenvectors of the moment of the inertia matrix, and the principal moments of inertia are the corresponding eigenvalues. Defining the total angular momentum relative to the body-fixed frame, $\Pi = \mathbb{I} \cdot \Omega$, where $\mathbb{I}$ is the moment of the inertia matrix, we find that Euler's equations may be represented in vector form as

$$\dot{\Pi} - \Pi \times \Omega = M.$$

The associated torque-free ($M = 0$) case is therefore determined by the nonlinear system

$$\dot{\Pi}_x = \eta_1\Pi_y\Pi_z, \tag{3.21}$$
$$\dot{\Pi}_y = -\eta_2\Pi_x\Pi_z, \tag{3.22}$$
$$\dot{\Pi}_z = \eta_3\Pi_x\Pi_y, \tag{3.23}$$

where the constants η_1, η_2, and η_3 are given by

$$\eta_1 = \frac{I_y - I_z}{I_yI_z}, \quad \eta_2 = \frac{I_x - I_z}{I_xI_z}, \quad \text{and} \quad \eta_3 = \frac{I_x - I_y}{I_xI_y}.$$

Without loss of generality, let us arrange the principal axes so that $I_x \geq I_y \geq I_z$, so that $\eta_1, \eta_2, \eta_3 \geq 0$. The resulting dynamics of the torque-free evolution of the body-fixed angular momentum vector (or, equivalently, angular velocity vector) are easily understood with the aid of a method first introduced in [242]. This construction utilizes two separate integrals of motion.

Proposition 3.1. *The* total angular momentum, *defined by*

$$L = \left(\Pi^T \Pi\right)^{1/2},$$

and the total energy, *defined by*

$$E = \frac{1}{2}\Omega^T \Pi = \frac{1}{2}\left(\frac{\Pi_x^2}{I_x} + \frac{\Pi_y^2}{I_y} + \frac{\Pi_z^2}{I_z}\right), \tag{3.24}$$

are integrals of motion for the torque-free ($M = 0$) motion of the rigid body.

Exercise 3.11. Prove Proposition 3.1.

Without loss of generality, we can normalize the angular momentum vector so that $\|\Pi\| = 1$, so that the ensuing dynamics takes place on unit sphere S^2. The nonlinear system (3.21)–(3.23), constrained to the sphere $\|\Pi\| = 1$, has exactly six fixed points, $\Pi = (\pm 1, 0, 0)$, $\Pi = (0, \pm 1, 0)$, and $\Pi = (0, 0, \pm 1)$.

Exercise 3.12. Linearize system (3.21)–(3.23) about each of its fixed points and show that the fixed points $\Pi = (\pm 1, 0, 0)$ and $\Pi = (0, 0, \pm 1)$ are (linear) centers, whereas the points $\Pi = (0, \pm 1, 0)$ are saddle points.

Meanwhile, constraint (3.24) may be recognized as an *ellipsoid* in angular momentum space, called the *ellipsoid of inertia*. The orbits of system (3.21)–(3.23) are therefore given by the intersection of the inertia ellipsoid (3.24) with sphere S^2. The path traced out by the angular velocity vector on sphere S^2 is known as a *polhode*. As an example, let us consider a rigid body with principal moments of inertia $I_x = 1.5$, $I_y = 1.0$, and $I_z = 0.5$. The angular momentum sphere S^2 (shaded) is plotted with the ellipsoid of inertia (outlined) for various values of energy in Figure 3.15. Individual orbits on the angular momentum sphere are therefore constrained to remain within these intersections. Several orbits at various values of energy are shown in Figure 3.16. The four heteroclinic orbits connecting the saddle points $\Pi = (0, \pm 1, 0)$ divide the angular momentum sphere into four sections, as they form the boundaries of the four center manifolds of the fixed points $\Pi = (\pm 1, 0, 0)$ and $\Pi = (0, 0, \pm 1)$, respectively.

Satellite Design

A word of caution, however, lingers for the expeditionary-hearted. We have perforce proven that the linear centers at the fixed points $\Pi = (\pm 1, 0, 0)$ and $\Pi = (0, 0, \pm 1)$ are in fact *nonlinear centers* and, hence, Lyapunov stable. It might therefore be tempting, if you work in the business of designing satellites, to design a satellite shaped in such a way that it rotates about its axis of symmetry with the smallest moment of inertia. Rotations about this axis are Lyapunov stable as they correspond to the fixed point

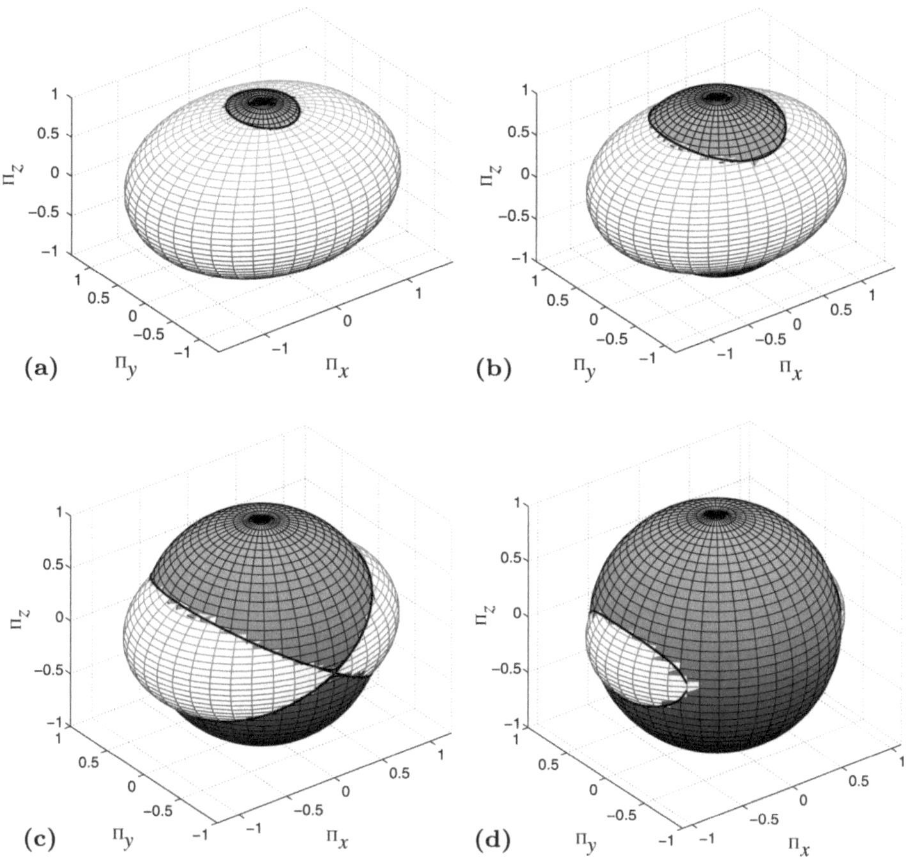

Figure 3.15: Intersection of inertia ellipsoid with angular momentum sphere for various values of energy for a rigid body with principal moments of inertia $I_x = 3$, $I_y = 2$, and $I_z = 1$. (a) $E = 0.9$, (b) $E = 0.7$, (c) $E = 0.5$, and (d) $E = 0.4$.

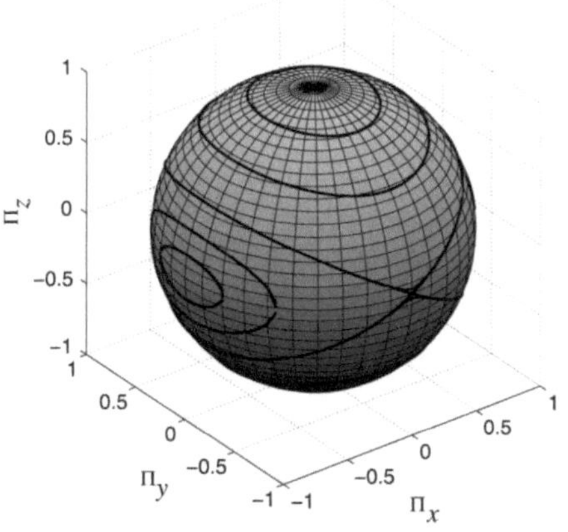

Figure 3.16: Various orbits on the angular momentum sphere.

$\Pi = (0, 0, +1)$. However, an interesting thing happens if the satellite is designed so as to rotate about Π_z. It loses control and ends up rotating about Π_x! The reason for this transition is that our analysis has not yet taken into account *energy dissipation*. As it turns out, in practice, angular momentum is always conserved, though energy ends up being dissipated (via friction, heat, etc.), as much as possible, into the vacuum of space. Let us see what happens in the presence of dissipative forces (torques).

Exercise 3.13. (a) Use Euler's equations to show that, in the presence of an external moment M, the total angular momentum changes according to

$$L\dot{L} = M \cdot \Pi.$$

Conclude that torques acting orthogonal to angular momentum vector Π do not affect the total angular momentum $L^2 = \Pi \cdot \Pi$.

(b) Show that total energy changes at the rate

$$\dot{E} = M \cdot \Omega.$$

In what direction must one apply an external moment to conserve both the total energy and the angular momentum of the system?

(c) Show that the moment $M = \Pi \times (\Pi \times \Omega)$ conserves total angular momentum while dissipating energy.

As an illustration of this effect, we took the rigid body discussed previously, with principal moments of inertia $I_x = 1.5$, $I_y = 1.0$, and $I_z = 0.5$, and numerically integrated Euler's equations with initial conditions on the $E = 0.99$ ellipsoid of inertia (a point close to the north pole $\Pi_z = (0, 0, 1)$) with the energy dissipating moment $M = 0.5\Pi \times (\Pi \times \Omega)$. The resultant trajectory is shown in Figure 3.17. The angular momentum starts very close to the Π_z-axis, so that rotations are initially about the principal axis with the *smallest* moment of inertia. Due to energy dissipation, the ellipsoid of inertia begins

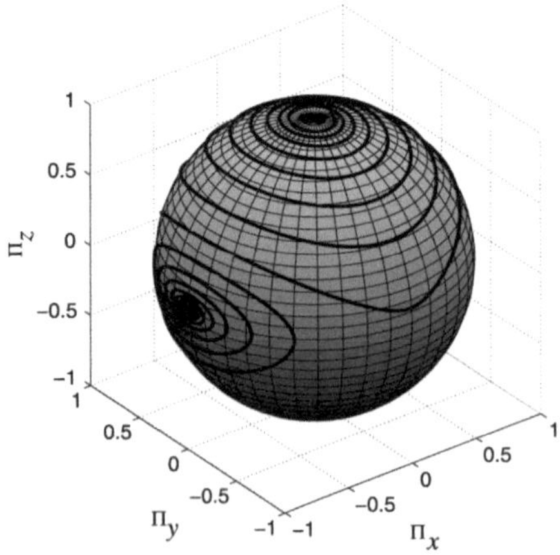

Figure 3.17: Orbit on angular momentum sphere with energy dissipation.

to shrink, causing the orbit to spiral away from the fixed point $\Pi_z = (0, 0, 1)$, eventually bounce away from the saddle points as the energy passes through $E = 0.5$, and spiral into the stable fixed point $\Pi_x = (1, 0, 0)$, corresponding to rotations about the axis with the greatest moment of inertia. Satellites should therefore be designed to rotate about the principal axis with the greatest moment of inertia.

As an interesting historical note, Explorer 1, the first Earth orbiting satellite launched by the United States, was designed to rotate precisely about its principal axis with the least moment of inertia. A photograph of the satellite is shown in Figure 3.18. As we have discussed, applying the torque-free Euler equations alone implies that this relative equilibrium should be Lyapunov stable. However, to the chagrin of its designers, the satellite quickly lost rotational stability and eventually approached a configuration in which it rotated about an axis perpendicular to its length (see [296] for details). This later spurred the study of *minimal energy configurations* in aeronautics and astrodynamics, which remains an active area of research today; see [78, 79] and [260].

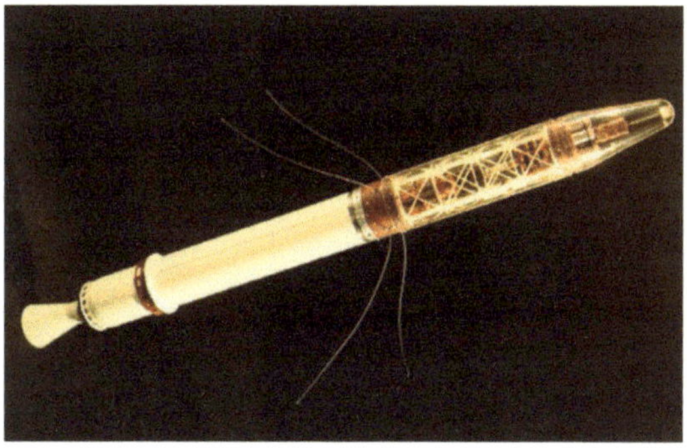

Figure 3.18: Photograph of Explorer 1 (courtesy of NASA).

4 Periodic Orbits

Previously we saw how fixed points in nonlinear systems possess invariant stable, unstable, and center manifolds, each of which lies tangent to the corresponding subspaces of a linearization. In this chapter, we will generalize this discussion to the theory of periodic orbits. In particular, we will look at families of periodic orbits, Poincaré maps, and Floquet theory.

4.1 Summation Notation

We first discuss Einstein's summation convention, a notational formalism for tensor multiplication first introduced by Albert Einstein during the early development of general relativity. The use of summation notation, as it is called, has its benefits as well as its drawbacks.

Let us begin by stating a number of rules that are universally employed when using summation notation. For our purposes, the only "tensors" we will encounter are scalars, vectors, and matrices; so you may simply think of a tensor as one of these.

Rules for summation notation

1. Each tensor is represented as having a certain number of "up-indices" (superscripted indices) as well as a certain number of "down-indices" (subscripted indices). The up-indices are called *contravariant* indices, whereas the down-indices are called *covariant* indices.
2. Whenever the same index is repeated in both an up and down position in a given term, summation over that index is implied.
3. Whenever an index is not repeated within a given term, that index is referred to as a *free index*.
4. Each term of an equation must possess the same free indices, appearing in the same positions (up or down).

By convention, we write the components of a vector with up-indices, and we write the components of a covector, i. e., a one-form, with down-indices. Thus, for example, V^μ represents the μth component of vector V, and ω_i represents the ith component of the one-form ω. As we will see in Chapter 6, the standard basis vectors of $\mathbb{R}^n$ are represented as

$$\left\{\frac{\partial}{\partial x^\mu}\right\}_{\mu=1}^{n},$$

so that vector V itself is literally

$$V = V^\mu \frac{\partial}{\partial x^\mu} = V^1 \frac{\partial}{\partial x^1} + \cdots + V^n \frac{\partial}{\partial x^n} = \sum_{\mu=1}^{n} V^\mu \frac{\partial}{\partial x^\mu}.$$

https://doi.org/10.1515/9783110597806-004

Similarly, the standard basis covectors of the dual space to $\mathbb{R}^n$ are represented as

$$\{dx^i\}_{i=1}^n,$$

so that the one-form ω is itself equivalent to

$$\omega = \omega_i dx^i = \omega_1 dx^1 + \cdots + \omega_n dx^n = \sum_{i=1}^{n} \omega_i dx^i.$$

Example 4.1. Let $A \in \mathbb{R}^{n\times n}$ and $v \in \mathbb{R}$. We will write the matrix product $x = Av$ using summation notation as follows:

$$x^\mu = A^\mu_\nu v^\nu.$$

This equation has a single free index that is a contravariant index. Hence this quantity represents the individual components of a vector. Moreover, by convention, the quantity A^μ_ν refers to the particular component of matrix A residing in the μth row and νth column. □

Example 4.2. Let $\omega = \omega_\mu dx^\mu$ be a one-form and $v = v^\mu \frac{\partial}{\partial x^\mu}$ a vector. Then ω acting on v is given by

$$\omega(v) = \omega_\mu dx^\mu \left(v^\nu \frac{\partial}{\partial x^\nu} \right) = \omega_\mu v^\nu dx^\mu \left(\frac{\partial}{\partial x^\nu} \right) = \omega_\mu v^\nu \delta^\nu_\mu = \omega_\mu v^\mu,$$

where δ^μ_ν is the Kronecker delta. Both μ and ν serve as summation indices in the preceding equation. □

Example 4.3. Given coordinates $(x^1, \ldots, x^n)$ on $\mathbb{R}^n$, we have

$$\frac{\partial x^\mu}{\partial x^\nu} = \delta^\mu_\nu,$$

where δ^μ_ν is the Kronecker delta. Thus, the Jacobian matrix of the identity transformation is the identity matrix. Moreover, given two smooth coordinate transformations $f : x \to y$ and $g : y \to z$, the chain rule yields

$$\frac{\partial z^\mu}{\partial y^\lambda} \frac{\partial y^\lambda}{\partial x^\nu} = \frac{\partial z^\mu}{\partial x^\nu},$$

which represents the components of the Jacobian matrix of the transformation $g \circ f$. □

4.2 Poincaré Maps

To study the stability and invariant manifolds of periodic orbits, we will make use of a Poincaré map, which is transversal to the periodic orbit at a particular point.

A Poincaré map is a way of converting a continuous dynamical system into a *discrete* dynamical system on a lower-dimensional set called the *Poincaré section*. Typically, the Poincaré section is taken to be a plane. In particular, we have the following definitions.

Definition 4.1. A *Poincaré section*, or *surface of section*, of a vector field $f : \mathbb{R}^n \to \mathbb{R}^n$ is an $(n-1)$-dimensional, orientable manifold Σ, which is submerged in $\mathbb{R}^n$ and transversal to vector field f at each $p \in \Sigma$. An *orientation* on Σ is a continuous unit normal vector field $n : \Sigma \to \mathbb{R}^n$, i. e., $|n| = 1$ and $n(x) \perp T_x\Sigma$ for all $x \in \Sigma$.

Definition 4.2. A *Poincaré map* of the nonlinear system $\dot{x} = f(x)$, associated with a given Poincaré section Σ, is a discrete map $P : \Sigma \to \Sigma$ defined by

$$Px_0 = \varphi(T; x_0),$$

where $\varphi : \mathbb{R} \times \mathbb{R}^n \to \mathbb{R}^n$ is the flow of vector field f and

$$T = \min\{t > 0 : \varphi(t, x_0) \in \Sigma \text{ and } f(\varphi(t, x_0)) \cdot n(\varphi(t, x_0)) > 0\}$$

for all $x_0 \in \Sigma$. If no orientation on Σ is specified, we instead define T by

$$T = \min\{t > 0 : \varphi(t, x_0) \in \Sigma\}.$$

A given Poincaré map may then be composed with itself any number of times, so that we may define the maps

$$P^n = \underbrace{P \circ \cdots \circ P}_{n \text{ times}}$$

for all $n \in \mathbb{N}$. For this reason, we sometimes refer to P as the *first return map*, to the map $P^2 : \Sigma \to \Sigma$ as the *second return map*, and so forth. An illustrative sketch of a Poincaré map is shown in Figure 4.1.

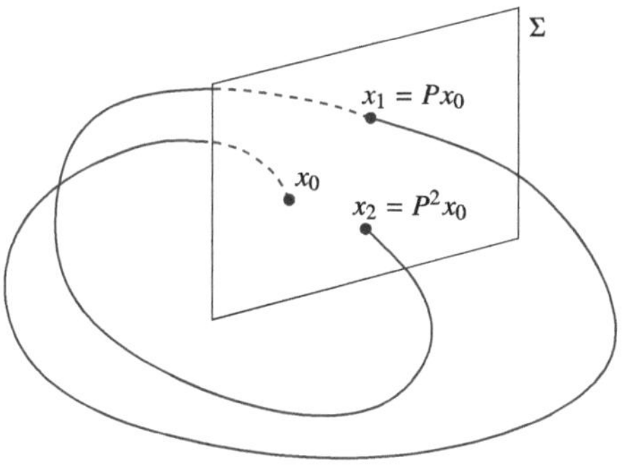

Figure 4.1: Illustrative sketch of a Poincaré map.

Example 4.4. As an example, we will consider the Hénon–Heiles system, which was proposed in [131] to describe the nonlinear planar motion of a star about the center of its galaxy. For further discussion, also see, for example, [23], [188], [249], and [280]. The equations of motion that govern this system are given by

$$\ddot{x} = -\frac{\partial U}{\partial x} \quad \text{and} \quad \ddot{y} = -\frac{\partial U}{\partial y},$$

where the potential $U(x, y)$ is given by

$$U(x,y) = \frac{1}{2}(x^2 + y^2) + x^2 y - \frac{y^3}{3}.$$

Following the discussion of §2.6, we may introduce the velocity variables $u = \dot{x}$ and $v = \dot{y}$ and write the Hénon–Heiles system as the coupled system of four first-order ordinary differential equations as follows:

$$\frac{d}{dt}\begin{bmatrix} x \\ y \\ u \\ v \end{bmatrix} = \begin{bmatrix} u \\ v \\ -x - 2xy \\ -y - x^2 + y^2 \end{bmatrix}. \tag{4.1}$$

This system has a first integral, which may be interpreted as the total mechanical energy, that is given by

$$E = \frac{1}{2}(u^2 + v^2) + U(x, y),$$

so that individual solution trajectories live on the 3-D constant-energy manifolds

$$\mathcal{M}_c = \{(x, y, u, v) \in \mathbb{R}^4 : E(x, y, u, v) = c\},$$

which foliate $\mathbb{R}^4$ (i. e., every point in $\mathbb{R}^4$ is contained in one of the level sets $\mathcal{M}_c$). We will take the planar surface

$$\Sigma = \{(x, y, u, v) \in \mathbb{R}^4 : x = 0\}$$

as our Poincaré section and consider the Poincaré map on $P : \Sigma \cap \mathcal{M}_c \to \Sigma \cap \mathcal{M}_c$. For our first trajectory, we choose initial conditions $x_0 = 0$, $y_0 = -0.1$, $v_0 = 0$, and u_0 such that the total energy of the orbit equals $E = 1/8$. The first 1,500 Poincaré returns to the surface $x = 0$ are plotted in Figure 4.2.

One clearly sees the ergodicity of the dynamics in this region. However, the bounded region to which the trajectories return has several "holes" through which this single trajectory never passes. Selecting initial conditions at the same energy in these vacuums, one obtains a more complete picture of the dynamics by plotting the Poincaré returns of several trajectories concurrently (Figure 4.3).

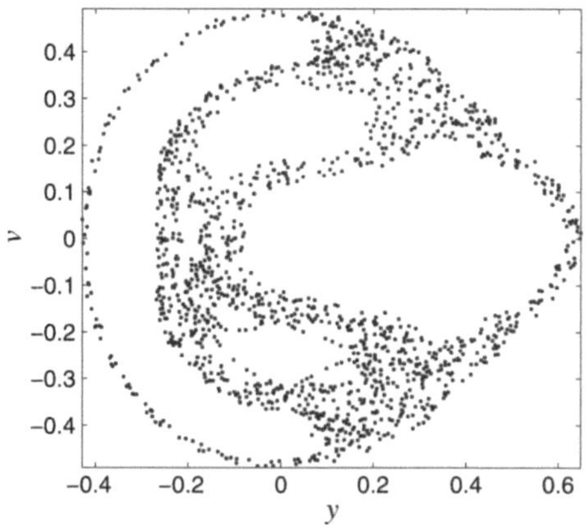

Figure 4.2: First 1,500 returns of a Poincaré map for a single trajectory in the stochastic region of the Hénon–Heiles system, with $E = 1/8$.

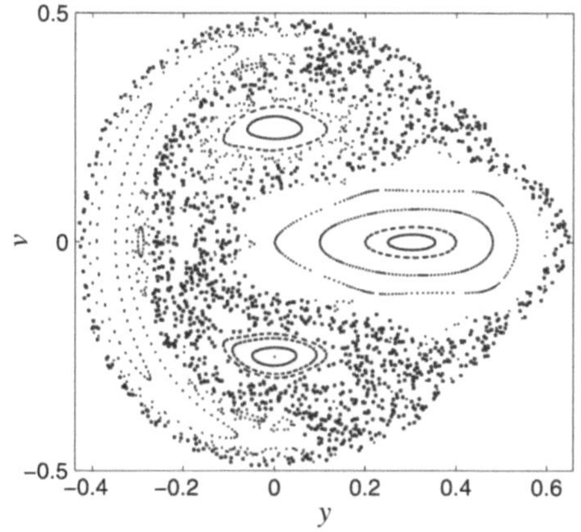

Figure 4.3: Poincaré maps for several trajectories in the Hénon–Heiles system, with $E = 1/8$.

To generate Poincaré maps using MATLAB, one integrates the equations of motion using `ode45.m`, making use of the built-in "Events" option. Sample code is provided below. Before calling `ode45.m` in the main file `henonmain.m`, we set the Events option to on by storing the name of our event file, `henonevents.m`, for Events in `odeset`, as illustrated below. Execution of `ode45` then returns additional variables `TE`, `XE`, `IE`, which correspond to the time of each event, the system state at each event, and whether or not the cataloged event was terminal, respectively. The event file, `henonevents.m`, is a function m-file that inputs the current time and state of the system and outputs the value of the event function, whether or not numerical integration should terminate when the system's state passes through a zero of the event function, and the direction, if any, in which the state passes through the zero of the event function. In the context of our current example, the event function is simply the variable $x(t)$, and the direction is taken as +1, so that an event (crossing the Poincaré section in the positive sense) is cataloged every time the solution trajectory passes through the 3-plane $x = 0$ in the positive sense, i. e., with x increasing. In the event function, the variable

ISTERMINAL can take the values 0 (for no) or 1 (for yes), and the variable DIRECTION may take the values 1, −1, or 0, depending on whether an event is cataloged when the state passes through a zero contour of the event function in the positive sense, the negative sense, or both senses, respectively. For more information, type `help ode45` at the command prompt in your open MATLAB window.

henonmain.m

```
clear
hold on
E = 1/8;
x0 = 0;
y0 = -0.1;
v0 = 0;
u0 = sqrt(2*E - y0^2 + 2*y0^3/3 - v0^2);
IC = [x0 y0 u0 v0];
Tspan = [0 10000];
OPTIONS = odeset('RelTol',1e-6,'AbsTol',1e-9);
OPTIONS = odeset('Events',@henonevents);
[T,X,TE,XE,IE] = ode45(@henonfun, Tspan, IC, OPTIONS);
plot(XE(:,2),XE(:,4),'b.')
xlabel('y'), ylabel('v')
```

henonfun.m

```
function F = henonfun(T,X)
x=X(1); y=X(2); u=X(3); v=X(4);
F = [u; v; -x - 2*x*y; -y -x^2 + y^2];
```

henonevents.m

```
function [VALUE,ISTERMINAL,DIRECTION] = henonevents(T,X)
VALUE = X(1);
ISTERMINAL = 0;
DIRECTION = 1;
```

□

Exercise 4.1. Plot the level contours of the potential function $U(x, y)$ on the x-y plane for $U = 1/6, 1/8, 1/12, 1/24$, and $1/100$.

Exercise 4.2. Use MATLAB to create a Poincaré map of the bounded region of the y-v plane on the energy surface $E = 1/12$.

4.3 Poincaré Reduction of the State-Transition Matrix

In this section, we describe how one may use the continuous-dynamics state-transition matrix in conjunction with a Poincaré map and energy integral to describe the local linearization of a periodic orbit. This is achieved by computing the local linearization of the corresponding discrete map on the intersection of the constant-energy manifold with the Poincaré section.

Poincaré Reduction

In this paragraph, we seek to determine first-order information about the flow near a periodic orbit. To achieve this, we will first derive a formula for the Jacobian of a general Poincaré map. To begin, let us fix an $\alpha \in \{1, \dots, n\}$ such that the flow of the system $\dot{x} = f(x)$ is transversal to the Poincaré section $\Sigma = \{x \in \mathbb{R}^n : x^\alpha = 0\}$ and let us consider a point $x_0 \in \Sigma$ that corresponds to an initial point on a trajectory.

Suppose that $T > 0$ is the amount of time required for the trajectory $x^\mu(t) = \varphi^\mu(t; x_0)$ to return to the surface Σ. Consider now a trajectory with a perturbed initial condition $x_0 + \delta x_0 \in \Sigma$, so that $\delta x_0^\alpha = 0$, and so that $|\delta x_0| \ll 1$. (Note that α is the fixed integer between 1 and n that corresponds to the coordinate defining the surface of section Σ.) In general, however, it is not the case that the perturbed trajectory returns to the Poincaré section after precisely a single period of the fiducial periodic orbit, i. e., in general, $\varphi^\alpha(T; x_0 + \delta x_0) \neq 0$. However, we may assume that $\varphi^\alpha(T + \delta T; x_0 + \delta x_0) = 0$ for some small δT.

We wish to construct a reduced, $(n-1) \times (n-1)$ state-transition matrix $\tilde{\Phi}^I_J$ $(I, J = 1, \dots, n; \neq \alpha)$ that maps an initial perturbation about the fixed point x_0, on the surface Σ, into the linearization of its first return, so that

$$P^I(x_0 + \delta x_0) = P^I(x_0) + \tilde{\Phi}^I_J \delta x_0^J + O(|\delta x_0|^2). \tag{4.2}$$

To do this, we require an estimate for the time required for the orbit $\varphi(t; x_0 + \delta x_0)$ to make its first return to Σ. Define the time-varying, trajectory displacement vector by $\xi(t) = \varphi(t; x_0 + \delta x_0) - \varphi(t; x_0)$ (so that, for example, $\xi(0) = \delta x_0 \in T_{x_0}\Sigma$). We may expand equation (2.12) about T, to the first order in δT, to obtain

$$\xi^\mu(T + \delta T) = \Phi^\mu_J \delta x_0^J + \dot{x}^\mu(T)\delta T,$$

where δT is defined such that $\xi^\alpha(T + \delta T) = 0$. Please note the utility of our summation convention. The free index μ runs from $\mu = 1, \dots, n$; however, the capital roman letters run only through $I, J = 1, \dots, n; \neq \alpha$. Since we are constraining $\xi^\alpha(T + \delta T) = 0$, we can readily solve for δT to find

$$\delta T = -\frac{1}{\dot{x}^\alpha(T)} \Phi^\alpha_J \delta x_0^J.$$

There is an exception in the summation notation. Since we have declared α a fixed constant, summation over the repeated αs is not implied. Combining the above two equations, we see that, to the first order,

$$P^I(x_0 + \delta x_0) - P^I(x_0) = \xi^I(T + \delta T) = \left(\Phi^I_J - \frac{\dot{x}^I(T)}{\dot{x}^\alpha(T)} \Phi^\alpha_J \right) \delta x^J_0.$$

We thus have obtain a reduced $(n-1) \times (n-1)$ state-transition matrix

$$[DP]^I_J = \tilde{\Phi}^I_J = \Phi^I_J - \frac{\dot{x}^I(T)}{\dot{x}^\alpha(T)} \Phi^\alpha_J, \tag{4.3}$$

so that, to the first order, we obtain (4.2). If we further require that x_0 is a point on a first-order periodic orbit with period T, so that $P^I(x_0) = x^I_0$, the preceding equation reduces to

$$P^I(x_0 + \delta x_0) = x^I_0 + \tilde{\Phi}^I_J \delta x^J_0.$$

Energy Reduction

We next suppose that our system is further in possession of an integral of motion $J : \mathbb{R}^n \to \mathbb{R}$, which often corresponds to an energy integral for the system. Trajectories will therefore be constrained to lie on a constant-energy manifold

$$\mathcal{M}_c = \{x \in \mathbb{R}^n : J(x) = c\}. \tag{4.4}$$

Suppose now that we would like to further restrict our choice of initial perturbation δx^I_0 so that it lies on this constant-energy manifold. We obtain a variation in the value of J with respect to variations in the initial condition by the equation

$$\delta J = \frac{\partial J}{\partial x^I} \delta x^I_0,$$

where the sum over I skips the index α, since by definition $\delta x^\alpha_0 = 0$. Constraining the variation to lie on the constant-energy manifold therefore implies that the $(n-1)$ initial variations are not linearly independent. We may therefore select a fixed index $A \in \{1, \ldots, \alpha - 1, \alpha + 1, \ldots, n\}$, subject to the condition $J_A \neq 0$, so that

$$\delta x^A_0 = -\frac{J_j(x_0)}{J_A(x_0)} \delta x^j_0,$$

where the indices i, j run from $i, j = 1, \ldots, n; \neq \alpha, A$. (Note the subscript J_μ denotes the partial derivative of J with respect to x^μ.) Equation (4.2) therefore implies that

$$P^i(x_0 + \delta x_0) = P^i(x_0) + \left(\tilde{\Phi}^i_j - \tilde{\Phi}^i_A \frac{J_j(x_0)}{J_A(x_0)} \right) \delta x^j_0.$$

We hence define a twice reduced $(n-2) \times (n-2)$ state-transition matrix for the reduced Poincaré map by the relation

$$M^i_j = \tilde{\Phi}^i_j - \tilde{\Phi}^i_A \frac{J_j(x_0)}{J_A(x_0)}, \tag{4.5}$$

where $\tilde{\Phi}$ is given by (4.3), so that $\delta x^i(T) = M^i_j \delta x^j_0$. A similar construction was used in [262] in the analysis of the dynamic environment of the asteroid 433 Eros; see also [1].

4.4 Invariant Manifolds of Periodic Orbits

As we have seen, the Poincaré map is an effective tool for reducing by one dimension the dimensionality of a nonlinear system in the vicinity of a periodic orbit, which manifests itself as a fixed point on the Poincaré section. As with fixed points in continuous nonlinear systems, periodic orbits possess certain invariant structures; these structures may be better understood in terms of the local, discrete maps they induce in the vicinity of the nominal orbit on transversal sections to the flow. Before we dissect such structures, let us first examine the invariant subspaces of linear discrete maps.

Theorem 4.1. *Consider the discrete map*

$$x_{n+1} = Ax_n, \tag{4.6}$$

where $x_n \in \mathbb{R}$, $A \in \mathrm{GL}(n;\mathbb{R})$, and suppose that A has eigenvalues $\lambda_j \in \mathbb{C}$ and associated, generalized eigenvectors $w_j = u_j + iv_j \in \mathbb{C}^n$. Then the subspaces defined by

$$\begin{aligned} E^s &= \mathrm{span}\{u_j, v_j : |\lambda_j| < 1\}, \\ E^u &= \mathrm{span}\{u_j, v_j : |\lambda_j| > 1\}, \\ E^c &= \mathrm{span}\{u_j, v_j : |\lambda_j| = 1\}, \end{aligned}$$

such that $\mathbb{R}^n = E^s \oplus E^u \oplus E^c$, are invariant with respect to action (4.6). Moreover, for all $x_0 \in E^s$, the sequence defined recursively by (4.6) has the property that $\lim_{n\to\infty} x_n = 0$. Similarly, for all $x_0 \in E^u$, the sequence defined recursively by (4.6) has the property that $\lim_{n\to\infty} x_n = \infty$. The subspaces E^s, E^u, and E^c are referred to, respectively, as the stable, unstable, and center subspaces of discrete map (4.6).

Invariance of subspaces E^s, E^u, and E^c follows from Lemma 1.2. It should be self-evident why the stability is determined by the moduli of the eigenvalues. In §4.6, we will explain why this differs from continuous systems, for which the stable, center, and unstable subspaces are defined based on the polarity of the real part of λ_j rather than the value of its magnitude in comparison to unity.

Remark. Suppose that coefficient matrix A of linear discrete system (4.6) has a unit eigenvalue $\lambda = 1$ with associated eigenvector u. Then every point on the subspace span$\{u\}$ is a fixed point of mapping (4.6). Similarly, if coefficient matrix A has an eigenvalue $\lambda = -1$ with associated eigenvector u, then every point on the subspace span$\{u\}$ is a periodic orbit with period 2 of mapping (4.6).

We next state the invariant manifold theorem for discrete maps, which is directly analogous to the development of invariant manifolds of fixed points in continuous dynamical systems, discussed in Chapter 3.

Theorem 4.2 (Invariant Manifold Theorem for Discrete Maps). *Let x^* be a fixed point of the nonlinear, discrete map $x_{n+1} = f(x_n)$ and let E^s, E^u, and E^c be, respectively, the stable, unstable, and center subspaces of the linearized discrete map $\xi_{n+1} = Df(x^*)\xi_n$, where $\xi_n = x_n - x^*$. Then there exist invariant stable, unstable, and center manifolds of x^*, which we will denote by M^s, M^u, and M^c, respectively. Moreover, the stable, unstable, and center manifolds are of equal dimension and lie tangent to the correspondingly named subspaces of the linearization.*

For discrete mappings, the invariance of a set, let us say the stable manifold M^s, implies that $f(x_n) \in M^s$ for all $x_n \in M^s$. Hence the corresponding trajectory of an initial condition is comprised of a discrete set of points contained within the invariant manifold that houses the initial condition. Development of the stable and center manifold theorem for discrete mappings follows analogously from the corresponding development of the stable and center manifold theorems for continuous dynamical systems, as discussed in Chapter 3.

Definition 4.3. Let γ be a periodic orbit of the continuous system $\dot{x} = f(x)$ that corresponds to a period-1 orbit of the Poincaré map $P : \Sigma \to \Sigma$ for some surface of section Σ. Let M^s, M^u, and M^c be respectively the stable, unstable, and center manifolds of the fixed point $x^* \in \Sigma$ corresponding to the periodic orbit γ. Then the flows $W^s(\gamma) = \varphi(t; M^s)$, $W^u(\gamma) = \varphi(t; M^u)$, and $W^c(\gamma) = \varphi(t; M^c)$ are referred to as the stable, unstable, and center manifolds of the periodic orbit γ, respectively.

Our next theorem shows that the stable, unstable, and center manifolds of a periodic orbit have the properties and dimensions that one would expect. It follows directly from the invariant manifold theorem for discrete maps.

Theorem 4.3 (Invariant Manifold Theorem for Periodic Orbits). *Let γ be a periodic orbit that corresponds to the fixed point $x^* \in \Sigma$ of the Poincaré map $P : \Sigma \to \Sigma$. If the linearization of the Poincaré map, $DP = \tilde{\Phi}$, has a k-dimensional stable subspace, an $(n - k - p - 1)$-dimensional unstable subspace, and a p-dimensional center subspace, then the sets $W^s(\gamma)$, $W^u(\gamma)$, and $W^c(\gamma)$ are $(k + 1)$-, $(n - k - p)$-, and $(p + 1)$-dimensional invariant manifolds, respectively. Moreover, they intersect at γ transversally and possess the stability properties $\omega(x_0) = \gamma$ for all $x_0 \in W^s(\gamma)$ and $\alpha(x_0) = \gamma$ for all $x_0 \in W^u(\gamma)$.*

In the case where the preceding formula yields a 1-D invariant manifold, this manifold is understood to correspond to the periodic orbit itself and is usually ignored by convention. For example, if the linearized Poincaré map DP does not have a center subspace, then the preceding theorem indicates that there is a 1-D center manifold for the periodic orbit γ. In this degenerate case, this 1-D manifold refers to the periodic orbit γ itself, so we say that γ does not have a center manifold.

As we will see in the next section, one may use the linearization of the Poincaré map (4.3) to locate 1-D, natural families of periodic orbits.

4.5 Families of Periodic Orbits

In this paragraph, we show how one may use the reduced state-transition matrices (4.3) and (4.5) to locate natural families of periodic orbits. To aid our notation, let us define the reduced coordinates $q \in \Sigma \cong \mathbb{R}^{n-1}$ and $r \in \Sigma \cap \mathcal{M}_c \cong \mathbb{R}^{n-2}$, so that we may describe the variational equations as $\delta q(T) = \tilde{\Phi}\delta q$ and $\delta r(T) = M\delta r$. A similar predictor–corrector method for computing natural families of periodic orbits is discussed in [67].

Predictor Step

We begin by assuming that we already know the location of a single fixed point $q^* \in \Sigma$, so that $P(q^*) = q^*$. This corresponds to a periodic orbit in a continuous system, with period T, so that $\varphi(T; q^*) = q^*$. We now wish to determine whether or not there is a neighboring fixed point on Σ, corresponding to a periodic orbit in the continuous system, with a slightly varied value of total energy. Instead of using the twice reduced state-transition matrix M and varying energy, we will use the once reduced state-transition matrix $\tilde{\Phi}$. We seek a $\delta q \in T_{q^*}\Sigma$ such that $q^* + \delta q$ is a fixed point of P, i. e.,

$$\begin{aligned} q^* + \delta q &= P(q^* + \delta q) \\ &= P(q^*) + \tilde{\Phi}\delta q. \end{aligned}$$

But since q^* is a fixed point of P, the preceding equation simplifies to

$$\delta q = \tilde{\Phi}\delta q, \tag{4.7}$$

which we immediately recognize as an eigenvalue problem. We have thus obtained the following theorem.

Theorem 4.4. *Let $q^* \in \Sigma$ be a fixed point of the Poincaré map $P : \Sigma \to \Sigma$ and $\tilde{\Phi} = DP$ the variation of the Poincaré map at q^*, as defined in (4.3). If $\tilde{\Phi}$ has an eigenvalue $\lambda = 1$ with geometric multiplicity m and if E^p is the associated eigenspace, then there exists* *an m-dimensional invariant manifold $M^p \subset \Sigma$ that is tangent to E^p at q^* and that is comprised entirely of fixed points of Poincaré map P.*

In particular, if the reduced state-transition matrix $\tilde{\Phi}$ has a single eigenvalue equal to $\lambda = 1$, then its associated eigenvector points tangent to the direction of a one-parameter family of fixed points of P that passes through the fixed point q^*. Since each of these fixed points corresponds to a periodic orbit of the full-dimensional system, one has thereby uncovered a one-parameter family of periodic orbits.

Corrector Step

In the previous section, we showed how one may use the linearization of the Poincaré map to estimate the location of a periodic orbit of different energy in the vicinity of a nominal periodic orbit. Due to the nonlinearities, the predicted location of the new periodic orbit will only be an estimate, which should then be refined with a corrector step. To correct for the error in our prediction, let us begin by assuming that we have an initial approximation $r_0 \in \Sigma \cap \mathcal{M}_c$ to a fixed point of P, where $\mathcal{M}_c$ is the constant-energy manifold defined in (4.4) and c is the energy of the orbit passing through r_0. We assume that there exists a displacement vector δr_0, with $|\delta r_0| \ll 1$, such that the corrected point $r_0 + \delta r_0$ is a fixed point of the Poincaré map with energy c, i. e., we assume that

$$r_0 + \delta r_0 = P(r_0 + \delta r_0).$$

Using the twice reduced state-transition matrix (4.5), we find that, to the first order,

$$r_0 + \delta r_0 = P(r_0) + M\delta r_0.$$

Rearranging and assuming that the matrix M does not possess a unity eigenvalue, we solve for the displacement vector, obtaining

$$\delta r_0 = (I - M)^{-1}(P(r_0) - r_0), \tag{4.8}$$

where I is understood to be the $(n-2) \times (n-2)$ identity matrix.

4.6 Floquet Theory

Suppose now that the trajectory $y(t) = \varphi(t; x_0)$, for some $x_0 \in \Sigma$, is a periodic orbit with period $T > 0$. Recall that the state-transition matrix satisfies the initial value problem

$$\dot{\Phi}(t) = Df(y(t))\Phi(t), \quad \Phi(0) = I, \tag{4.9}$$

i. e., the state-transition matrix is the principal matrix solution of the linearized equations (2.13). Thus, if y represents a periodic orbit with period T, the coefficient matrix $Df(y(t))$ is itself T-periodic since it is evaluated along the nominal trajectory. Of particular distinction is the value of the state-transition matrix after a single orbit period; this value merits its own name.

Definition 4.4. Let $\Phi(t)$ be the state-transition matrix corresponding to a periodic orbit y with period T. Then the matrix $\Phi(T)$ is referred to as the *monodromy matrix* of the periodic orbit.

The monodromy matrix of a periodic orbit may be used in conjunction with an important theorem due to [94] to make long-term predictions regarding the linearized flow.

Theorem 4.5 (Floquet). *Any fundamental matrix solution to the nonautonomous, linear system*

$$\dot{x} = A(t)x, \tag{4.10}$$

where $A : \mathbb{R} \to \mathbb{R}^{n\times n}$ *is a* T*-periodic coefficient matrix, takes the form*

$$\Psi(t) = Z(t)e^{Rt}, \tag{4.11}$$

where $R \in \mathbb{R}^{n\times n}$ *is a constant matrix and* $Z : \mathbb{R} \to \mathrm{GL}(n;\mathbb{R})$ *is a nonsingular, differentiable,* T*-periodic matrix-valued function of time. Moreover, if* $\Phi(t)$ *is the principal matrix solution, then* $Z(0) = I$.

Proof. Let $\Psi : \mathbb{R} \to \mathrm{GL}(n;\mathbb{R})$ be any fundamental matrix solution for (4.10). Since coefficient matrix $A(t)$ is T-periodic, it follows that $\Psi(t+T)$ is also a fundamental matrix solution for (4.10), and thus $\Psi(t+T) = \Psi(t)C$, for some nonsingular $C \in \mathrm{GL}(n;\mathbb{R})$, since any two fundamental matrix solutions are connected in this way (see [126] or [56] for details). Therefore, there exists a matrix R, not necessarily unique, such that $C = e^{RT}$. Now define

$$Z(t) = \Psi(t)e^{-Rt},$$

as in (4.11). To check whether $Z(t)$ is T-periodic, let us compute

$$Z(t+T) = \Psi(t+T)e^{-RT}e^{-Rt} = \Psi(t)e^{-Rt} = Z(t).$$

Hence, the theorem follows. □

This theorem naturally applies to the linearization about a periodic orbit in a nonlinear system, as we will see in the following corollary.

Corollary 4.1. *Let γ be a periodic orbit of a nonlinear system $\dot{x} = f(x)$. Then the state-transition matrix along γ, determined by the initial value problem (4.9), takes the form*

$$\Phi(t) = Z(t)e^{Rt}, \tag{4.12}$$

where $Z : \mathbb{R} \to \mathrm{GL}(n; \mathbb{R})$ is a nonsingular, differentiable, T-periodic matrix-valued function of time that satisfies the boundary conditions $Z(0) = Z(T) = I$.

We next define characteristic exponents and multipliers of a periodic orbit, making an important distinction between the eigenvalues of R and the eigenvalues of the monodromy matrix e^{RT}.

Definition 4.5. Let γ be a periodic orbit in a nonlinear system $\dot{x} = f(x)$ and $\Phi(t)$ the state-transition matrix along γ, which is of the form given by (4.12). Then the *characteristic exponents* of the periodic orbit γ are the eigenvalues λ_i of the constant matrix R. The *characteristic multipliers* (also called *Floquet multipliers*) of the periodic orbit γ are the eigenvalues μ_i of the monodromy matrix e^{RT}.

As a corollary to this definition, we have the following obvious proposition.

Proposition 4.1. *The characteristic multipliers and characteristic exponents of a periodic orbit are related by the equation $\mu_i = e^{\lambda_i T}$. Moreover, we have the statements*
- $\mathbb{R}\{\lambda_i\} < 0$, *if and only if* $|\mu_i| < 1$;
- $\mathbb{R}\{\lambda_i\} = 0$, *if and only if* $|\mu_i| = 1$;
- $\mathbb{R}\{\lambda_i\} > 0$, *if and only if* $|\mu_i| > 1$.

Example 4.5. Let x^* be a fixed point of the nonlinear system $\dot{x} = f(x)$. The solution trajectory through x^* is the constant trajectory $\gamma(t) = \varphi(t; x^*) = x^*$. For each $T > 0$, we might consider regarding γ as a periodic orbit with period T. In this case, the T-periodic coefficient matrix $Z(t)$ found in (4.12) must identically equal the identity matrix, $Z(t) = I$, so that the state-transition matrix is given by $\Phi(t) = e^{Rt}$. It therefore follows that $\dot{\Phi} = R\Phi$ and, by comparison with (4.9), that

$$R = Df(x^*).$$

Given the correspondences between the characteristic multipliers and characteristic exponents laid out in Proposition 4.1, we immediately see that the stable and center manifold theorems for fixed points are a special case of the stable and center manifold theorems for periodic orbits. □

Next, we will see how the characteristic exponents (and hence multipliers) of a periodic orbit are related to its Lyapunov exponents.

Proposition 4.2. *Let γ be a periodic orbit of the nonlinear system $\dot{x} = f(x)$. Then the Lyapunov exponents, ν_i, of the trajectory γ are simply the characteristic exponents, so that $\nu_i = \lambda_i$.*

Proof. Since γ is a periodic orbit, its state-transition matrix must be of the form given by (4.12). Let μ_i be the eigenvalues of e^{RT} (i. e., the characteristic multipliers) and let ξ_i be the associated eigenvectors, normalized so that $|\xi_i| = 1$. Next, let us use definition (2.17) to calculate the Lyapunov exponents along γ in the directions ξ_i for $i = 1, \dots, n$. We obtain

$$\chi(\xi_i) = \lim_{t\to\infty} \frac{1}{t} \log \|\Phi(t)\xi_i\|.$$

We can take the limit to exist by Oseledec's multiplicative ergodic theorem [229]. Therefore, for any monotonically increasing sequence (t_n), we obtain the limit

$$\chi(\xi_i) = \lim_{n\to\infty} \frac{1}{t_n} \log \|\Phi(t_n)\xi_i\|.$$

In particular, let us choose $t_n = nT$, where T is the period of the orbit γ. Since $Z(nT) = I$, for all $n \in \mathbb{N}$, we therefore obtain

$$\chi(\xi_i) = \lim_{n\to\infty} \frac{1}{nT} \log \|e^{nRT}\xi_i\| = \lim_{n\to\infty} \frac{1}{nT} \log \|\mu_i^n \xi_i\| = \lim_{n\to\infty} \frac{1}{T} \ln \mu_i.$$

Since $\mu_i = e^{\lambda_i T}$, it follows that $\chi(\xi_i) = \lambda_i$. Moreover, since the full spectrum of Lyapunov exponents is obtained, we have $\nu_i = \chi(\xi_i)$ for each of the eigenvalues ξ_i of the matrix e^{RT}, thereby completing the proof. □

Example 4.6. As an example of Floquet theory, we again consider the chaotic regime (energy $E = 1/8$) of the Hénon–Heiles system (4.1). Two periodic orbits for this system were recently published in [1]: the first has a period $T = 32.040$ and initial condition $y_0 = -0.0170$; the second has a period $T = 32.377$ and initial condition $y_0 = 0.5729$. Both orbits initially pass through the planes $x = 0$ and $v = 0$ and lie on the constant-energy surface $\mathcal{M}_{1/8}$, so that the initial value u_0 may be obtained using

$$u_0 = \sqrt{2E - y^2 + \frac{2y^3}{3}},$$

where $E = 1/8$. The orbits corresponding to these initial conditions are shown in Figure 4.4.

Exercise 4.3. Write a program that reproduces the trajectories shown in Figure 4.4 for the Hénon–Heiles system (4.1). Compute the linearized equations (2.11) that determine the state-transition matrix for the Hénon–Heiles system. Numerically integrate these with the two periodic orbits for a single-orbit period to determine the Floquet multipliers and Lyapunov exponents for each trajectory.

This example illustrates the importance of the theory of periodic orbits in dynamical systems. Being asymptotic quantities, Lyapunov exponents are typically estimated

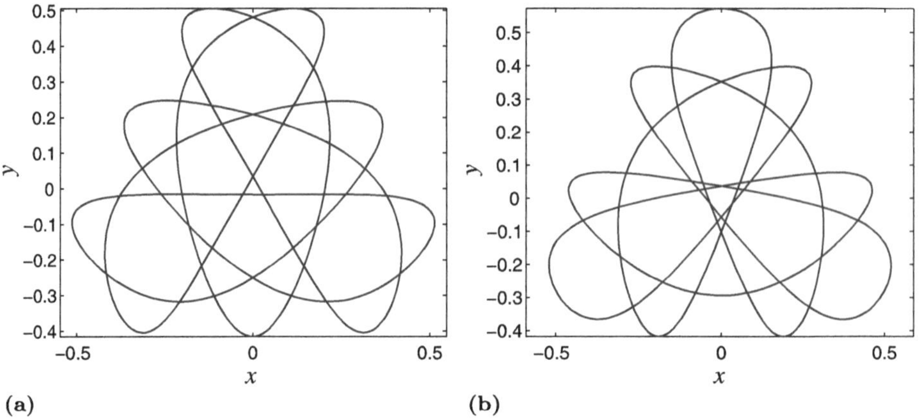

Figure 4.4: An unstable (a) and stable (b) periodic orbit in the Hénon–Heiles system.

using long-term integrations that must be handled delicately to achieve accurate results. On the other hand, Lyapunov exponents for periodic orbits are much simpler to compute, as one only needs to carry out a numerical simulation for a fixed interval of time, corresponding to the period of the orbit. In addition to this computational benefit, periodic orbits are often of direct practical value of themselves. In astrodynamics, the natural place to park a spacecraft or satellite is on a periodic orbit. Determining the stability of such orbits is therefore pertinent to real-life applications. □

Exercise 4.4. Using the linearization of the fixed point L_1 in the circular restricted three-body problem (2.36)–(2.39), estimate the location of a periodic orbit (on the center manifold) of the fixed point L_1, numerically verify and correct the orbit using (4.8), and determine its stability. Such orbits are similar to *halo orbits* (see, for example, [88]). More than a mere mathematical curiosity, halo orbits have been used during many satellite missions to the libration points in the Earth–Sun system; see, for example, [76] and [149].

4.7 Application: Periodic Orbit Families in the Hill Problem

In this section, we discuss a limiting case of the circular restricted three-body problem discussed in §2.8, which describes motion in the vicinity of the second body as $\mu \to 0$. The resulting system is known as the *Hill restricted three-body problem* and was first proposed by Hill [136]. Early references to periodic orbits discovered in the Hill problem are [153], [207, 208], and [283]. The first comprehensive study of the families of periodic orbits in the Hill problem, a topic we will consider in this section, is described in [128]. New families of periodic orbits were more recently found by Hénon [129, 130] and Perko [234]. The term "Hill's equation," however, has been given a broader meaning and may also refer to an entire class of differential equations, with applications to science, engineering, physics, astrodynamics, and electrical engineering (see [194]).

Hill's equations are not obtained by simply substituting $\mu = 0$ into the equations governing the circular restricted three-body problem (2.34)–(2.35): such an endeavor can only lead to the classically understood motions of the two-body problem. Instead, for each small, nonzero value of μ, we enlarge the region immediately surrounding the second body by an appropriate coordinate change. Within this region, the gravitational effects due to the second body are of the same order or greater than the gravitational effects due to the first body. (For definiteness, think of the first body as the Sun and the second body as the Earth.) Such a scaling must depend upon the mass parameter μ. Finally, we let $\mu \to 0$ while simultaneously enlarging the vicinity of the second body via our one-parameter family of coordinate transformations. The resulting equations determine the dynamics of Hill's circular restricted three-body problem.

Exercise 4.5. By making the substitutions

$$x = 1 - \mu + \mu^{1/3}\xi \quad \text{and} \quad y = \mu^{1/3}\eta$$

in the second-order system (2.34)–(2.35) and by subsequently taking the limit as $\mu \to 0$, show that one obtains the resulting system

$$\ddot{\xi} - 2\dot{\eta} = \frac{\partial U}{\partial \xi}, \tag{4.13}$$

$$\ddot{\eta} + 2\dot{\xi} = \frac{\partial U}{\partial \eta}, \tag{4.14}$$

where

$$U(\xi, \eta) = \frac{1}{\rho} + \frac{3\xi^2}{2} \tag{4.15}$$

and where $\rho = \sqrt{\xi^2 + \eta^2}$.

Exercise 4.6. Show that the function

$$J(\xi, \eta, \dot{\xi}, \dot{\eta}) = \frac{1}{2}(\dot{\xi}^2 + \dot{\eta}^2) - U(\xi, \eta), \tag{4.16}$$

where $U(\xi, \eta)$ is defined in (4.15), is an integral of motion for system (4.13)–(4.14). This function is the *Jacobi integral* for the Hill problem.

By defining the state variable $x = (\xi, \eta, u, v)$, we can rewrite system (4.13)–(4.14) as the first-order system

$$\frac{d}{dt}\begin{bmatrix} \xi \\ \eta \\ u \\ v \end{bmatrix} = \begin{bmatrix} u \\ v \\ 2v - \xi/\rho^3 + 3\xi \\ -2u - \eta/\rho^3 \end{bmatrix}. \tag{4.17}$$

The state-transition matrix satisfies matrix differential equation (2.11) and initial condition (2.10). For this example, the Jacobian of $f(x)$ is given by

$$Df(x) = \frac{\partial f}{\partial x} = \begin{bmatrix} 0 & 0 & 1 & 0 \\ 0 & 0 & 0 & 1 \\ A & B & 0 & 2 \\ C & D & -2 & 0 \end{bmatrix},$$

where

$$\begin{aligned} A &= 3 - \frac{1}{\rho^3} + \frac{3\xi^2}{\rho^5}, \\ B = C &= \frac{3\xi\eta}{\rho^5}, \\ D &= -\frac{1}{\rho^3} + \frac{3\eta^2}{\rho^5}. \end{aligned}$$

Hence one must solve the system

$$\begin{aligned} \dot{x}(t) &= f(\varphi(t; x_0)), \\ \dot{\Phi}(t) &= Df(\varphi(t; x_0)) \cdot \Phi(t) \end{aligned}$$

of 20 coupled, nonlinear differential equations to solve for a given solution trajectory with its time-varying state-transition matrix.

Exercise 4.7. Plot the forbidden regions of the Hill restricted three-body problem, (4.13)–(4.15), for various values of the Jacobi constant, (4.16). The following MATLAB code, which was used to generate one of the plots in Figure 2.14, might be useful (with modification for the Hill problem).

```
mu = 0.15;
[X,Y] = meshgrid(linspace(-2,2), linspace(-2,2));
r1 = sqrt((X+mu).^2+Y.^2);
r2 = sqrt((X-1+mu).^2+Y.^2);
C = -3;
F = -C - 0.5*(X.^2+Y.^2) - (1-mu)./r1 - mu./r2;
contourf(X,Y,F,[0 0],'linewidth',3)
hold on
plot(-mu,0,'kx','linewidth',2)
plot(1-mu,0,'kx','linewidth',2)
xlabel('x'), ylabel('y')
hold off
```

Exercise 4.8. In this exercise, you will use the predictor–corrector method that was presented earlier in the chapter to locate a periodic orbit in the Hill problem.

(a) Write a MATLAB program that uses ode45 to integrate Hill's equations (4.17) and the linearized equations for the state-transition matrix. Use initial conditions

$$\xi_0 = -0.36577, \quad \eta_0 = 0, \quad u_0 = 0,$$

and v_0 so that the Jacobi integral (4.16) takes the value $J = 0$. Use the Events option to terminate the numerical integration when the trajectory returns to the plane $\eta = 0$, passing in the positive sense, with $\xi < 0$. (The condition $\xi < 0$ can be taken into account with an IF THEN when determining the value of ISTERMINAL in the EVENTS function.)

(b) Determine the twice reduced state-transition matrix upon the first return of the trajectory and use it to correct for the initial condition of the trajectory.

(c) Use the corrected initial condition and rerun the program. Plot the resulting trajectory on the ξ-η plane, thereby reproducing the trajectory in Figure 4.5.

(d) Determine the error $\|P(r_0) - r_0\|$ using both the initial and corrected trajectory. Here, $r = (\xi, u, v)$.

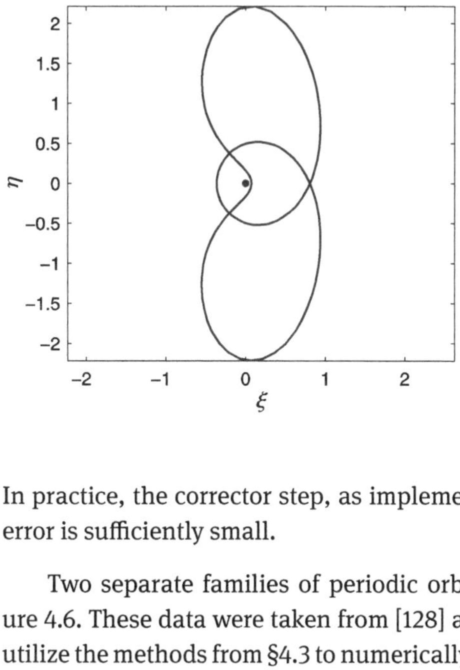

Figure 4.5: A periodic orbit in the Hill system.

In practice, the corrector step, as implemented in (b), is normally repeated until the error is sufficiently small.

Two separate families of periodic orbits in the Hill problem are shown in Figure 4.6. These data were taken from [128] and the references therein, though one can utilize the methods from §4.3 to numerically integrate along these families, thereby reconstructing these plots. Each point on the two curves g and g' represents a symmetric periodic orbit in the Hill problem. The constant Γ is a multiple of the Jacobi constant, $\Gamma = -2J$, that is used due to convention. Since these periodic orbits are symmetric, one takes $u_0 = 0$. One finally determines v_0 using the Jacobi constant together with the value of ξ_0. Several periodic orbits from the g-family are plotted in Figure 4.7.

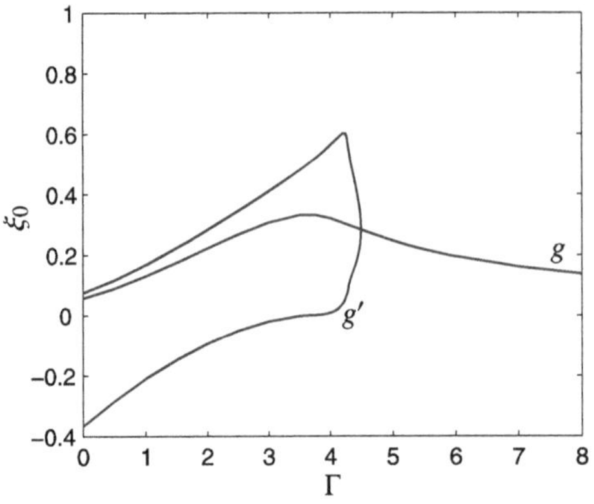

Figure 4.6: Two families of symmetric periodic orbits on the (Γ, ξ_0) plane. Each point represents a periodic orbit.

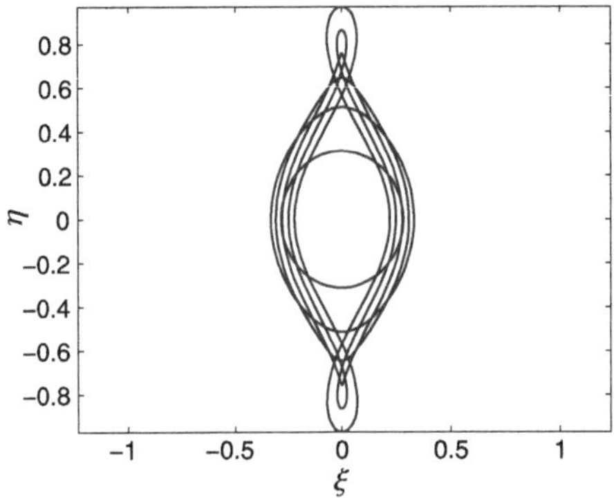

Figure 4.7: Several members of a symmetric periodic orbit family (family g).

It is interesting to observe the eigenvalues of the 3×3 reduced state-transition matrix $\tilde{\Phi}$ as one traces these families of periodic orbits. As it turns out, $\det(\tilde{\Phi}) = 1$, so the product of the eigenvalues equals 1. Each of these periodic orbits has a unity eigenvalue whose associated eigenvector gives the direction of the nearby periodic orbit in (ξ, u, v)-space. As one follows the g-family, the eigenvalues of $\tilde{\Phi}$ go from a unity eigenvalue and two reciprocal real eigenvalues (left-half of curve) to a unity eigenvalue and two complex conjugate eigenvalues with unit modulus (right-half of curve). The transition occurs precisely at the intersection of the g- and g'-families, at which point $\tilde{\Phi}$ has a single repeated eigenvalue $\lambda = 1$ with algebraic multiplicity 3 and geometric multiplicity 1. At this intersection point, the single linearly independent eigenvector lies tangent to the g'-family, and one of the generalized eigenvectors lies tangent to the g-family.

5 Bifurcations and Chaos

In this final chapter on nonlinear systems, we introduce an important result for 2-D phase flows, known as the *Poincaré–Bendixson theorem*. In addition, we discuss bifurcations and hysteresis in planar dynamical systems and present a brief overview of chaos theory in higher-dimensional systems. The Poincaré–Bendixson theorem provides a complete enumeration of the possible fates of an orbit that enters a trapping region in two dimensions: it can either settle down to a fixed point or approach a limit cycle. This theorem rules out chaos from ever taking hold in 2-D autonomous systems. Next, we provide a brief overview of *bifurcations* in nonlinear systems and a related effect known as *hysteresis*. These are relevant for systems whose behavior depends on a parameter. A *bifurcation* is a sudden qualitative change in the behavior of a system. *Hysteresis* is a powerful consequence of nonlinearity: systems that undergo a bifurcation do not necessarily return to their original state, even if the parameter returns to its nominal levels. We will see why the greatest deficit of modern society is our inability, on the whole, to think nonlinearly. Finally, we will conclude the chapter with *period doubling bifurcations* and a brief introduction to the theory of chaos.

5.1 Poincaré–Bendixson Theorem

The main subject of this paragraph is the *Poincaré–Bendixson theorem*, first introduced in [237] and [19]. This theorem applies to planar autonomous systems. Theoretically it is important for two fundamental reasons: it can be used as a tool to prove the existence of a periodic orbit in planar flows, and it also proves that *chaos* is impossible in smooth, 2-D autonomous flows.

For a generalization of the Poincaré–Bendixson theorem to closed 2-D manifolds, see [266]. An interesting application to the Poincaré–Bendixson theorem includes oscillatory phenomena that occur during the biochemical process of *glycolysis* in yeast cells and muscle extracts; see [48] and [101]. A simple model of this is proposed in [267] and further discussed in [274]. Limit cycles also appear in circuit analysis.

Another application area in which the Poincaré–Bendixson theorem proves useful is the *Van der Pol oscillator*, introduced in [286] by Dutch electrical engineer and physicist Balthasar van der Pol while he was studying electric circuits. He referred to limit cycles as *relaxation oscillations*. See [110] and [115] for additional discussion and analysis on the Van der Pol oscillator. The Van der Pol oscillator itself has been used in a variety of interdisciplinary applications, including modeling of action potentials of neurons (see [93], [220]), geological faults in seismology (see [45]), and aeroelastic flutters (see [71, 72], [146], [147]).

https://doi.org/10.1515/9783110597806-005

Without further ado, we now state the theorem.

Theorem 5.1 (Poincaré–Bendixson). *Let $R \subset \mathbb{R}^2$ be a trapping region for the smooth autonomous system $\dot{x} = f(x)$ and let $x_0 \in R$. Then the omega limit set $\omega(x_0)$ is either*
1. *a fixed point;*
2. *a periodic orbit (i. e., limit cycle); or*
3. *the union of a finite number of fixed points p_i, $i = 1, \ldots, k$, and heteroclinic orbits $(y_i)_{i=1}^k$, with $\alpha(y_i) = p_i$ and $\omega(y_i) = p_{(i \bmod k)+1}$.*

Proof. See [231] or [294]. □

Recall from §3.2 that a *trapping region* is a closed, compact, positively invariant set. Theorem 5.1 therefore states a complete enumeration of possible ultimate motions for an orbit that enters a trapping region: the orbit *must* settle down to either a fixed point or a periodic orbit as $t \to \infty$. It is therefore impossible for an orbit that enters a trapping region to wander around without settling down to a periodic orbit or fixed point. The dimensionality assumption, i. e., the fact that the theorem is stated for 2-D flows, is crucial. The theorem does not hold in systems whose dimension is greater than two. It also does not hold for *nonautonomous systems* since time dependency can be thought of as an extra dimension.

Example 5.1. Consider the system

$$\begin{aligned}\dot{x} &= y + x(\lambda - x^2 - y^2),\\ \dot{y} &= -x + y(\lambda - x^2 - y^2).\end{aligned}$$

Notice that this system may be cast in polar coordinates to yield

$$\begin{aligned}\dot{\theta} &= -1,\\ \dot{r} &= r(\lambda - r^2).\end{aligned}$$

The polar-coordinate formulation reveals, for the case $\lambda > 0$, a limit cycle at $r = \sqrt{\lambda}$, since we directly observe

$$r^2 > \lambda \text{ implies } \dot{r} < 0, \quad r^2 < \lambda \text{ implies } \dot{r} > 0, \quad r^2 = \lambda \text{ implies } \dot{r} = 0.$$

The limit cycle $r = \sqrt{\lambda}$ is considered a *stable limit cycle*, as when r is initially outside of this periodic orbit (i. e., $r_0 > \sqrt{\lambda}$), the trajectory must asymptotically approach the limit cycle as $t \to \infty$. This follows since the radius is decreasing, and since trajectories cannot cross. Similarly, if $0 < r_0 < \sqrt{\lambda}$, trajectories must spiral outward and asymptotically approach the limit cycle. The time-reversed system exhibits an *unstable* limit cycle. □

Example 5.2. Prove that the nonlinear system $(\dot{x}, \dot{y}) = f(x, y)$, given by

$$\dot{x} = x - y - x^3 - xy^2,$$
$$\dot{y} = x + y - x^2y - y^3,$$

has a periodic orbit in the region $R = \{(x, y) \in \mathbb{R}^2 : \sqrt{x^2 + y^2} \leq 2\}$.

This system has a single equilibrium point at the origin $(x, y) = (0, 0)$.

Exercise 5.1. Show that the eigenvalues of the linearization at $(0, 0)$ are $\lambda = 1 \pm i$. Conclude that the origin is an *unstable spiral point.*

As a result of the preceding exercise, we may conclude that there exists an $\varepsilon > 0$ such that all trajectories that cross the boundary $\partial B_\varepsilon(0)$ are moving *away* from the origin. Note that $\partial B_\varepsilon(0)$ is literally a circle of radius ε centered at the origin.

Next, consider the simple closed contour $\Gamma = \partial B_2(0) = \partial R$, consisting of a circle of radius 2. Let the *outward* normal n be oriented *away* from the origin.

Exercise 5.2. Show that $f(x, y) \cdot n(x, y) < 0$ for all $(x, y) \in \Gamma$.

As a result of the preceding computations, we may conclude that the region $R' = R \setminus B_\varepsilon(0)$ is a trapping region for our nonlinear system. Since it does not contain any fixed points, all orbits that enter R' must have a limit cycle. Therefore, by the Poincaré–Bendixson theorem, we may conclude that R' must contain a periodic orbit. In fact, it can be shown that the *basin of attraction* for this periodic orbit, i. e., the set of all points whose omega limit set consists of the periodic orbit, is given by $\mathbb{R}^2 \setminus (0, 0)$. In other words, the trajectories emanating from all points except for the origin approach this periodic orbit in the limit as $t \to \infty$. □

Exercise 5.3 (Square limit cycle). Consider the system

$$\dot{x} = (y + 0.5x)(1 - x^2),$$
$$\dot{y} = -x(1 - y^2).$$

Compute a linearization for each of the five fixed points $p_1 = (-1, -1)$, $p_2 = (-1, 1)$, $p_3 = (1, 1)$, $p_4 = (1, -1)$, and $p_5 = (0, 0)$, showing that the first four are saddles and the origin is an unstable spiral. Explain why $\gamma_1 = \{-1\} \times (-1, 1)$, $\gamma_2 = (-1, 1) \times \{1\}$, $\gamma_3 = \{1\} \times (1, -1)$, and $\gamma_4 = (1, -1) \times \{-1\}$ are images of heteroclinic orbits of the system.

Let S be the square $S = [-1, 1] \times [-1, 1]$ and note that its boundary is given by

$$\partial S = \bigcup_{i=1}^{4} (p_i \cup \gamma_i).$$

Explain why the fact that $\omega(x_0) = \partial S$, for all $x_0 \in S \setminus \partial S \setminus (0, 0)$, does not contradict the Poincaré–Bendixson theorem.

As with Lyapunov functions, the Poincaré–Bendixson theorem is a fundamental tool in the toolbox of any dynamicist; however, construction of a trapping region, as with the construction of a Lyapunov function, is often a difficult task and may only be tractable for the simplest of problems.

5.2 Bifurcation and Hysteresis

Bifurcation theory is an important subdiscipline of dynamical systems. Many texts talk in great depth about the subject of bifurcations, such as [110], [115], [188], [235], [274], and [294]. In this text, we will content ourselves with paying a passing homage to bifurcation theory, getting across the main idea, without enumerating the various types of bifurcations that exist or providing a detailed analysis of individual types.

Definition 5.1. Let $\dot{x} = f(x; \mu)$ be a smooth dynamical system that depends on a parameter μ. Then the specific parameter value $\mu = \mu_0$ is a *bifurcation point* of the system if there exists an $\varepsilon > 0$ such that the system exhibits dynamics for values of $\mu \in (\mu_0, \mu_0+\varepsilon)$ that are *qualitatively different* from the dynamics exhibited by the system for values of $\mu \in (\mu_0 - \varepsilon, \mu_0)$.

Keep in mind that, in applications, dynamical systems are *models* of real-world phenomena. Hence, often various simplifying assumptions are made. When modeling real-world systems, a given dynamical system often depends on one or more *parameters*—numbers that depend on physical constants and dimensions but that are regarded as constant for the purpose of solving the corresponding system of ordinary differential equations. One often wishes to see how real systems behave *qualitatively* under different values of these parameters. This is the first reason for studying bifurcation theory—to gain an understanding of the qualitative behavior that a system can exhibit under various values of the system's parameters. The second reason is more subtle and goes back to the earlier statement that dynamical systems are models and not laws of nature. Since they only *model* physical phenomena, it is possible that the parameter values, which are treated as constants during the mathematical analysis of the system, might, actually, be slowly changing in time. For example, when modeling the Earth's seasonal weather patterns, one might make the simplifying assumption that the CO_2 level in the atmosphere is a constant, although over time it is slowly changing. Hence the CO_2 level might represent a *parameter* for the system one is studying. Now, if one of a system's *slowly varying parameters* happens to creep up and pass through a *bifurcation point*, the physical system will exhibit a *sudden* qualitative change in its dynamics; a *bifurcation* is said to have occurred. Although one might encounter many different types of bifurcations, we will focus our attention on a type of bifurcation that always involves a limit cycle and a fixed point. These bifurcations are known as *Hopf* bifurcations.

Definition 5.2. Let $\dot{x} = f(x; \mu)$ be a smooth dynamical system that depends on a parameter μ. Then a *Hopf bifurcation* is a type of bifurcation in which a periodic orbit coalesces with an equilibrium point at the bifurcation point, leaving an equilibrium point with the opposite stability.

Hopf bifurcations, sometimes also known as *Poincaré–Andronov–Hopf bifurcations*, were originally introduced in [238] and studied extensively later on by Andronov and Witt [6]. These types of bifurcations were generalized to higher dimensions in [148]. Hopf bifurcations are discussed in each of the references (above) on bifurcation theory and in [111] and [195]. We will restrict our focus to Hopf bifurcations in 2-D systems. Hopf bifurcations are further classified into several types.

Definition 5.3. A *supercritical Hopf bifurcation* is a Hopf bifurcation in which a stable limit cycle and an unstable spiral point coalesce, leaving only a single stable spiral point.

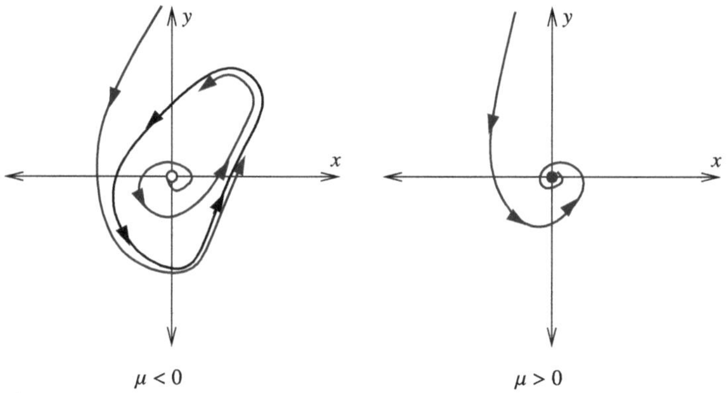

Figure 5.1: Illustration of a supercritical Hopf bifurcation.

An illustration of a supercritical Hopf bifurcation is shown in Figure 5.1. We see that, for values of the parameter $\mu < 0$, there exists an *unstable* spiral equilibrium point at the origin, which is centered inside of a *stable* limit cycle. Orbits from far away are attracted to the limit cycle, and orbits initially close to the origin are repelled from the origin and attracted to the limit cycle. As the value of μ increases toward the bifurcation point $\mu = 0$, the size of the limit cycle decreases until finally, at $\mu = 0$, the limit cycle degenerates to a single point. Since the limit cycle is stable and attracts orbits that lie outside the limit cycle, when it degenerates to a single equilibrium point, it becomes a *stable* equilibrium point. Hence, for values of $\mu > 0$, orbits are attracted to the origin.

The other type of Hopf bifurcation that arises is called a *subcritical Hopf bifurcation*. Subcritical Hopf bifurcations have been found to occur in the dynamics of brain neurons (see [139], [250]), aeroelastic flutter (see [73]), and unstable fluid flow (see [74]). As duly noted in [274], "The subcritical case is always much more dramatic, and potentially dangerous in engineering applications." This statement is unavoidable due to the phenomenon of *hysteresis*, which may occur during a subcritical Hopf bifurcation, as we will explain below.

Definition 5.4. A *subcritical Hopf bifurcation* is a Hopf bifurcation in which an unstable limit cycle and a stable spiral point coalesce, leaving only a single unstable spiral point.

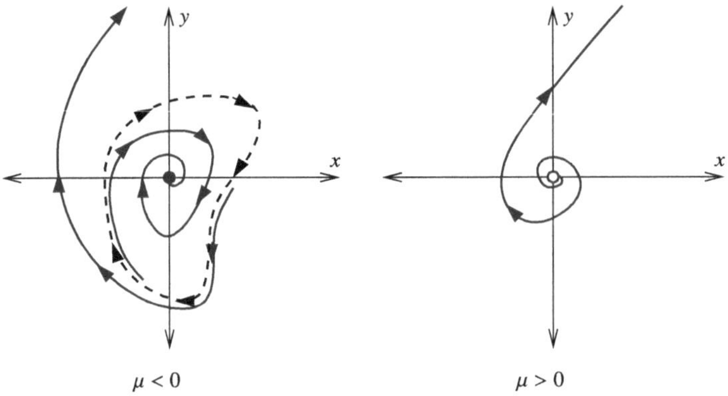

Figure 5.2: Illustration of a subcritical Hopf bifurcation.

An illustration of a subcritical Hopf bifurcation is shown in Figure 5.2. For parameter values $\mu < 0$, we find a *stable* spiral equilibrium point at the origin, which is centered inside of an *unstable* limit cycle. Hence, orbits with initial conditions slightly away from the origin *relax* back to the origin. In fact, for any initial condition interior to the region bounded by the limit cycle, the resulting orbit will exhibit *damped oscillations* that eventually settle down to the equilibrium point as $t \to \infty$. Naturally, initial conditions on the limit cycle itself follow the path of a periodic orbit. For initial conditions just outside of the region bounded by the limit cycle, the corresponding orbits are repelled from the limit cycle and venture off into some other region of phase space. They might be attracted to some other equilibrium point, another limit cycle, or infinity, or, in the case of three or more dimensions, they might wind up falling into the grasp of a *strange attractor*, i. e., a compact, fractal, chaotic region of phase space. This time, as the value of the parameter μ slowly increases toward the bifurcation point $\mu = 0$, the unstable limit cycle shrinks, engulfing the origin, until it degenerates to a single point. That single point is now an *unstable* equilibrium point.

To understand why Strogatz refers to subcritical Hopf bifurcations as both *dramatic* and *dangerous*, consider the following scenario. Assume a real-world physical system whose model depends on some parameter μ. This parameter is slowly changing in time; how the system behaves, however, at any instant is computed as though that parameter were constant. Suppose now that our physical system is in a state corresponding to the stable equilibrium point depicted in Figure 5.2, with $\mu < 0$. If a perturbation shifts the system's state, the system relaxes back to its former, stable equilibrium. As time passes, the parameter μ slowly creeps up to the bifurcation point $\mu = 0$. Now, suddenly, the system is at an unstable equilibrium point! The system state starts changing dramatically, and the trajectory of the state starts departing from the vicinity of the origin and is attracted to some other region of phase space. As a consequence of the system's wandering off, the parameter μ quickly corrects itself and drifts back below the critical value $\mu = 0$, the equilibrium point becomes stable again, and the limit cycle reemerges. But now, the system state is on the *outside* of the limit cycle, and hence it can *never* return to its former equilibrium. The system state is *forever* lost to another region of phase space, with no possibility of ever returning to the equilibrium and stability that it once held so dear. This phenomenon is known as *hysteresis*. Hysteresis may be defined, for our needs, as follows.

Definition 5.5. *Hysteresis* is a phenomenon in which a nonlinear system that depends on a parameter does not return to its original state upon tracing a closed loop in parameter space.

Hence, nonlinear systems, especially ones that exhibit hysteresis, behave very differently than the limits that our *linear* intuition and human experience allow. They are not like your thermostat. If you set your thermostat too high, and the temperature gets too hot, you can just turn the thermostat back down and the temperature will return to its starting value. *Nonlinear* systems, on the other hand, behave *nonlinearly*.

This is why the greatest deficit of modern civilization may be our inability, on the whole, to think nonlinearly. For example, if you view the world with your *linear* intuition, you might naively think that if the carbon dioxide levels get too high and the Earth gets too warm, we can simply turn down the thermostat by burning less fossil fuels and the Earth will recover. However, as we demonstrated in this paragraph, with a single subcritical Hopf bifurcation, a dynamical system that has been stable for all of history can suddenly lose that stability forever. Hence, due to the nonlinear nature of weather patterns and climate change, one must act *preemptively* and not *retroactively* to protect the planet from forever losing the stability that it enjoys.

Exercise 5.4. Use `pplane8.m` to plot the phase portrait for the system

$$\dot{x} = \mu x - y + xy^2,$$
$$\dot{y} = x + \mu y + y^3$$

for $\mu = -0.2$, $\mu = 0$, and $\mu = 0.2$. What type of Hopf bifurcation do you observe?

5.3 Period Doubling Bifurcations

In this section, we will discuss another important type of bifurcation, known as *period doubling bifurcations*, which present when chaos is lurking nearby. We will codify our discussion in the context of a well-known example.

The simplest system known to exhibit *chaos* was introduced in [253] and is given by

$$\dot{x} = -y - z, \tag{5.1}$$

$$\dot{y} = x + ay, \tag{5.2}$$

$$\dot{z} = b + z(x - c), \tag{5.3}$$

where a, b, and c are parameters of the system. Note that, rather remarkably, this system possesses only a single nonlinear term. The Rössler system was studied extensively in [227], documenting a cascade of *period doubling bifurcations* that preceded the onset of chaos. We will proceed in a similar fashion in this chapter. We begin by stating a formal definition.

Definition 5.6. A *period doubling bifurcation* in a smooth dynamical system $\dot{x} = f(x;\mu)$, which depends on the parameter μ, is a bifurcation in which a periodic orbit spontaneously doubles its period as μ passes through the bifurcation point.

Period doubling bifurcations are well known in *discrete mappings*, a topic that is beyond the scope of this text. However, we would feel remiss if we did not state an obvious connection. Suppose that Π is a Poincaré section of the phase space transversal to a periodic orbit γ. Now let us follow changes in the periodic orbit γ as the parameter μ is varied. Let $n(\mu)$ be the number of discrete points in the intersection $\Pi \cap \gamma$. Then a *period doubling bifurcation* is said to have occurred at the parameter value $\mu = \mu_0$ (also referred to as the *bifurcation point*) if there exists an $\varepsilon > 0$ such that $n(\mu_0 + \kappa) = 2n(\mu_0 - \kappa)$ for all $\kappa \in (0, \varepsilon)$. In other words, the number of intersections of the periodic orbit γ with the surface Π *doubles* as the parameter passes through the bifurcation point $\mu = \mu_0$.

Definition 5.7. An *attractor* of a dynamical system $\dot{x} = f(x)$ is a set toward which solutions evolve over time. A *strange attractor* is an attractor of noninteger dimension.

See [77] or [87] for a discussion on noninteger (fractal) geometries. See also [230] for an introduction to chaos in dynamical systems. Examples of attractors include stable fixed points, stable limit cycles, and omega limit sets. *Strange attractors* often appear in chaotic systems and possess *fractal structures* based on their noninteger dimensionality. The culmination of a cascade of period doubling bifurcations is often a *strange attractor*, as we will see in the context of the Rössler system.

The attractor for the Rössler system (5.1)–(5.3) is a stable limit cycle for a range of various parameter values. We will consider only the case with $a = b = 0.2$ fixed and analyze the dynamics as we vary the parameter c.

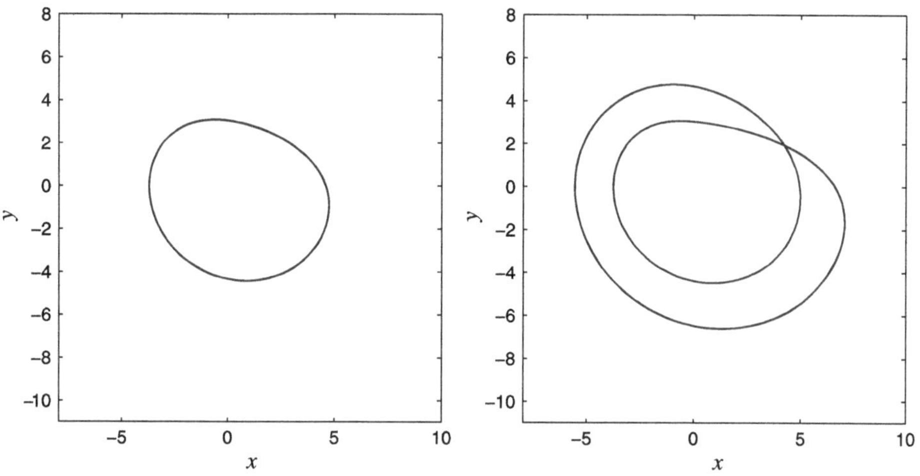

(a) $c = 2.5$. (b) $c = 3.5$.

Figure 5.3: Attractors of the Rössler system for two different values of c.

In Figure 5.3, we plot the attractors for the Rössler system and the parameter values $c = 2.5$ and $c = 3.5$. Technically, these are plots of the *projections* of the attractors onto the x-y plane, as the attractors themselves are orbits in 3-D space. Note that, for $c = 2.5$, the attractor is an order-1 periodic orbit, i. e., the number of intersection points of the periodic orbit and a given Poincaré section is 1. This orbit is a stable limit cycle (as are attracting, periodic orbits). We can play the same trick and plot the attractor for the Rössler system at a parameter value of $c = 3.5$, and we find that the limit cycle is now an *order-2* periodic orbit! Therefore, at some point between $c = 2.5$ and $c = 3.5$, a *period doubling* bifurcation must have occurred.

So far, we have seen that a period doubling bifurcation occurs between $c = 2.5$ and $c = 3.5$. Next, let us plot the x-y planar projection of the attractor for the parameter value $c = 4.5$. This is shown in Figure 5.4(a). We observe that, somewhere between $c = 3.5$ and $c = 4.5$, a second *period doubling* bifurcation must have occurred, as the attractor for $c = 4.5$ is an order-4 periodic orbit. So far all of this is noteworthy but not astonishing. But, rather suddenly, when you plot the x-y planar projection of the attractor for $c = 5$, something very different happens (Figure 5.4(b)). The attractor is no longer a fourth-order, eighth-order, or even sixteenth-order periodic orbit. The attractor is a fractal set of noninteger dimension—a *strange attractor*. The orbits bounce back and forth within the confines of the attractor, never settling down to any single periodic orbit or pattern. Their motion is now governed by the laws of *chaos*. And yet, we are studying a completely deterministic system, i. e., precise orbits are completely determined based on the initial conditions. However, trajectories in the chaotic region do possess strong *sensitivity* to initial conditions, as we will discuss in the Section 5.4.

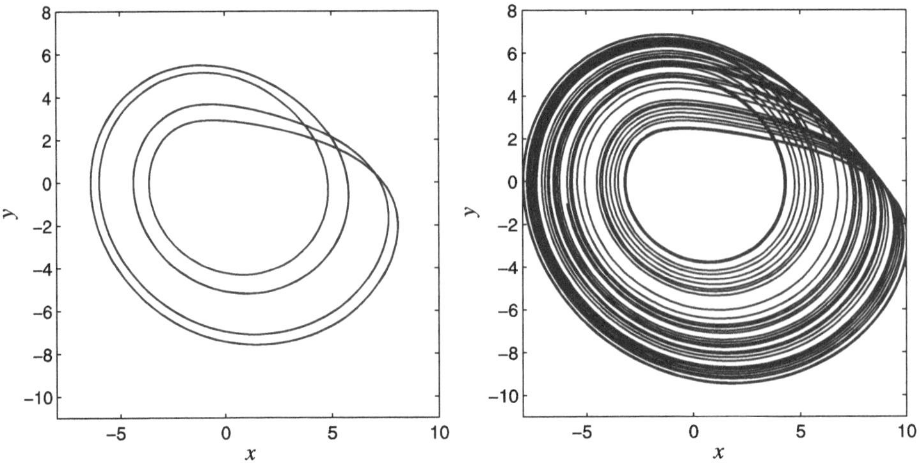

(a) $c = 4.5$. (b) $c = 5.0$.

Figure 5.4: Attractors of the Rössler system for two different values of c.

To find and plot an attractor, one must simply arbitrarily pick a point in phase space in the *basin of attraction* of the attractor, numerically integrate the resulting trajectory far into the future, and then use the trajectory's *final* endpoint as initial condition for the attractor. In Figures 5.3 and 5.4, I chose initial condition $(0, 0, 0)$ and then numerically integrated the resulting trajectory using `ode45.m` to several hundred time units. I then used the *final* point of my original trajectory as initial condition for the given plots, relying on the fact that my trajectory must have ended up indistinguishably close to its attracting set.

Exercise 5.5. Use MATLAB to reproduce the plots of the attractors shown in Figures 5.3 and 5.4.

To better visualize these *period doubling cascades*, one can represent the parameter space graphically using a *Feigenbaum diagram*, named in honor of Feigenbaum's seminal work [90] on discovering a fundamental constant related to period doubling cascades. The idea is that one can instead look at the entire *spectrum* of parameter values and graphically represent something about the attractor for each value. An example of this for the Rössler system is given in Figure 5.5. For 100 equally spaced values of c on the interval $c \in [2.5, 6]$, the attractor was computed using MATLAB, along with each of the *local maxima* of the variable x. These local maxima were then plotted for each given value of c. The number of local maxima then corresponds to the order of the periodic orbit. From the diagram we see that, for $c = 3$ and $c = 3.5$, the attractor is an order-2 periodic orbit and that, for $c = 4$, the attractor is an order-4 periodic orbit. Hence, the Feigenbaum diagram is a way to *visualize* the chaoticity of the system

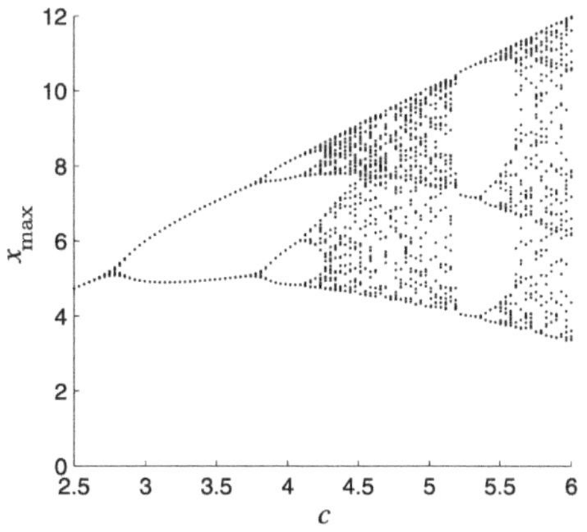

Figure 5.5: Period doubling cascade for the Rössler system.

and the nature of the attractor simultaneously throughout a continuum of parameter space.

Exercise 5.6. Reproduce the Feigenbaum diagram for the Rössler system shown in Figure 5.5. *Hints:* you should use a for-loop to run 100 equally spaced parameter values. For each iteration of the loop, first run the orbit with arbitrary initial conditions for a sufficient amount of time. Then use the final point of this orbit as the initial point on a new orbit (which approximates the attractor). Use `ode45`'s Events option to save local maxima in x and then plot the local maxima values against the given value of c. Clear your state variables, as outputted by `ode45`, after you run your initial setup orbit and at the end of each iteration of the for-loop.

Exercise 5.7. Based on Figure 5.5, of what order would you suspect the attracting periodic orbit to be for the parameter values $c = 5.3$ and $c = 5.45$? Compute and plot the attractor at these parameter values to confirm your hypothesis.

5.4 Chaos

We have reviewed several systems with very interesting behavior that we referred to as *chaotic*. These are systems in which orbits in compact sets do not settle down to either a periodic orbit or a fixed point. Rather, they hop about in an apparently random fashion as they approach an attracting set of noninteger dimension. In this final section of the chapter, we will provide a precise definition of *chaos*. Before we proceed to the definition, we will first define two distinguishing properties possessed by chaotic systems.

The first property is called *sensitivity to initial conditions*. Intuitively, this property means that two trajectories with similar initial conditions, no matter how alike those initial conditions are, will rapidly deviate from one another. In other words, even though a given system might be *deterministic*, in the absence of *precise* data, its long-term behavior might be largely unpredictable.

Definition 5.8. The flow $\varphi : \mathbb{R} \times \mathbb{R}^n \to \mathbb{R}^n$ of a smooth dynamical system $\dot{x} = f(x)$ is said to exhibit *sensitive dependence on initial conditions* in an invariant set $D \subset \mathbb{R}^n$ if there exists a fixed $r > 0$ such that, for every $x \in D$ and $\varepsilon > 0$, there exists a $y \in D \cap B_\varepsilon(x)$ and a time $t \in \mathbb{R}$ such that $|\varphi(t; x) - \varphi(t; y)| > r$.

There exists a single $r > 0$ that works for every initial condition x on the invariant set D. Also, point y can be made *arbitrarily close* to point x. The definition states that *for every* $x \in D$ there is always a point *arbitrarily close* to x whose orbit leaves the r-neighborhood of the orbit $\varphi(t; x)$ in finite time.

A common test for sensitivity involves computation of the Lyapunov exponents of the flow. Flows with positive Lyapunov exponents are always sensitive to initial conditions, as a positive Lyapunov exponent indicates exponential divergence of nearby trajectories.

The next property is known as *topological transitivity*, which is sometimes referred to as *mixing*. There is a famous example, due to [9], in which one takes a shaker with 20 % rum and 80 % cola and finds, upon shaking it, that every region of "phase space," no matter how small, contains approximately 20 % rum and 80 % cola. We state the definition more formally as follows.

Definition 5.9. An invariant set $D \subset \mathbb{R}$ is said to be *topologically transitive* with respect to the phase flow φ if, for any two open sets $U, V \subset D$, there exists a t such that $\varphi(t; U) \cap V \neq \emptyset$.

In regard to the *rum shaker analogy*, if you fix any open set V in the rum shaker and consider any other open set U, then after a sufficient amount of time, some of the rum and cola that were initially in the open set U will pass through set V.

Finally, we define a *chaotic set*, or a set that exhibits *chaos*, as a compact, invariant set with the preceding properties.

Definition 5.10. A compact, invariant set $D \subset \mathbb{R}^n$ is said to be *chaotic* with respect to the phase flow φ if the flow is both sensitive to initial conditions and topologically transitive in D.

Some authors also require periodic orbits to be *dense* in D as part of the definition of chaos. We will follow [294] in not requiring this as part of the definition but as a property possessed by chaotic sets.

Exercise 5.8. Show that the flow of $\dot{x} = x$ on $\mathbb{R}$ is *topologically transitive* and possesses *sensitivity to initial conditions* on the set $\mathbb{R}^+ = (0, \infty)$. Explain why this counterexample necessitates the requirement that a set must be *compact* to be classified as chaotic.

Exercise 5.9. Write a MATLAB program that numerically integrates the Lorenz system of (2.5)–(2.7) using `ode45.m` for a time span of 50 time units. Create a triple for-loop that runs the simulation with initial conditions

$$x_0 = -10 + 0.01i,$$
$$y_0 = -10 + 0.01j,$$
$$z_0 = 25 + 0.01k,$$

where $i, j, k = 1, \ldots, 10$. Plot the final point, $(x(50), y(50), z(50))$, for each of these 1,000 orbits concurrently on the same graph. Plot the initial conditions for each orbit on the same graph using a different color. Explain in detail how the concepts of *topological transitivity* and *sensitivity to initial conditions* are relevant to your graph. You should explain what you see in the context of the actual mathematical definitions for these two concepts.

Advice: This simulation might take 15 to 20 minutes. It is advisable to use a waitbar (type `help waitbar` in the MATLAB command prompt for more information) and run the simulation with $i, j, k = 1, \ldots, 3$ first to make sure your program works before running it with $i, j, k = 1, \ldots, 10$. Store the final point of each orbit as a row in a matrix during each step of the for-loops. At the end of the triple for-loop, clear your state and time variable (using, for example, `clear T X`), so that if the next simulation requires fewer steps, you do not keep saving the same point for your final position.

5.5 Application: Billiards

The game of billiards has long been an inspiration to students of dynamical systems; indeed, *Hadamard's billiards*, a particular type of billiard problem involving geodesics on a surface of constant negative curvature (a two-holed doughnut, in fact), was the first soluble dynamical system proven to be chaotic (see [113]). Other special billiard geometries were introduced in [12], [38, 39], and [270], each carrying the author's name. Sinai billiards, also known as a *Lorentz gas*, is of particular interest in the study of statistical thermodynamics. Other applications turn up in the study of nonequilibrium gases (see [68]) and quantum mechanical fiber lasers (see [184]).

Whereas dynamical systems are normally specified by a differential equation, a billiard problem is specified by a connected set (referred to as the "billiard table"), usually taken to be convex, but not necessarily (Sinai billiards are played within a rectangle minus a central circular hole). Both dynamical systems and billiard problems may be regarded either as initial value problems or as flow-map problems in which one visualizes the motion of all trajectories simultaneously. The dynamics of a point (the "billiard") in a billiard problem is very simple: the billiard satisfies Newton's first law of motion—that a body in motion will remain in motion unless acted upon by an external force—while within the interior of the billiard table, and it satisfies Newton's third law of motion—that to every action there is an equal and opposite reaction—when it encounters the boundary of the billiard table (these laws of motion are due to [224]).

Elliptic Billiards

We will consider the case in which the billiard table is a convex set $\Omega \subset \mathbb{R}^2$ with boundary $\Gamma = \partial\Omega$. In particular, we will take Ω to be the closed ellipse $x^2/a^2 + y^2/b^2 \leq 1$, as studied in [182, 183]. The velocity of a billiard is taken as constant between successive collisions with the boundary of the billiard table. When the billiard strikes the boundary, an impulse is applied to the billiard by the edge of the billiard table in an amount equal to twice the normal component of the billiard's momentum and in the direction of the inward-pointing normal to the table's boundary: the billiard's velocity is instantaneously reflected about the line tangent to the boundary.

To create an algorithm that may be used to implement a computer simulation of the dynamics of the billiard problem, we first suppose that (x_n, y_n) and (u_n, v_n) are the position and velocity of the billiard immediately following its nth impact with the boundary. The position (x_n, y_n) must satisfy the boundary condition, and the velocity (u_n, v_n) must be directed "inward," i. e., we require that $x_n^2/a^2 + y_n^2/b^2 = 1$ and $x_n u_n/a^2 + y_n v_n/b^2 < 0$, respectively. Now, the path traced by the billiard between its nth and $(n+1)$st collision with the boundary may be described parametrically by

$$x(s) = x_n + u_n s,$$
$$y(s) = y_n + v_n s.$$

Substituting these equations into the constraint equation that defines the boundary and further recognizing that the point (x_n, y_n) satisfies the same constraint equation (since it, too, was a boundary point), we obtain

$$[(b^2 u_n^2 + a^2 v_n^2)s + 2(b^2 x_n u_n + a^2 y_n v_n)]s = 0.$$

Solving for this equation's nonzero root, we find that the billiard's $(n+1)$st collision will occur at the position $(x_{n+1}, y_{n+1}) = (x_n + u_n s_f, y_n + v_n s_f)$, where

$$s_f = -\frac{2(b^2 x_n u_n + a^2 y_n v_n)}{b^2 u_n^2 + a^2 v_n^2}.$$

Similarly, the updated velocity following the $(n+1)$st collision is given (in vector notation) by

$$\mathbf{v}_{n+1} = \mathbf{v}_n - 2\mathbf{v}_n \cdot \hat{\mathbf{n}}(x_{n+1}, y_{n+1}),$$

where $\mathbf{v}_n = (u_n, v_n)$ and $\hat{\mathbf{n}}(x_{n+1}, y_{n+1})$ is an outward-pointing unit vector at the point (x_{n+1}, y_{n+1}), i. e., a unit vector in the direction of $(\frac{x_{n+1}}{a^2}, \frac{y_{n+1}}{b^2})$. This algorithm can be used to plot trajectories in the elliptic billiard problem; two such trajectories are shown in Figure 5.6.

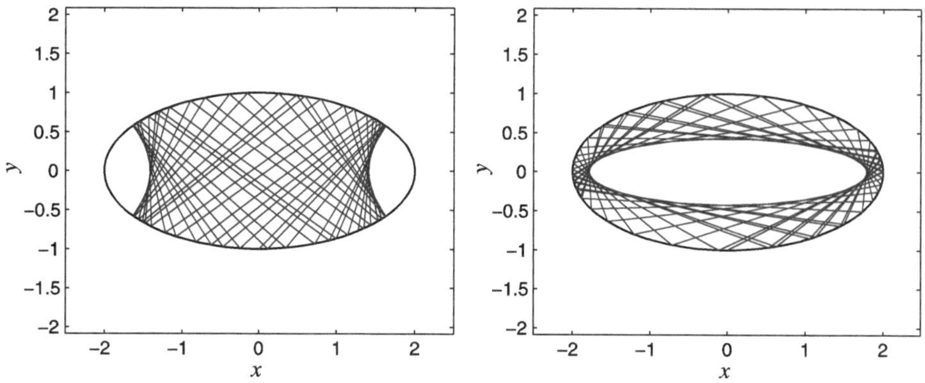

Figure 5.6: Sample orbits in elliptic billiard problem.

Exercise 5.10. Write a MATLAB code to generate trajectories in the elliptic billiard problem for the first 50 collisions. Reproduce the trajectories shown in Figure 5.6 by using $a = 2$, $b = 1$, $x_0 = 0$, $y_0 = -1$, $u_0 = \cos(\alpha)$, and $v_0 = \sin(\alpha)$ for both $\alpha = \pi/4$ and $\alpha = 3\pi/20$.

Elliptic billiards possess two integrals of motion: total kinetic energy and the product of the angular momenta about the ellipse's foci. Proving that a quantity is an integral of motion in a billiard problem requires two steps: proving it is conserved along the straight-line paths traversed between successive collisions with the boundary and proving it is conserved during the collisions.

Exercise 5.11. Prove that the kinetic energy $K = \frac{1}{2}\mathbf{v} \cdot \mathbf{v}$ is an integral of motion.

Exercise 5.12. Let $\mathbf{f} = (\sqrt{a^2 - b^2}, 0)$ be the position of one of the ellipse's foci and let $-\mathbf{f}$ be the other. Define the relative position vectors

$$\mathbf{r}_1 = \mathbf{r} - \mathbf{f} \quad \text{and} \quad \mathbf{r}_2 = \mathbf{r} + \mathbf{f}$$

of a billiard relative to the ellipse's foci, where $\mathbf{r} = (x, y)$.

(a) Show that the scalar quadruple product $F = (\mathbf{r}_1 \times \mathbf{v}) \cdot (\mathbf{r}_2 \times \mathbf{v})$, which represents the product of the angular momenta of the billiard about the two foci, is conserved along straight-line trajectories.

(b) Show that the scalar quadruple product F that was defined in part (a) is conserved during collisions of the billiard with the ellipse's boundary.

This demonstrates that F is an integral of motion under the billiard dynamics.

A planar billiard problem is a 4-D one. In general, a Poincaré map can be used to reduce the dimensionality by one; however, we can further use the conservation of kinetic energy to further reduce the dimensionality by one. It is natural to make the section surface the boundary of the billiard table itself, so that the Poincaré map is simply given by the mapping that sends the nth collision to the $(n+1)$st collision. To aid in our visualization, we will introduce two new variables to describe the state space "information" of each collision: an angle φ that describes the location of the collision along the boundary (consider points on the boundary parameterized by $(a\cos\varphi, b\sin\varphi)$) and a direction $p = \cos\alpha$, where α is the angle between the billiard's new velocity immediately following the collision and the forward-pointing tangent vector of the boundary. (The shape of this reduced space has the topology of a torus.) The Poincaré maps for the systems $a = 2$, $b = 1$, $x_0 = 0$, $y_0 = -1$, $u_0 = \cos\alpha$, and $v_0 = \sin\alpha$ for $\alpha = \kappa\pi/2$, $\kappa = 0, 0.1, 0.2, \ldots, 1.0$ and for the first 500 collisions are plotted concurrently in Figure 5.7.

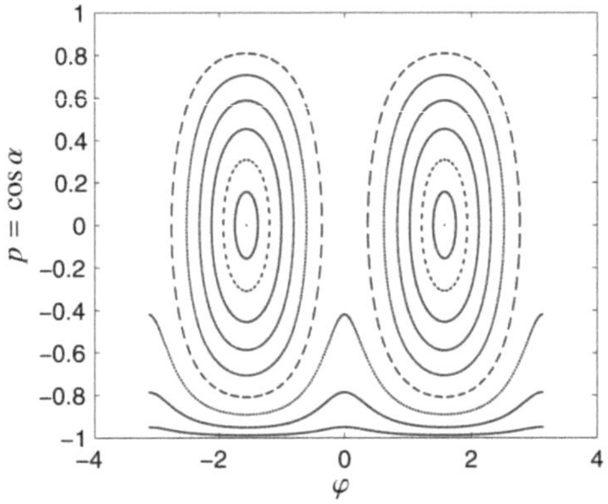

Figure 5.7: Poincaré map for various trajectories in the elliptic billiard problem.

Exercise 5.13. Write a program that reproduces the Poincaré maps shown in Figure 5.7. Explain how the level contours that appear in these plots are related to the function F defined in Exercise 5.12.

Stadium Billiards

Not all billiard problems are so well behaved: imagine a game of *stadium billiards*. The billiard table for stadium billiards is comprised of a rectangular region with a semicircle joined at either end. The dynamics of the stadium billiard game was introduced and proven to be chaotic by Bunimovich [38, 39]. For definiteness, our stadium will be comprised of a square with side length 2 that is sandwiched between two semicircles with radius 1.

To demonstrate the mixing properties of this ergodic problem, consider a 100×100 array of billiards that are equally spaced inside a square of side length 0.1 that is centered at the origin. Suppose each billiard has initial velocity $(\cos(\alpha), \sin(\alpha))$, where $\alpha = \arctan(1/\pi)$. (Note that the irrational angle of the velocity guarantees that no two

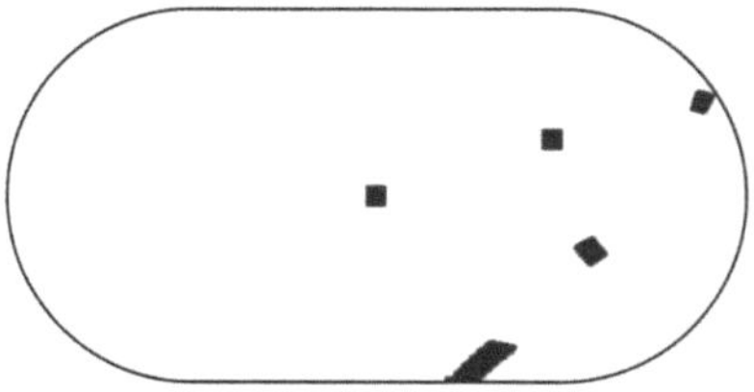

(a) $t = 0$, 1, 2, 3, and 4 (concurrent).

(b) $t = 5$, 6, 7, 8, and 9 (concurrent).

(c) $t = 10$, 11, and 12 (concurrent).

(d) $t = 13$.

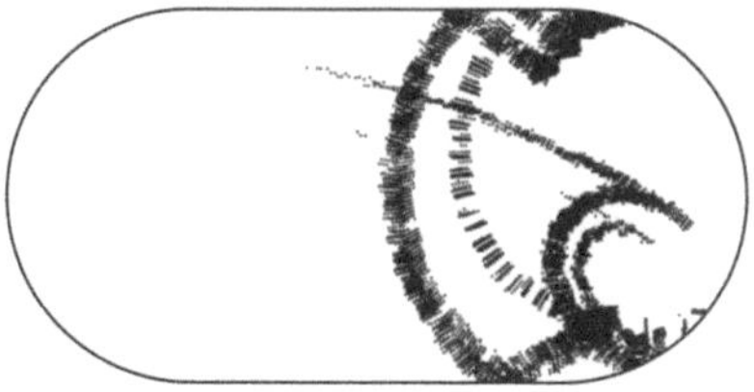

(e) $t = 14$.

(f) $t = 15$.

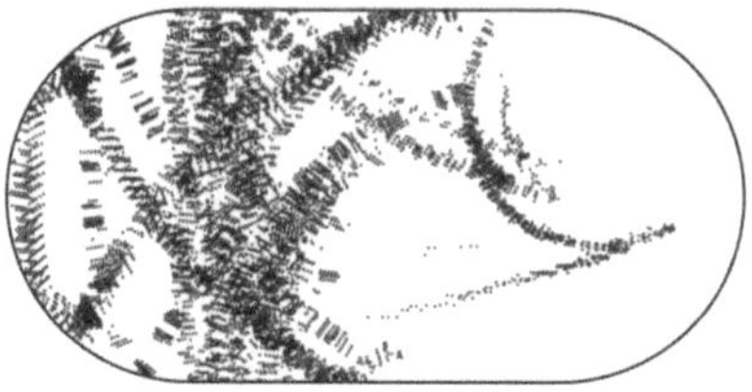

(g) $t = 16$.

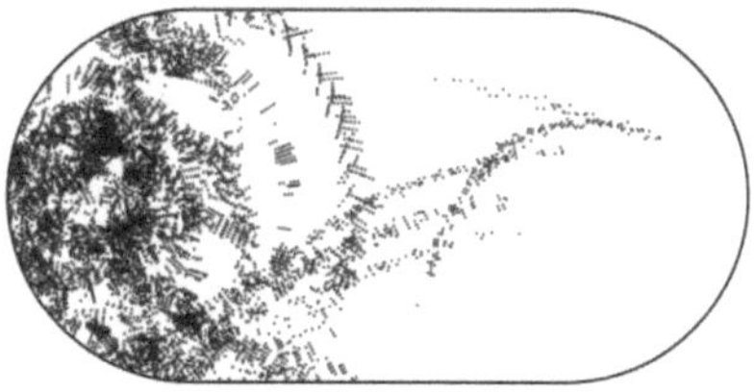

(h) $t = 18$.

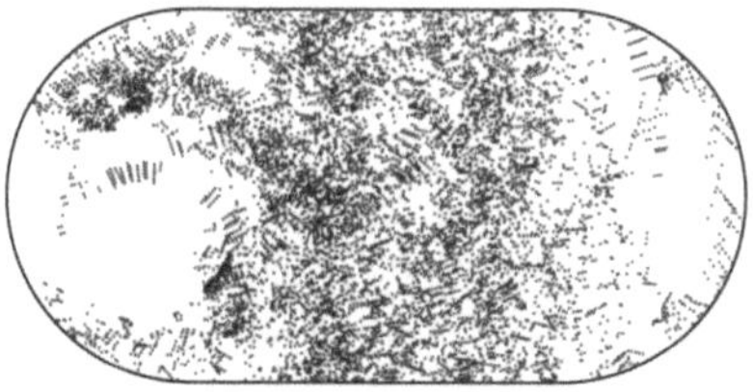

(i) $t = 20$.

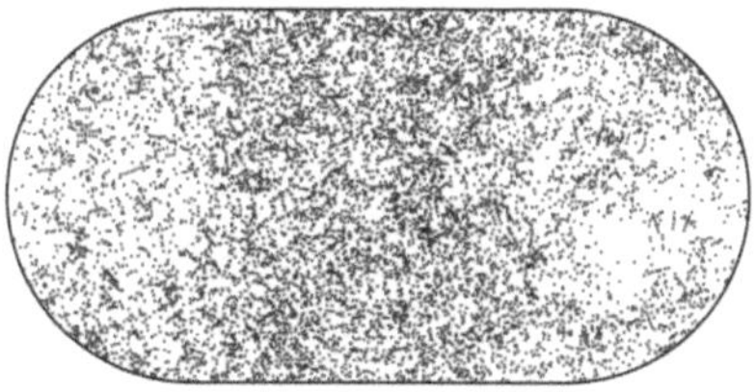

(j) $t = 25$.

Figure 5.8: Mixing in the stadium billiard problem.

virtual billiards in the 100×100 array lie on the same trajectory.) Each billiard interacts solely with the stadium's boundary and not with other billiards. The evolution of this tightly packed group of billiards over the first 25 time units is shown in Figure 5.8. The upper-left plot shows the position of all 10,000 billiards at the times $t = 0, 1, 2, 3, 4$; the upper-right plot shows the positions at times $t = 5, 6, 7, 8, 9$. In the second row, the plot on the left shows the positions at times $t = 10, 11, 12$; the plot on the right shows the positions at time $t = 13$. The third row has plots of the positions at times $t = 14$ (left) and $t = 15$ (right), the fourth row at times $t = 16$ (left) and $t = 18$ (right), and the fifth row at times $t = 20$ (left) and $t = 25$ (right).

Exercise 5.14. Write a computer code that approximately reproduces the plots in Figure 5.8.

We see that, after about 25 units of time, a small, initially tightly packed collection of virtual billiard balls has become almost uniformly distributed within the entire stadium. This demonstrates sensitivity to the initial conditions as well as to the mixing properties of the stadium billiard system.

Part II: **Geometric Mechanics**

6 Differentiable Manifolds

In this chapter, we will lay out some of the basic concepts of modern differential geometry. In particular, we properly introduce the concept of differentiable manifolds, usurping our previous working definitions, which we used in the context of Chapters 3 and 4. An important hallmark of the more modern definitions presented in this chapter is the intrinsic nature of the definitions: manifolds will be presented in such a way that they must no longer lean on the crutch of reference to a higher-dimensional Euclidean space. Similarly, we will present notions of vectors on manifolds that rely only upon the manifold's intrinsic structure. Finally, we will discuss the flows of vector fields on manifolds and bracket operators.

This chapter further constitutes the mathematical foundation of geometric mechanics, our focus in the remainder of the text. The reason for exploring the country of geometric mechanics is akin to the reason Captain Kirk gave for climbing the mountain El Capitan at Yosemite National Park in Star Trek V: *because it is there*. The award for our survey quickly turns profitable, however, as our pursuit will reveal hidden structures that complement our understanding of and analytical capacity for the classical techniques of mechanics.

6.1 Differentiable Manifolds

One must perforce begin with a discussion of *topological* manifolds. Please see [216] for a review of basic topological notions, as needed.

Definition 6.1. An *n-dimensional, topological manifold* M is a Hausdorff space[1] such that every point in M has an open neighborhood that is homeomorphic to an open set in $\mathbb{R}^n$.

A *coordinate chart* of the topological manifold M is a pair (U_i, φ_i), where U_i is an open subset of M and $\varphi_i : U_i \to V_i \subset \mathbb{R}^n$ is a homeomorphism onto an open set V_i of $\mathbb{R}^n$. For all $p \in U_i$, the numbers

$$\varphi_i(p) = (x^1(p), \dots, x^n(p))$$

are called *local coordinates* for the point p relative to the coordinate chart U_i.

Given a topological manifold M, one may cover M with a collection of open sets $\{U_i\}$ such that

$$M = \bigcup_i U_i.$$

A collection of coordinate charts that covers M is referred to as an *atlas* for M.

1 A *Hausdorff space* is a topological space in which distinct points can be separated by disjoint neighborhoods.

https://doi.org/10.1515/9783110597806-006

To escalate from a topological to a differential manifold, one requires a *differentiable structure* on the topological manifold M. This structure is precisely a set of *compatibility conditions* between the various coordinate charts of M, which we define precisely as follows.

Definition 6.2. Two coordinate charts, $(U_\alpha, \varphi_\alpha)$ and (U_β, φ_β), are called *compatible* if either
1. $U_\alpha \cap U_\beta = \emptyset$ or
2. $U_\alpha \cap U_\beta \neq \emptyset$, and the mapping $\varphi_\alpha \circ \varphi_\beta^{-1} : \varphi_\beta(U_\alpha \cap U_\beta) \to \varphi_\alpha(U_\alpha \cap U_\beta)$ is a smooth homeomorphism (i. e., a diffeomorphism).

The compatibility condition is illustrated graphically in Figure 6.1. Since the mapping $\varphi_\alpha \circ \varphi_\beta^{-1}$ is a mapping from a set in $\mathbb{R}^n$ to another set in $\mathbb{R}^n$, one may require that such a mapping be differentiable and define derivatives in the usual sense when dealing with mappings in Euclidean space. We therefore say that two coordinate charts on a topological manifold M are *compatible* if the corresponding mapping $\varphi_\alpha \circ \varphi_\beta^{-1}$ is a smooth mapping from $\varphi_\beta(U_\alpha \cap U_\beta)$ onto $\varphi_\alpha(U_\alpha \cap U_\beta)$. In Figure 6.1, the intersection $U_\alpha \cap U_\beta$ is represented by the crosshatch region in M and the images $\varphi_\alpha(U_\alpha \cap U_\beta)$ and $\varphi_\beta(U_\alpha \cap U_\beta)$ by the crosshatch regions of $V_\alpha = \varphi_\alpha(U_\alpha)$ and $V_\beta = \varphi_\beta(U_\beta)$, respectively.

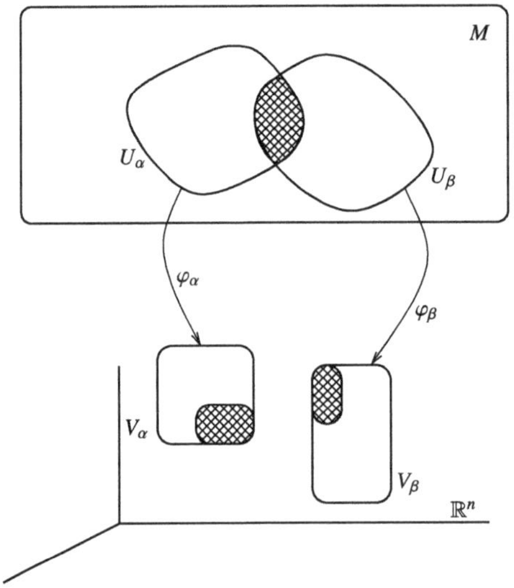

Figure 6.1: Illustration of compatibility condition.

Definition 6.3. A *differentiable structure* on a topological manifold M is a *maximal atlas* $\mathcal{A}$ consisting exclusively of mutually compatible coordinate charts. An atlas $\mathcal{A}$ is a *maximal atlas* if the coordinate chart $(\tilde{U}, \tilde{\varphi}) \in \mathcal{A}$ whenever $(\tilde{U}, \tilde{\varphi})$ is compatible with the various coordinate charts of $\mathcal{A}$.

In practice, one typically defines a compatible atlas and then extends it to a maximal atlas theoretically by simply including all charts that are compatible with the present atlas.

Definition 6.4. An *n-dimensional differentiable manifold* is an n-dimensional topological manifold with a differentiable structure.

Exercise 6.1. Prove that $\mathbb{R}^n$, $S^n = \{x \in \mathbb{R}^{n+1} : x \cdot x = 1\}$, and the n-torus

$$T^n = \underbrace{S^1 \times \cdots \times S^1}_{n\text{–times}}$$

are n-dimensional differentiable manifolds.

6.2 Vectors on Manifolds

In §3.2, we presented a classical view in which manifolds are regarded as sets of points, in a higher-dimensional Euclidean space, which may be locally represented as the graph of a smooth function about each point. In such a context, erection of tangent vectors is natural: one simply takes vectors in the higher-dimensional space that lie tangent to the graph at a given point. The path to defining tangent vectors, however, which is fundamental to our discussion of dynamical systems on manifolds, in the context of the modern, *intrinsic* definition of a manifold, is at first a bit more opaque than the view of its classical counterpart. After all, how is one to conceive of a tangent vector to a sphere when limited *only* to the sphere's *intrinsic* geometry and without viewing it as a subset of 3-D Euclidean space? We endeavor to answer such a question in this paragraph. We first define an appropriate equivalence class for curves on manifolds.

Definition 6.5. Let M be an n-dimensional, differentiable manifold and $I \subset \mathbb{R}$ an open interval containing the point $0 \in I$. The smooth curves $c_1, c_2 : I \to M$ are said to be *equivalent* at $p \in M$, denoted $c_1 \sim c_2$, if
- $c_1(0) = c_2(0) = p$ and
- $(\varphi \circ c_1)'(0) = (\varphi \circ c_2)'(0)$ for some coordinate chart (U, φ), with $p \in U$.

This notion of equivalency among curves on M is illustrated in Figure 6.2. Basically, two curves passing through a point $p \in M$ are equivalent if the velocity vectors in $\mathbb{R}^n$ of the curves $\varphi \circ c_1$ and $\varphi \circ c_2$ are identical in some set of local coordinates in a neighborhood U containing p. This definition is an *intrinsic* one. We make no reference to the possibility that manifold M is embedded in a higher-dimensional Euclidean space.

Exercise 6.2. Show that if the curves $c_1, c_2 : I \to M$ are equivalent at $p \in M$ relative to the coordinate chart $(U_\alpha, \varphi_\alpha)$, then they are equivalent relative to any compatible chart (U_β, φ_β).

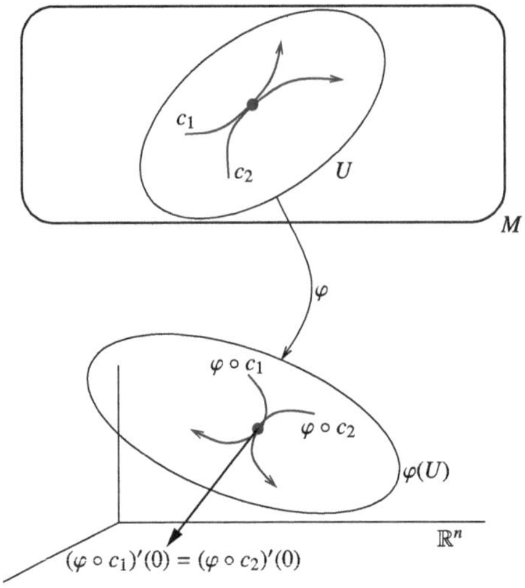

Figure 6.2: Illustration of equivalent curves on M.

The preceding exercise shows that our definition of equivalency is independent of the choice of coordinate chart. Hence, this definition of equivalency forms an equivalence class on the set of curves passing through a fixed point $p \in M$. Given a curve $c : I \to M$ passing through $p \in M$, we denote its equivalence class by $[c]_p$, or simply $[c]$ when point p is understood.

Definition 6.6. A *tangent vector* v to a manifold M at the point $p \in M$ is an equivalence class of curves passing through p, as defined in Definition 6.5.

Every curve in a given equivalency class, when mapped into $\mathbb{R}^n$ by the homeomorphism $\varphi : U \to \mathbb{R}^n$, has the same velocity vector at the point $\varphi(p) \in \mathbb{R}^n$.

Exercise 6.3. Prove that the set of tangent vectors at a point $p \in M$ forms a vector space. Show that this vector space is isomorphic to $\mathbb{R}^n$.

Definition 6.7. The set of tangent vectors at a point $p \in M$ is referred to as the *tangent space* to M at p and is denoted T_pM.

Definition 6.8. Let U be a coordinate neighborhood of $p \in M$ and $v \in T_pM$ a tangent vector to M at p. The *components* of the tangent vector v relative to the coordinate chart (U, φ) are the numbers $v^1, \ldots, v^n$, defined by

$$v^i = (\varphi \circ c)'(0) \cdot e_i,$$

where $(e_1, \ldots, e_n)$ are the basis vectors for $\mathbb{R}^n$ and $c : I \to M$ is any representative curve in the equivalence class defined by v.

The components of a vector $v \in T_pM$ relative to a coordinate chart (U, φ) are literally the components of the tangent vector $(\varphi \circ c)'(0)$ of any representative curve c in the equivalency class defined by v, i. e.,

$$(\varphi \circ c)'(0) = (v^1, \dots, v^n) \in \mathbb{R}^n.$$

Definition 6.9. Let M be an n-dimensional, differentiable manifold. The *tangent bundle* of M, denoted TM, is the disjoint union of all the tangent spaces to M, i. e.,

$$TM = \mathop{\dot{\bigcup}}_{p \in M} T_pM = \{(p, v) : p \in M,\ v \in T_pM\}.$$

Exercise 6.4. Prove that the tangent bundle TM of an n-dimensional differentiable manifold M is a $2n$-dimensional differentiable manifold.

Remark. Although each tangent space T_pM is isomorphic to the vector space $\mathbb{R}^n$, one cannot view the tangent bundle as a direct product of M with $\mathbb{R}^n$. While this is true *locally*, it may not accurately reflect the *global* topology of TM. As an example of this, consider the Klein bottle. A Klein bottle may *locally* be viewed as a direct product $S^1 \times S^1$; however, its global topology is different from that of the 2-torus $T^2 = S^1 \times S^1$.

Next we discuss the connection between vectors on manifolds and directional differentiation. Because of this link, modern geometers have adopted a certain convention for denoting vectors in terms of their components, which we will present momentarily.

Definition 6.10. Let $v \in T_pM$ and let $f : M \to \mathbb{R}$. Then the *directional derivative* of f in the direction v, denoted D_vf, is given by

$$D_vf = \sum_{i=1}^{n} v^i \frac{\partial f}{\partial x^i}, \tag{6.1}$$

where $(x^1, \dots, x^n)$ are local coordinates relative to some coordinate chart (U, φ) and $(v^1, \dots, v^n)$ are the components of vector v relative to the same coordinate chart (U, φ).

Exercise 6.5. Show that the directional derivative, as defined in Definition 6.10, is independent of the choice of coordinate chart used to perform the calculation.

Another way to think of the directional derivative is as follows. Let $c : I \to M$ be a representative curve in the equivalence class of curves defined by the vector v. Then the directional derivative of $f : M \to \mathbb{R}$ in the direction $v \in T_pM$ is equivalent to the ordinary time derivative of the function $f \circ c : I \to \mathbb{R}$ evaluated at $t = 0$, when the curve passes through $c(0) = p$, i. e.,

$$D_vf = \left.\frac{d}{dt}(f(c(t)))\right|_{t=0}.$$

One may not use the chain rule when evaluating this composite function, as derivatives of functions on manifolds are by themselves undefined. The utility of this composition is to circumvent differentiation on the manifold altogether, using a representative curve for v to compose a real-valued function of a real variable. This relation is independent of the choice of representative curve as the derivative mentioned in the preceding line depends only on the velocity of the curve at the instant it passes through point p.

In this regard, vectors on manifolds may themselves be viewed as differentiations. Vectors are machines that act on scalar functions and output the time rate of change of a given scalar function along any representative curve in the equivalence class of curves defined by the given vector. Due to the perspective of vectors as operators on functions, modern geometers have initialized the following notational convention linking vectors with their components relative to a given coordinate chart. Let $(x^1, \dots, x^n)$ be a set of local coordinates of a coordinate chart (U, φ), so that

$$\varphi(p) = (x^1(p), \dots, x^n(p))$$

for all $p \in U$ and let the numbers $(v^1, \dots, v^n)$ represent the components of the vector $v \in T_pM$ relative to the coordinate chart (U, φ). Then one may represent v in local coordinates by

$$v = \sum_{i=1}^{n} v^i \frac{\partial}{\partial x^i} = v^i \frac{\partial}{\partial x^i}. \tag{6.2}$$

The final equality holds due to the summation convention outlined in §4.1, i. e., summation over the repeated index is implied. By comparing (6.1) with (6.2), one may equivalently represent the directional derivative of function f in the direction v in the following equivalent ways:

$$D_v f = v[f] = v^i \frac{\partial}{\partial x^i}[f] = v^i \frac{\partial f}{\partial x^i}.$$

The quantities $\frac{\partial}{\partial x^i}$ themselves represent the basis vectors for $T_pM \cong \mathbb{R}^n$ relative to the local coordinates $(x^1, \dots, x^n)$ of the coordinate chart (U, φ), so that

$$(e_1, \dots, e_n) = \left(\frac{\partial}{\partial x^1}, \dots, \frac{\partial}{\partial x^n}\right).$$

Proposition 6.1. *Let (U, φ) and (V, ψ) be two coordinate charts on M, with local coordinates $(x^1, \dots, x^n) = \varphi(p)$ for all $p \in U$ and $(y^1, \dots, y^n) = \psi(p)$ for all $p \in V$, respectively. Then*

$$\frac{\partial}{\partial x^i} = \frac{\partial y^j}{\partial x^i} \frac{\partial}{\partial y^j}$$

at each $p \in U \cap V$.

Proof. The vectors $\frac{\partial}{\partial y^1}, \dots, \frac{\partial}{\partial y^n}$ form a basis for T_pM, and hence for each $i = 1, \dots, n$ there exists a set of scalars a_i^j such that

$$\frac{\partial}{\partial x^i} = a_i^k \frac{\partial}{\partial y^k}.$$

Now let this vector field act on the coordinate functions $y^j : V \to \mathbb{R}$. We obtain

$$\frac{\partial y^j}{\partial x^i} = a_i^k \frac{\partial y^j}{\partial y^k} = a_i^k \delta_k^j = a_i^j,$$

thereby completing the proof. □

Exercise 6.6. Let (U, φ) and (V, ψ) be two coordinate charts on M, with local coordinates $(x^1, \dots, x^n) = \varphi(p)$ for all $p \in U$ and $(y^1, \dots, y^n) = \psi(p)$ for all $p \in V$. Suppose that $u \in T_pM$ has components $(u^1, \dots, u^n)$ relative to the chart (U, φ) and components $(v^1, \dots, v^n)$ relative to the chart (V, ψ) and that $p \in U \cap V$. Show that

$$v^j = u^i \frac{\partial y^j}{\partial x^i}.$$

Another useful concept in geometric mechanics is that of the cotangent space.

Definition 6.11. Let M be an n-dimensional differentiable manifold. The *cotangent space* to M at p, denoted T_p^*M, is the dual vector space to T_pM. Vectors in T_p^*M are referred to as *covectors*, *differential forms*, or *one-forms*.

In other words, T_p^*M is the set of mappings $\omega : T_pM \to \mathbb{R}$. The cotangent space T_p^*M is also a vector space that is isomorphic to T_pM, which in turn is isomorphic to $\mathbb{R}^n$; see [62] for additional details.

The basis for the dual space, relative to the local coordinates $(x^1, \dots, x^n)$, is denoted by the symbols $dx^1, \dots, dx^n$. The basis vectors for the dual space share the property

$$dx^i \left(\frac{\partial}{\partial x^j} \right) = \delta_j^i,$$

where δ_j^i is the Kronecker delta. (Recall that each dx^i is a linear functional on the tangent space, i. e., $dx^i : T_pM \to \mathbb{R}$.)

As an example instance of usage of this notation, we might rewrite the ith component of the vector $v \in T_pM$ as

$$v^i = dx^i \left((\varphi \circ c)'(0) \right) = dx^i \left(v^j \frac{\partial}{\partial x^j} \right) = v^j dx^i \left(\frac{\partial}{\partial x^j} \right) = v^j \delta_j^i = v^i.$$

Finally, one may collect each of the cotangent spaces to a manifold into a single bundle structure in a similar fashion as we did with the tangent spaces.

Definition 6.12. Let M be an n-dimensional differentiable manifold. The *cotangent bundle* of M, denoted T^*M, is the disjoint union of all the cotangent spaces to M, i. e.,

$$T^*M = \mathop{\dot{\bigcup}}_{p\in M} T_p^*M = \{(p,\omega) : p \in M, \omega \in T_p^*M\}.$$

Also in analog with the tangent bundle, one may show that the cotangent bundle itself has the structure of a $2n$-dimensional differentiable manifold. Cotangent bundles will prove useful later in our study of Hamiltonian mechanics.

6.3 Mappings

Next, we will take up the topic of mappings from one manifold into another. Such a concept is obviously crucial to representing dynamical systems on manifolds, as the flow of a vector field itself may be considered a one-parameter family of smooth mappings of a manifold onto itself, passing through the identity mapping at the initial time. Before we embark upon such a discussion, however, the more rudimentary idea of mappings from manifolds to manifolds must first be considered.

Definition 6.13. Let $f : M \to N$ be a smooth mapping from the m-dimensional differentiable manifold M into the n-dimensional differentiable manifold N. The *linearization* of f at the point $x \in M$, denoted by the symbol $Df(x)$, is the vector space transformation $Df(x) : T_xM \to T_{f(x)}N$, which sends $v = [c] \in T_xM$ to $Df(x) \cdot v = [f(c)] \in T_{f(x)}N$, where $c : I \to M$ is a representative curve for the equivalence class $v \in T_xM$.

We now show how the linearization of a mapping $f : M \to N$ transforms the vector v when written out in terms of its components with respect to a given set of local coordinates. Let (U,φ) be a coordinate chart about the point $x \in M$ and (U',ψ) a coordinate chart about the point $f(x) \in N$. Now let $V = f(U) \cap U'$ and consider the restricted chart (V,ψ) about the point $f(x)$.

Now let $v \in T_xM$. In the local coordinates $(x^1,\dots,x^m)$ defined by the chart (U,φ), we may express

$$v = v^i \frac{\partial}{\partial x^i},$$

where the summation runs through $i = 1,\dots,m$. Let $c : I \to M$ be a representative curve in the equivalence class defined by v, so that

$$(\varphi \circ c)'(0) = (v^1,\dots,v^m).$$

According to Definition 6.13, the mapping $Df(x)$ sends the vector $v \in T_xM$ to the vector $Df(x) \cdot v$ defined by the equivalence class of curves represented by the curve $f \circ c : I \to N$. Now if we use the local chart (V,ψ), we may express the components of the vector $Df(x) \cdot v$ relative to the local coordinates $(y^1, \ldots, y^n)$ by differentiating the curve $\psi \circ f \circ c : I \to \mathbb{R}^n$ at $t = 0$, so that

$$Df(x) \cdot v = (\psi \circ f \circ c)'(0).$$

See Figure 6.3 for an illustration of this mapping.

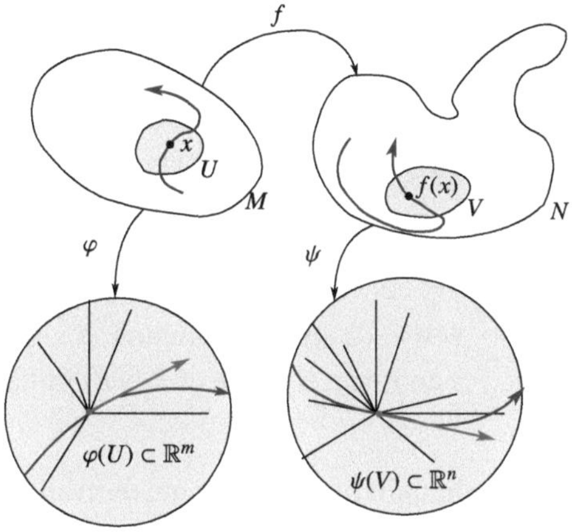

Figure 6.3: Illustration of the linearization of a smooth mapping.

In local coordinates, the matrix for the linear transformation $Df(x)$ may be represented by the $n \times m$ matrix

$$Df(x) = \begin{bmatrix} \frac{\partial y^1}{\partial x^1} & \cdots & \frac{\partial y^1}{\partial x^m} \\ \vdots & \ddots & \vdots \\ \frac{\partial y^n}{\partial x^1} & \cdots & \frac{\partial y^n}{\partial x^m} \end{bmatrix},$$

known as the *Jacobian matrix*.

6.4 Vector Fields and Flows

We next generalize the concept of differentiable equations and solution flows to their geometric analogs on manifolds. Recall that in Euclidean space, an autonomous, nonlinear differential equation is defined by a vector field $f : \mathbb{R}^n \to \mathbb{R}^n$. The corresponding differential equation is then given by (2.1). On differentiable manifolds, therefore, a system of differential equations is again defined by specification of a vector field on M. To be precise, consider the following.

Definition 6.14. A *vector field* X on a differentiable manifold M is a map $X : M \to TM$ that assigns a vector $X(x) \in T_xM$ to every point $x \in M$. We will denote the set of smooth vector fields on M by the symbol $\mathfrak{X}(M)$.

Remark. We will say that a vector field X is r times continuously differentiable, denoted $X \in \mathfrak{X}^r(M)$, whenever each of the components $(X^1, \dots, X^n)$ of X relative to a given coordinate chart are r times continuously differentiable functions of the local coordinates $x^1, \dots, x^n$. Due to the differentiable structure on M, this requirement is independent of the given chart.

Once a vector field is defined on M, one seeks "solution curves" to the implied system of differential equations, i. e., curves whose tangent vectors at all points coincide with the prescribed vector field. Such curves are known as *integral curves.*

Definition 6.15. An *integral curve* of the vector field $X \in \mathfrak{X}(M)$ with initial condition $x_0 \in M$ at $t = 0$ is a smooth curve $c : I \to M$ such that $c(0) = x_0$ and $c'(t) = X(c(t))$.

Finally, one may define the *flow* of a vector field in an entirely analogous fashion as its Euclidean counterpart.

Definition 6.16. The *flow* of the vector field $X \in \mathfrak{X}(M)$ is a smooth function $\varphi : \mathbb{R} \times M \to M$ such that, for fixed x_0, the curve $\varphi(t; x_0)$ is an integral curve of X with initial condition x_0.

We previously examined how vectors may be thought of as directional derivatives of functions at the point to which they are attached. One may also view vector fields as derivations. If $X \in \mathfrak{X}^r(M)$, then we can view X as a mapping $X : C^{r+1}(M) \to C^r(M)$.

Definition 6.17. A mapping $X : C^r(M) \to C^{r-1}(M)$ is a *derivation* if it satisfies
1. $X(af + bg) = aXf + bXg$ and
2. $X(fg) = fX(g) + gX(f)$

for all $a, b \in \mathbb{R}$ and $f, g \in C^r(M)$.

Proposition 6.2. *An r times continuously differentiable vector field $X \in \mathfrak{X}^r(M)$ is a derivation $X : C^{r+1}(M) \to C^r(M)$.*

Proof. Let X be given in local coordinates by

$$X = X^i \frac{\partial}{\partial x^i}.$$

Then, for all $f \in C^{r+1}(M)$, we have

$$X[f] = X^i \frac{\partial f}{\partial x^i} \in C^r(M).$$

It follows that, for all $a, b \in \mathbb{R}$ and $f, g \in C^{r+1}(M)$, we have, relative to local coordinates,

$$X[af + bg] = X^i \frac{\partial}{\partial x^i}[af + bg] = aX^i \frac{\partial f}{\partial x^i} + bX^i \frac{\partial g}{\partial x^i} = aX[f] + bX[g].$$

Similarly,

$$X[fg] = X^i \frac{\partial(fg)}{\partial x^i} = f(x)X^i \frac{\partial g}{\partial x^i} + g(x)X^i \frac{\partial f}{\partial x^i} = fX[g] + gX[f].$$

Hence the proposition follows. □

The preceding definitions and concepts provide us with the correct technology to define and solve systems of differential equations on manifolds. We will take up the topic of how such dynamical systems arise in the next chapter. But first we must introduce an important operation on vector fields, known as *Jacobi–Lie brackets.*

6.5 Jacobi–Lie Bracket

Let us consider the mapping $X(Yf) - Y(Xf)$ defined by commuting the order in which the vector fields X and Y act upon a given scalar function f. This mapping is one of importance, so we give it a special name.

Definition 6.18. Let $X, Y \in \mathfrak{X}^2(M)$. Then the *Jacobi–Lie bracket* of X and Y, or simply the *Lie bracket* or *commutator* of X and Y, denoted $[X, Y]$, is the C^1-vector field

$$[X, Y] = XY - YX.$$

To better understand the quantity $[X, Y]$, it will be beneficial to compute how it acts on a given scalar function.

Proposition 6.3. *Let $X, Y \in \mathfrak{X}^2(M)$ be twice continuously differentiable vector fields that are represented in local coordinates as*

$$X = X^i \frac{\partial}{\partial x^i}, \quad Y = Y^i \frac{\partial}{\partial x^i}.$$

Then the Jacobi–Lie bracket of X and Y, denoted $[X, Y] \in \mathfrak{X}^1(M)$, may be represented in terms of local coordinates by

$$[X, Y] = \left(X^i \frac{\partial Y^j}{\partial x^i} - Y^i \frac{\partial X^j}{\partial x^i} \right) \frac{\partial}{\partial x^j}.$$

Proof. By Proposition 6.2, we may consider the Lie bracket $[X, Y] \in \mathfrak{X}^1(M)$ to be a derivation on the set of twice continuously differentiable functions $C^2(M)$. To see how the bracket $[X, Y]$ acts on a given C^2 function f, we compute

$$\begin{aligned}
[X, Y]f &= X(Yf) - Y(Xf) \\
&= X^i \frac{\partial}{\partial x^i} \left(Y^j \frac{\partial f}{\partial x^j} \right) - Y^i \frac{\partial}{\partial x^i} \left(X^j \frac{\partial f}{\partial x^j} \right) \\
&= X^i Y^j \frac{\partial^2 f}{\partial x^i x^j} + X^i \frac{\partial Y^j}{\partial x^i} \frac{\partial f}{\partial x^j} - Y^i X^j \frac{\partial^2 f}{\partial x^i x^j} - Y^i \frac{\partial X^j}{\partial x^i} \frac{\partial f}{\partial x^j}
\end{aligned}$$

$$= \left(X^i \frac{\partial Y^j}{\partial x^i} - Y^i \frac{\partial X^j}{\partial x^i} \right) \frac{\partial f}{\partial x^j}.$$

The proposition follows. In the preceding calculation, we rewrote the terms

$$X^i Y^j \frac{\partial^2 f}{\partial x^i \partial x^j} - Y^i X^j \frac{\partial^2 f}{\partial x^i \partial x^j} = X^i Y^j \left(\frac{\partial^2 f}{\partial x^i \partial x^j} - \frac{\partial^2 f}{\partial x^j \partial x^i} \right)$$

by manipulating the index of summation on the second term. These terms then cancel each other by Clairaut's theorem since f is twice continuously differentiable. □

Proposition 6.4. *The bracket of two vector fields satisfies the following properties:*
(i) *linearity in the first and second arguments:*

$$[aX + bZ, Y] = a[X, Y] + b[Z, Y],$$
$$[X, aY + bZ] = a[X, Y] + b[X, Z];$$

(ii) *antisymmetry:*

$$[Y, X] = -[X, Y];$$

(iii) *the Jacobi identity:*

$$[[X, Y], Z] + [[Y, Z], X] + [[Z, X], Y] = 0; \quad \textit{and}$$

(iv) *the Leibniz rule:*

$$[fX, gY] = fg[X, Y] + fX(g)Y - gY(f)X$$

for all $X, Y, Z \in \mathfrak{X}(M)$, $a, b \in \mathbb{R}$, *and* $f, g \in C^\infty(M)$.

Proof. Properties (i) and (ii) follow immediately from the definition of the bracket. To prove the Jacobi identity, it suffices to consider the quantities

$$[[X, Y], Z] = XYZ - YXZ - ZXY + ZYX,$$
$$[[Y, Z], X] = YZX - ZYX - XYZ + XZY,$$
$$[[Z, X], Y] = ZXY - XZY - YZX + YXZ.$$

Clearly these sum to zero. Finally, to show (iv), we carry out the computation in local coordinates. We have

$$\begin{aligned}
[fX, gY] &= fX(gY) - gY(fX) \\
&= fX^i \frac{\partial}{\partial x^i}\left(gY^j \frac{\partial}{\partial x^j} \right) - gY^i \frac{\partial}{\partial x^i}\left(fX^j \frac{\partial}{\partial x^j} \right) \\
&= \left(fX^i \frac{\partial g}{\partial x^i} Y^j + fgX^i \frac{\partial Y^j}{\partial x^i} - gY^i \frac{\partial f}{\partial x^i} X^j - fgY^i \frac{\partial X^j}{\partial x^i} \right) \frac{\partial}{\partial x^j} \\
&= fg[X, Y] + fX(g)Y - gY(f)X,
\end{aligned}$$

which completes our proof. □

We have from the preceding proposition all the elements of an important concept.

Definition 6.19. A *Lie algebra* is a vector space with binary operation $[\cdot,\cdot]$ satisfying conditions (i)–(iii) of Proposition 6.4.

We will discuss Lie algebras and their relation to Lie groups and geometric mechanics on Lie groups further in Chapter 9. Notice, however, that the set of all vector fields $\mathfrak{X}(M)$ on a manifold M constitutes a Lie algebra with respect to the bracket operation.

Next we examine the following geometric interpretation of the Lie bracket, which relates the flow along the bracket direction $[X, Y]$ with a concatenation of forward and reverse flows along the original vector fields X and Y. This geometric interpretation is discussed, for example, in [3]. Its relation to control theory is examined in [4], [37], and [225].

Proposition 6.5. *Let $X, Y \in \mathfrak{X}(M)$ and let $\varphi_t^X : M \to M$ be the flow along X for a duration t. Then, for any fixed $x \in M$, the curve $c : I \to M$ defined by*

$$c(t) = \varphi_{\sqrt{t}}^{-Y} \circ \varphi_{\sqrt{t}}^{-X} \circ \varphi_{\sqrt{t}}^{Y} \circ \varphi_{\sqrt{t}}^{X}(x)$$

is differentiable and $c'(0) = [X, Y](x)$; here $I \subset \mathbb{R}$ is an open interval containing the origin.

Proof. To see this, let us pass to a set of local coordinates $(x^1, \ldots, x^n)$ for the given chart (U, ψ). Without loss of generality, we may impose the condition that $\psi(x) = 0$, where $x \in M$ is the point at which we are computing the bracket $[X, Y]$. We may also take the open interval I to be sufficiently small so that the image of curve c lies entirely within our open neighborhood U. We will consider the curve's trajectory in local coordinates, $\psi \circ c : I \to \mathbb{R}^n$, as depicted in Figure 6.4, and denote its tangent vector by $R = (\psi \circ c)'(0) \in \mathbb{R}^n$.

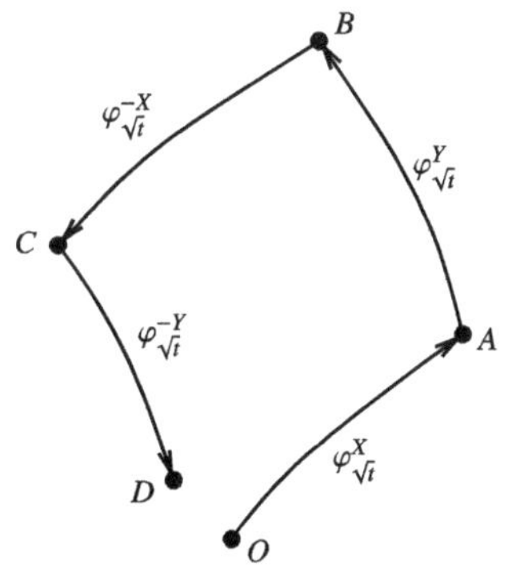

Figure 6.4: Geometric interpretation of the bracket $[X, Y]$.

Let us begin by considering the curve $\psi \circ \varphi^X_{\sqrt{t}}$ connecting the origin to point A. Expressing the location of point A using a second-order Taylor series in $\sqrt{t}$, we obtain the asymptotic expansion

$$x^i_A = X^i\sqrt{t} + \frac{1}{2}X[X^i]t + O(t^{3/2})$$

as $t \to 0$. Here, $X[X^i]$ denotes the vector field X acting on the ith component X^i, i. e.,

$$X[X^i] = X^j \frac{\partial X^i}{\partial x^j}.$$

Also, whenever we denote a vector field or one of its components without a subscript, we will assume that such a quantity is to be considered as being evaluated at the origin. When a subscript appears, we assume that the given quantity is evaluated at the given location; for example, X^i_A represents the ith component (relative to our local coordinates) of vector field X evaluated at point A.

To complete the proof, we must approximate the locations x^i_B, x^i_C, and x^i_D using a second-order Taylor series expansion in $\sqrt{t}$. To achieve this, we will need an approximation to the values of the vector fields X and Y at the points A, B, and C; as it turns out, however, we will only require a first-order approximation in $\sqrt{t}$ for these vector fields. At the point A, to the first order in $\sqrt{t}$, we have

$$\begin{aligned} X^i_A &= X^i + X[X^i]\sqrt{t} + O(t), \\ Y^i_A &= Y^i + X[Y^i]\sqrt{t} + O(t). \end{aligned}$$

Similarly, at the points B and C we obtain

$$\begin{aligned} X^i_B &= X^i_A + Y_A[X^i_A]\sqrt{t} + O(t) \\ &= X^i + (X[X^i] + Y[X^i])\sqrt{t} + O(t), \\ Y^i_B &= Y^i + (X[Y^i] + Y[Y^i])\sqrt{t} + O(t), \\ Y^i_C &= Y^i_B - X_B[Y^i_B]\sqrt{t} + O(t) \\ &= Y^i + Y[Y^i]\sqrt{t}. \end{aligned}$$

Now that we have obtained first-order approximations to the vector fields X and Y at the points A, B, and C, we can systematically approximate the locations of these points. To the second order in $\sqrt{t}$, the point B is located at

$$\begin{aligned} x^i_B &= x^i_A + Y^i_A\sqrt{t} + \frac{1}{2}Y_A[Y^i_A]t + O(t^{3/2}) \\ &= (X^i + Y^i)\sqrt{t} + \left(\frac{1}{2}X[X^i] + X[Y^i] + \frac{1}{2}Y[Y^i]\right)t + O(t^{3/2}). \end{aligned}$$

Notice that, since the vector field Y only appears in the terms of order $\sqrt{t}$ or higher, we only needed the terms in the expansion for Y^i_A up to order $\sqrt{t}$. Continuing in a similar fashion, we next obtain

$$\begin{aligned} x_C^i &= x_B^i - X_B^i\sqrt{t} + \frac{1}{2}X_B[X_B^i]t + O(t^{3/2}) \\ &= Y^i\sqrt{t} + \left(X[Y^i] - Y[X^i] + \frac{1}{2}Y[Y^i]\right)t + O(t^{3/2}), \\ x_D^i &= x_C^i - Y_C^i\sqrt{t} + \frac{1}{2}Y_C[Y_C^i]t + O(t^{3/2}) \\ &= (X[Y^i] - Y[X^i])t + O(t^{3/2}). \end{aligned}$$

However, since $(\psi \circ c)(t) = x_D(t) + O(t^{3/2})$, it follows that, as $t \to 0$,

$$R^i = (\psi \circ c)'(0)^i = X[Y^i] - Y[X^i],$$

which equals the Lie bracket $[X, Y]$ expressed in local coordinates, thereby completing the proof. □

The geometric interpretation for the Lie bracket lends itself immediately to the following corollary.

Corollary 6.1. *Let $X, Y \in \mathfrak{X}(M)$ and $x \in M$. Then*

$$\varphi_t^Y \circ \varphi_t^X(x) = \varphi_t^X \circ \varphi_t^Y(x)$$

if and only if $[X, Y] = 0$.

Exercise 6.7. Use Proposition 6.5 to prove Corollary 6.1.

More importantly, we have the following observation.

Remark. Whenever $[X, Y] \neq 0$, one may obtain a new, independent direction of motion by concatenating infinitesimal paths along the X and Y vector fields.

This observation is extremely important in control theory. It has implications especially for sub-Riemannian, or underactuated, control systems, in which the span of the control vector fields does not equal the whole of the tangent space. For instance, the geometric interpretation of the Lie bracket presented in Proposition 6.5 is relevant to the parallel parking of an automobile, which we will discuss in our next example.

Example 6.1. In this example, we consider a simplified model for an automobile. The configuration manifold is given by $Q = S^1 \times \mathbb{R}^2$. We choose (θ, x, y) as our local coordinates, where (x, y) represent the location of the center of mass and θ represents the angle between the car and the x-axis, as depicted in Figure 6.5. See [223] for a slightly more realistic version of the parallel parking problem, in which an additional degree of freedom is added by including the directional angle of the steering column and modeling the configuration manifold as $Q' = T^2 \times \mathbb{R}^2$.

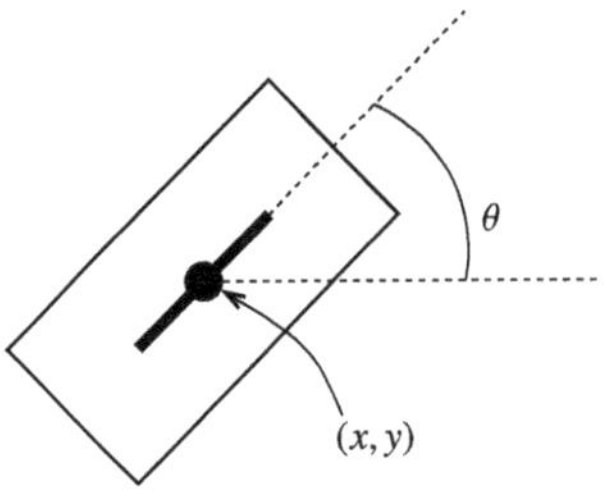

Figure 6.5: Model for an automobile.

Now suppose we are given the control vector fields

$$X_1 = \frac{\partial}{\partial \theta},$$
$$X_2 = \cos\theta \frac{\partial}{\partial x} + \sin\theta \frac{\partial}{\partial y}.$$

Vector field X^1 represents rotations of the vehicle about its vertical axis of inertia, whereas vector field X_2 represents motion in the direction of the frame of the car (Figure 6.6). We will suppose that we have controls in these two independent directions, i. e., we can drive the car in any fashion such that its velocity is contained within the span of vector fields X_1 and X_2. We now would like to ask the question, *How does one perform parallel parking of the vehicle?* In other words, suppose we would like to actualize a net displacement in the direction

$$X_3 = -\sin\theta \frac{\partial}{\partial x} + \cos\theta \frac{\partial}{\partial y},$$

which is orthogonal to the control direction X_2.

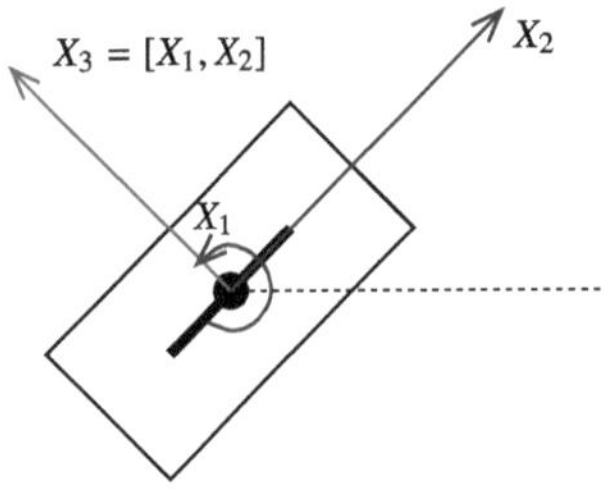

Figure 6.6: Control vector fields for the automobile.

To answer this question, let us compute the bracket of vector fields X_1 and X_2. We obtain

$$\begin{aligned}[X_1, X_2] &= X_1 X_2 - X_2 X_1 \\ &= \frac{\partial}{\partial \theta}\left(\cos\theta \frac{\partial}{\partial x} + \sin\theta \frac{\partial}{\partial y}\right) - \left(\cos\theta \frac{\partial}{\partial x} + \sin\theta \frac{\partial}{\partial y}\right)\left(\frac{\partial}{\partial \theta}\right)\end{aligned}$$

$$= -\sin\theta\frac{\partial}{\partial x} + \cos\theta\frac{\partial}{\partial y} = X_3.$$

We see that the Jacobi–Lie bracket of control vector fields X_1 and X_2 yields a third, independent direction X_3, which lies perpendicular to the alignment of the vehicle, i. e., vector X_2. Hence, by concatenating forward and reverse motion with infinitesimal rotations, one obtains a new direction of motion that is perpendicular to the vehicle (Figure 6.7). We therefore see that by "wiggling," one may steer the vehicle in a direction perpendicular to the body of the car itself, allowing one to *parallel park*. □

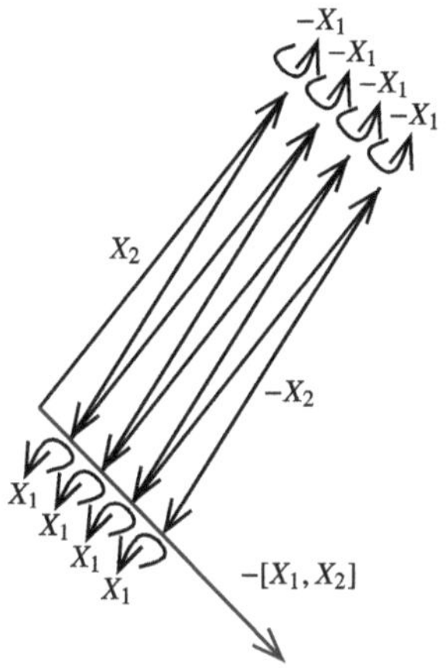

Figure 6.7: Concatenation of control vector fields X_1 and X_2 yields a new, independent direction of motion.

We will return again to the notion of the bracket of two vector fields in later chapters. What we actually broached in the previous example of the parallel parking of an automobile was our first example of a *nonholonomic* system. Such systems possess nonintegrable constraints on their velocity. For example, a nonholonomic constraint for the parallel parking problem is given by requiring that at no time may the velocity of the system have a component along the X_3 direction that lies perpendicular to the forward direction of the car. This specific constraint is an example of what is called a *knife edge* constraint. We will return to this idea at the end of Chapter 7 and revisit it once again in Chapter 10 for a more detailed discussion on approaches to handling such systems.

6.6 Differential Forms

In this section, we lay out some basic definitions and results regarding differential forms. For more details on the subject, see, for instance, [214]. We will pay particular attention to differential two-forms, as they will be the most complicated form encountered in this text.

Definition 6.20. A *differential k-form* on a manifold Q is a smooth assignment of an antisymmetric, k-multilinear mapping $\alpha_q : T_qQ \times \cdots \times T_qQ \to \mathbb{R}$ on each tangent space T_qQ. The set of all differential k-forms on Q is denoted $\mathcal{A}^k(Q)$.

If we omitted the antisymmetry requirement, such a mapping would be called a $(0\ k)$-tensor. Differential *one*-forms (sometimes simply called *differential forms*) are just smooth covector fields on the cotangent bundle T^*Q and therefore may locally be represented by $\alpha = \alpha_i dq^i$. Similarly, differential two-forms also have *components* relative to sets of local coordinates. In particular, if α is a differential two-form, then its *components* α_{ij}, relative to a local set of coordinates $(q^1, \dots, q^n)$, are defined by

$$\alpha_{ij} = \alpha\left(\frac{\partial}{\partial q^i}, \frac{\partial}{\partial q^j}\right).$$

Due to the antisymmetry requirement, it follows that $\alpha_{ij} = -\alpha_{ji}$. Also, bilinearity of two forms implies that, for arbitrary $v, w \in T_qQ$, we may write

$$\alpha(u, v) = \alpha_{ij} v^i w^j.$$

Definition 6.21. If α and β are two differential one-forms on a manifold Q, then the *wedge product* (or *exterior product*) of α and β, denoted $\alpha \wedge \beta$, is the differential two-form defined by the relation

$$\alpha \wedge \beta(u, v) = \alpha(u)\beta(v) - \alpha(v)\beta(u)$$

for all $u, v \in TQ$.

Wedge products can be further defined for higher-order differential forms (i. e., the wedge product of a k-form and an l-form). We refer the reader to [3] for details.

The definition of the wedge product is equivalent to the determinant

$$\alpha \wedge \beta(u, v) = \det \begin{bmatrix} \alpha[u] & \alpha[v] \\ \beta[u] & \beta[v] \end{bmatrix},$$

so that it has the geometric interpretation of equaling the (signed) area of the parallelogram defined by the vectors $(\alpha[u], \beta[u])$ and $(\alpha[v], \beta[v])$. See [8] for additional details.

Exercise 6.8. Show that any differential two-form α, with components α_{ij}, may be locally expressed in terms of the basis dq^i for the cotangent bundle as follows:

$$\alpha = \frac{1}{2}\alpha_{ij} dq^i \wedge dq^j.$$

The exterior product of two one-forms can further be generalized to higher-order exterior monomials as follows.

Definition 6.22. If $\alpha_1, \dots, \alpha_k$ are differential one-forms, then the differential k-form $\alpha_1 \wedge \cdots \wedge \alpha_k$ is defined by the relation

$$(\alpha_1 \wedge \cdots \wedge \alpha_k)(v_1, \dots, v_k) = \det \begin{bmatrix} \alpha_1[v_1] & \cdots & \alpha_1[v_k] \\ \vdots & & \vdots \\ \alpha_k[v_1] & \cdots & \alpha_k[v_k] \end{bmatrix}$$

for all $v_1, \dots, v_k \in T_qQ$.

Given the set of k differential one-forms $\alpha_1, \dots, \alpha_k$, we may therefore define a mapping $\alpha : T_qQ \to \mathbb{R}^k$ by the relation $\alpha(v) = (\alpha_1[v], \dots, \alpha_k[v])$. The preceding definition states that the value of the exterior product $\alpha_1 \wedge \cdots \wedge \alpha_k$ on the vectors $v_1, \dots, v_k \in T_qQ$ is equal to the oriented volume in $\mathbb{R}^k$ of the parallelepiped formed by the image of these vectors under the mapping α.

Definition 6.23. Let α be a differential k-form on a manifold N and $\varphi : M \to N$ a smooth mapping from manifold M to manifold N. Then the *pullback* $\varphi^*\alpha$ of α by φ is the differential k-form on M defined by

$$(\varphi^*\alpha)_q(v_1, \dots, v_k) = \alpha_{\varphi(q)}\left(d_q\varphi \cdot v_1, \dots, d_q\varphi \cdot v_k\right)$$

for all $v_1, \dots, v_k \in T_xM$.

Definition 6.24. Let α be a differential k-form and X a vector field on Q. Then the *interior product* $\mathbf{i}_X\alpha$ is the differential $(k-1)$-form defined by

$$(\mathbf{i}_X\alpha)_q(v_1, \dots, v_{k-1}) = \alpha_q(X(q), v_1, \dots, v_{k-1})$$

for all $v_1, \dots, v_k \in T_qQ$.

Definition 6.25. Let α be a differential k-form on Q that is given locally by the expression

$$\alpha = \alpha_{i_1 \cdots i_k} dx^{i_1} \wedge \cdots \wedge dx^{i_k}.$$

Then the *exterior derivative* of α is the $(k+1)$-form $d\alpha$ defined locally by

$$d\alpha = \frac{\partial \alpha_{i_1 \cdots i_k}}{\partial x^j} dx^j \wedge dx^{i_1} \wedge \cdots \wedge dx^{i_k}.$$

Exercise 6.9. Prove that if $F = Pdx + Qdy + Rdz$ is a one-form on $\mathbb{R}^3$ for given smooth functions $P, Q, R : \mathbb{R}^3 \to \mathbb{R}$, then the exterior derivative is given by

$$dF = \left(\frac{\partial R}{\partial y} - \frac{\partial Q}{\partial z}\right) dy \wedge dz + \left(\frac{\partial P}{\partial z} - \frac{\partial R}{\partial x}\right) dz \wedge dx + \left(\frac{\partial Q}{\partial x} - \frac{\partial P}{\partial y}\right) dx \wedge dy.$$

How is this expression related to the curl of the covector field F?

Exercise 6.10. Prove the identity

$$d\alpha(X, Y) = X[\alpha(Y)] - Y[\alpha(X)] - \alpha([X, Y])$$

that is satisfied for any one-form α and vector fields X, Y.

Integration of Differential Forms on Manifolds

The definitions we discussed in the previous paragraph immediately lend themselves to the concept of integration on manifolds. The theory is based on the fact that all differential k-forms in $\mathbb{R}^k$ are proportional to $dx^1 \wedge \cdots dx^k$.

Definition 6.26. If $\alpha = f(x^1, \ldots, x^k) dx^1 \wedge \cdots \wedge dx^k$ is a differential k-form in $\mathbb{R}^k$ and $\sigma \subset \mathbb{R}^k$ is a compact set, then we define the *integral of α over σ* in terms of the ordinary Riemann integral by the relation

$$\int_\sigma \alpha = \int_\sigma f(x^1, \ldots, x^k) dx^1 \cdots dx^k.$$

Definition 6.27. Suppose Ω is a k-dimensional connected region of the n-dimensional manifold Q, parameterized by local coordinates $(x^1, \ldots, x^k)$ via the smooth parameterization mapping $\varphi : (\sigma \subset \mathbb{R}^k) \to \Omega \subset Q$. Then the *integral of the k-form α over the set Ω* is given by

$$\int_\Omega \alpha = \int_\sigma \varphi^* \alpha,$$

where $\varphi^* \alpha$ is the pullback of α by the mapping φ.

Theorem 6.1 (Stokes' Theorem). *If Ω is a compact, oriented k-dimensional submanifold of Q with boundary $\partial\Omega$ and α is a $(k-1)$ form on Ω, then*

$$\int_\Omega d\alpha = \int_{\partial\Omega} \alpha.$$

Exercise 6.11. Use Stokes' theorem and the result from Exercise 6.9 to prove the classical form of Stokes' theorem: for a smooth 2-D surface $\Sigma \subset \mathbb{R}^3$ with boundary $\partial\Sigma$ and differentiable vector field $\mathbf{F}$,

$$\int_\Sigma \operatorname{curl} \mathbf{F} \cdot d\mathbf{S} = \int_{\partial\Sigma} \mathbf{F} \cdot d\mathbf{r}.$$

6.7 Riemannian Geometry

In this paragraph, we introduce two important concepts from Riemannian geometry: affine connections and covariant differentiation of vector fields. The need for such creatures arises out of the fundamental geometric fact that the ordinary coordinate derivatives of vector fields fail to form invariant quantities. See, for example, [70] and [75] for details. To define differentiation of vector fields on manifolds, one requires a structure known as a *connection*.

Connections and Covariant Differentiation

Given a vector field on a manifold, one naturally desires a way to construct a coordinate-invariant way of computing derivatives. This operation is provided by defining an *affine connection* on the manifold.

Definition 6.28. An *affine connection* ∇ on a differentiable manifold Q is a mapping $\nabla : \mathfrak{X}(Q) \times \mathfrak{X}(Q) \to \mathfrak{X}(Q)$, denoted by $(X, Y) \mapsto \nabla_X Y$, such that

1. $\nabla_{fX+gY} Z = f\nabla_X Z + g\nabla_Y Z$;
2. $\nabla_X(Y + Z) = \nabla_X Y + \nabla_X Z$; and
3. $\nabla_X(fY) = f\nabla_X Y + X[f]Y$

for all $X, Y, Z \in \mathfrak{X}(Q)$ and $f, g \in C^1(Q)$.

Since the covariant derivative $\nabla_X Y$ depends on the value of X only at the point $q \in Q$ where it is computed, whereas it depends on the values of vector field Y along the curves passing through q with tangent vector X, one can immediately use this operation to define derivatives of vector fields along curves.

Theorem 6.2. *Let Q be a manifold with an affine connection ∇. Then there exists a unique correspondence that associates to a vector field V along the smooth curve $c : [a, b] \to Q$ another vector field $\frac{DV}{Dt}$ along c, called the covariant derivative of V along c, such that*

1. $\frac{D}{Dt}(V + W) = \frac{DV}{Dt} + \frac{DW}{Dt}$;
2. $\frac{D}{Dt}(fV) = \frac{df}{dt}V + f\frac{DV}{Dt}$; *and*
3. *if V is induced by a vector field $Y \in \mathfrak{X}(Q)$, i. e., $V(t) = Y(c(t))$, then $\frac{DV}{Dt} = \nabla_{c'(t)}Y$*

for all $V, W \in \mathfrak{X}(Q)$ and $f \in C^1(Q)$.

Proof. First, suppose there exists a correspondence satisfying the three criteria of Theorem 6.2. Let $X_i = \frac{\partial}{\partial q^i}$ be the moving frame corresponding to the coordinate basis $\frac{\partial}{\partial q^i}$. Then we can write $V = v^i X_i$, where $v^i = v^i(t)$ and $X_j = X_j(c(t))$. By criteria 1 and 2, we have

$$\frac{DV}{Dt} = \frac{dv^i}{dt}X_i + v^i\frac{DX_i}{Dt}.$$

By criterion 3 of the theorem and item 1 of Definition 6.28, it follows that

$$\frac{DX_i}{Dt} = \nabla_{dc/dt}X_i = \nabla_{\frac{dx^j}{dt}X_j}X_i = \frac{dx^j}{dt}\nabla_{X_j}X_i.$$

Combining the two preceding equations, we find

$$\frac{DV}{Dt} = \frac{dv^i}{dt}X_i + \frac{dx^j}{dt}v^i\nabla_{X_j}X_i. \tag{6.3}$$

Therefore, if such a correspondence exists, then $\frac{DV}{Dt}$ is unique. To prove existence, define $\frac{DV}{Dt}$ by (6.3). □

We next define a set of quantities that are useful when performing coordinate-based calculations.

Definition 6.29. Given an affine connection $\nabla : \mathfrak{X}(Q) \times \mathfrak{X}(Q) \to \mathfrak{X}(Q)$ and a set of local coordinates $(q^1, \ldots, q^n)$ in the open set $U \subset Q$, the *Christoffel symbols* Γ^k_{ij} of ∇ relative to the coordinates $(q^1, \ldots, q^n)$ are the quantities defined by the relation

$$\nabla_{\partial_i} \partial_j = \Gamma^k_{ij} \partial_k,$$

where $\partial_i = \partial/\partial q^i$ denotes the standard coordinate basis vector for the tangent space in the direction q^i.

Exercise 6.12. Suppose that Γ^k_{ij} are the components of a connection ∇ relative to coordinates $(q^1, \ldots, q^n)$ in the neighborhood $U \subset Q$ and $\tilde{\Gamma}^{\gamma}_{\alpha\beta}$ are the components relative to coordinates $(x^1, \ldots, x^n)$ in the neighborhood $V \subset Q$. Show that in the coordinate overlap region $U \cap V$, we obtain

$$\tilde{\Gamma}^{\gamma}_{\alpha\beta} = \frac{\partial x^{\gamma}}{\partial q^k} \left(\frac{\partial q^i}{\partial x^{\alpha}} \frac{\partial q^j}{\partial x^{\beta}} \Gamma^k_{ij} + \frac{\partial^2 q^k}{\partial x^{\alpha} \partial x^{\beta}} \right). \tag{6.4}$$

Hint: compute $\nabla_{\frac{\partial}{\partial x^{\alpha}}} \frac{\partial}{\partial x^{\beta}}$ using 1 to 3 of Definition 6.28 and the relations

$$\frac{\partial}{\partial x^{\alpha}} = \frac{\partial q^i}{\partial x^{\alpha}} \frac{\partial}{\partial q^i} \quad \text{and} \quad \frac{\partial}{\partial x^{\beta}} = \frac{\partial q^j}{\partial x^{\beta}} \frac{\partial}{\partial q^j}.$$

This demonstrates that the Christoffel symbols are not tensorial quantities (i. e., invariant quantities that transform by the chain rule via multiple factors of the Jacobian matrix).

To see how one might express the covariant derivative of two vector fields in local coordinates, let us set $X = x^i \partial_i$ and $Y = y^j \partial_j$, where X and Y are vector fields. Then we compute

$$\nabla_X Y = \nabla_{x^i \partial_i} (y^j \partial_j) = x^i \nabla_{\partial_i} (y^j \partial_j) = x^i y^j \nabla_{\partial_i} \partial_j + x^i \partial_i [y^j] \partial_j.$$

Using the Christoffel symbols and recognizing that $x^i \partial_i [y^j] = X[y^j]$, we obtain

$$\nabla_X Y = (\Gamma^k_{ij} x^i y^j + X[y^k]) \frac{\partial}{\partial q^k}. \tag{6.5}$$

Note that the covariant derivative $\nabla_X Y$ depends on the value of X at point q and the values of vector field Y along a curve passing through q in the direction X, so that, in particular, quantities such as $\nabla_{\dot{c}} V$ and $\nabla_{\dot{c}} \dot{c}$, where V is a vector field along a particular curve c, are meaningful.

Definition 6.30. A vector field V defined along a smooth curve $c : [a, b] \to Q$ is called *parallel* if its covariant derivative $\nabla_{\dot{c}} V = 0$ along c. If $V(t)$ is such a vector field, the vector $V(b) \in T_{c(b)}Q$ is called the *result of parallel transport* of the vector $V(a) \in T_{c(a)}Q$ along curve c.

Covariant differentiation therefore introduces a notion of *parallelism* between neighboring tangent spaces. More accurately, parallelism and covariant differentiation should be thought of as two sides of the same basic structure, for, without parallelism, it is impossible to compare vectors belonging to neighboring tangent spaces to compute difference quotients when computing the derivative. Thus, an affine connection literally *connects* vectors together from neighboring tangent spaces.

Exercise 6.13. Let X and Y be vector fields and $q \in Q$ a fixed point on manifold Q. Let $c : [-\varepsilon, \varepsilon] \to Q$ be a smooth curve such that $c(0) = q$ and $c'(0) = X$, and let vector field $\tilde{Y}$ be the result of parallel transport of the vector $Y(q) \in T_qQ$ along curve c. Show that

$$\nabla_X Y(q) = \lim_{t \to 0} \frac{Y(c(t)) - \tilde{Y}(t)}{t}.$$

Note that both $Y(c(t)), \tilde{Y}(t) \in T_{c(t)}Q$. *Hint:* use the fact that $\nabla_X \tilde{Y} = 0$ to approximate the coordinates $\tilde{y}^k(t)$ of the vector-valued function $\tilde{Y}$ in a Taylor expansion about $t = 0$.

Definition 6.31. A smooth curve $c : [a, b] \to Q$ is called a *geodesic* of manifold Q relative to a given connection ∇ if and only if $\nabla_{\dot{c}} \dot{c} = 0$ along c, i. e., if and only if curve c in a parallel fashion transports its own tangent vector.

Thus, a curve $c(t) = (q^1(t), \ldots, q^n(t))$ is a geodesic if and only if it satisfies the system of second-order differential equations

$$\ddot{q}^k + \Gamma^k_{ij} \dot{q}^i \dot{q}^j = 0 \tag{6.6}$$

for $k = 1, \ldots, n$, known as the *geodesic equations*.

Finally, we introduce two additional tensor quantities of importance that are determined once one is given an affine connection.

Definition 6.32. Let ∇ be an affine connection on Q. Then the *torsion* of ∇ is a mapping $T : \mathfrak{X}(Q) \times \mathfrak{X}(Q) \to \mathfrak{X}(Q)$ defined by the relation

$$T(X, Y) = \nabla_X Y - \nabla_Y X - [X, Y], \tag{6.7}$$

where $X, Y \in \mathfrak{X}(Q)$ and $[\cdot, \cdot]$ is the ordinary Jacobi–Lie bracket of vector fields on Q. A *symmetric connection* is an affine connection whose torsion vanishes identically.

Exercise 6.14. Show that the Christoffel symbols of a symmetric connection are symmetric in their lower indices, i. e., $\Gamma^k_{ij} = \Gamma^k_{ji}$.

Definition 6.33. Let ∇ be an affine connection on Q. Then the *curvature* of ∇ is a mapping $R : \mathfrak{X}(Q) \times \mathfrak{X}(Q) \times \mathfrak{X}(Q) \to \mathfrak{X}(Q)$ defined by the relation

$$R(X,Y,Z) = \nabla_X \nabla_Y Z - \nabla_Y \nabla_X Z - \nabla_{[X,Y]} Z, \tag{6.8}$$

where $X, Y, Z \in \mathfrak{X}(Q)$ and $[\cdot,\cdot]$ is the ordinary Jacobi–Lie bracket of vector fields on Q.

Exercise 6.15. Let $\xi(\varepsilon)$ be the result of parallel transport of a vector $\xi \in T_q Q$ along the closed square of side length ε in coordinate space parallel to the x^i- and x^j-axes. Show that

$$\lim_{\varepsilon \to 0} \frac{\xi^k(\varepsilon) - \xi}{\varepsilon^2} = -R^k_{lij} \xi^l.$$

Thus, curvature may be viewed as a measure of the extent to which a vector does not return to its initial value upon being transported in a parallel fashion along a differential, closed curve.

Metrics on Manifolds

The theory of Riemannian geometry arises out of the desire to be able to measure distances on manifolds. The fundamental object that allows one to compute the distance of a curve is a *metric*. Once a manifold is endowed with such a structure, many additional concepts soon follow, such as the notions of *parallelism* between neighboring tangent spaces, *geodesics*, and *curvature*.

Definition 6.34. A *Riemannian manifold* is a differentiable manifold Q in which each tangent space $T_q Q$ is equipped with an inner product $\langle \cdot, \cdot \rangle_q$ (that is, a symmetric, bilinear, positive definite form), known as the *Riemannian metric* or *Riemannian structure*, which varies differentiably from point to point in the sense that, for any coordinate chart (U, φ) with local coordinates $(q^1, \dots, q^n)$, the functions $g_{ij}(q^1, \dots, q^n) = \langle \frac{\partial}{\partial q^i}, \frac{\partial}{\partial q^j} \rangle_q$, known as the *components of the metric* relative to the chart (U, φ), are differentiable functions of q in the set U.

Given a metric structure on a manifold Q, the *differential arclength ds* of a curve $c : [-a, b] \to Q$ at $c(t)$ is defined by the relation

$$ds^2 = \langle c'(t)dt, c'(t)dt \rangle_{c(t)}.$$

The *total arclength* $\mathcal{L}[c]$ of c is then given by

$$\mathcal{L}[c] = \int_a^b \sqrt{\langle c'(t), c'(t) \rangle} dt.$$

Exercise 6.16. Show that $\mathscr{L}[c]$ is independent of the parameterization of curve c, i. e., if $\tau : [a,b] \to [c,d]$ is a monotonic function of t and $\tilde{c}(\tau) = c(t(\tau))$, then

$$\mathscr{L}[c] = \int_c^d \sqrt{\langle \tilde{c}'(\tau), \tilde{c}'(\tau)\rangle} d\tau.$$

Definition 6.35. A connection ∇ on a Riemannian manifold Q is *compatible with the metric* if, for any vector fields $X, Y, Z \in \mathfrak{X}(Q)$, the following relation holds:

$$X\langle Y, Z\rangle = \langle \nabla_X Y, Z\rangle + \langle Y, \nabla_X Z\rangle.$$

Thus, connections that are compatible with the metric are those for which the product rule holds for directional derivatives of the inner product $\langle Y, Z\rangle$ in the direction of vector field X.

A Riemannian structure on Q endows each point $q \in Q$ with a natural vector space isomorphism $\flat : T_qQ \to T_q^*Q$, defined by the relation $X^\flat[Y] = \langle X, Y\rangle_q$, for all $X, Y \in T_qQ$. In local coordinates, $X^\flat = g_{ij}X^i dq^j$. The inverse of this transformation is denoted $\sharp = \flat^{-1}$. A *dual metric* is therefore given on the cotangent bundle by the inner product $\langle \cdot, \cdot \rangle_q^* : T_q^*Q \times T_q^*Q \to \mathbb{R}$ defined by

$$\langle u, v\rangle_q^* = \langle u^\sharp, v^\sharp\rangle_q$$

for every $u, v \in T_q^*Q$. One can then locally define the components of the dual metric by the relations $g^{ij} = \langle dq^i, dq^j\rangle_q^*$. Locally, one may therefore further write $u = u_i dq^i$ and

$$u^\sharp = g^{ij} u_i \frac{\partial}{\partial q^j}.$$

Exercise 6.17. Prove that $[g^{ij}] = [g_{ij}]^{-1}$, that is, $g^{ik}g_{kj} = \delta^i_k$.

The following theorem was introduced in [185].

Theorem 6.3 (Fundamental Theorem of Riemannian Geometry). *Given a Riemannian manifold Q, there exists a unique affine connection ∇ on Q, known as the Levi-Civita connection, that is symmetric and compatible with the Riemannian metric. Moreover, the Christoffel symbols of the Levi-Civita connection may be expressed relative to local coordinates by the relation*

$$\Gamma^k_{ij} = \frac{1}{2} g^{kr} \left(\frac{\partial g_{jr}}{\partial q^i} + \frac{\partial g_{ri}}{\partial q^j} - \frac{\partial g_{ij}}{\partial q^r} \right). \tag{6.9}$$

Proof. Let us first assume that there exists such a connection. Since this connection is compatible with the metric, we may write

$$X\langle Y, Z\rangle = \langle \nabla_X Y, Z\rangle + \langle Y, \nabla_X Z\rangle,$$

$$Y\langle Z,X\rangle = \langle\nabla_Y Z,X\rangle + \langle Z,\nabla_Y X\rangle,$$
$$Z\langle X,Y\rangle = \langle\nabla_Z X,Y\rangle + \langle X,\nabla_Z Y\rangle.$$

Assuming the connection is also symmetric, so that $[X,Y] = \nabla_X Y - \nabla_Y X$, it follows that $X\langle Y,Z\rangle + Y\langle Z,X\rangle - Z\langle X,Y\rangle$ is equal to

$$\langle [X,Z],Y\rangle + \langle [Y,Z],X\rangle + \langle [X,Y],Z\rangle + 2\langle Z,\nabla_Y X\rangle,$$

and hence

$$\begin{aligned}\langle Z,\nabla_Y X\rangle = \frac{1}{2}\,(&X\langle Y,Z\rangle + Y\langle Z,X\rangle - Z\langle X,Y\rangle \\ &- \langle [X,Z],Y\rangle - \langle [Y,Z],X\rangle - \langle [X,Y],Z\rangle).\end{aligned} \tag{6.10}$$

However, this equation then uniquely determines the connection $\nabla_Y X$.

Exercise 6.18. Show that the connection defined by (6.10) is symmetric and compatible with the metric. Conclude that such a metric exists.

Finally, to show that the Christoffel symbols are given by (6.9), let us apply (6.10) to the vector fields $X = \frac{\partial}{\partial q^i}$, $Y = \frac{\partial}{\partial q^j}$, and $Z = \frac{\partial}{\partial q^r}$. Recognizing the relations

$$\frac{\partial}{\partial q^r} = \delta_r^t \frac{\partial}{\partial q^t} \quad \text{and} \quad \nabla_{\frac{\partial}{\partial q^j}}\left(\frac{\partial}{\partial q^i}\right) = \Gamma_{ij}^s \frac{\partial}{\partial q^s},$$

we obtain

$$\langle Z,\nabla_Y X\rangle = g_{ts}\delta_r^t\Gamma_{ij}^s = g_{rs}\Gamma_{ij}^s.$$

Meanwhile, since our choice of vector fields X, Y, and Z commute, the right-hand side of (6.10) yields

$$g_{rs}\Gamma_{ij}^s = \frac{1}{2}\left(\frac{\partial g_{jr}}{\partial q^i} + \frac{\partial g_{ri}}{\partial q^j} - \frac{\partial g_{ij}}{\partial q^r}\right).$$

Contracting each side with g^{kr} yields the result. □

Remark. If $c : [a,b] \to Q$ is a geodesic of the Levi-Civita connection, then it must have constant speed. To see this, note that

$$\frac{d}{dt}\langle \dot c(t),\dot c(t)\rangle = \dot c(t)\langle \dot c(t),\dot c(t)\rangle = 2\langle \nabla_{\dot c(t)}\dot c(t),\dot c(t)\rangle = 0,$$

since time differentiation of a scalar along curve c is equivalent to action on the function by vector field $\dot c(t)$.

Exercise 6.19. Let Q be a Riemannian manifold. Show that if two vector fields V and W are parallel along a smooth curve $c : [a,b] \to Q$, relative to the Levi-Civita connection, then $\langle V(t),W(t)\rangle_{c(t)}$ is constant.

Exercise 6.20 (Motion in a Rotating Frame). Consider extended Euclidean space $Q = \mathbb{R} \times \mathbb{R}^3$ with local coordinates (t, q^1, q^2, q^3) and metric $g_{ij} = \delta_{ij}$. Show that under the coordinate transformation

$$\begin{aligned} t &= t, \\ x &= \cos(\omega t)q^1 + \sin(\omega t)q^2, \\ y &= -\sin(\omega t)q^1 + \cos(\omega t)q^2, \\ z &= q^3, \end{aligned}$$

which represents a uniformly rotating reference frame, with angular speed ω, the covariant acceleration of a particle may be expressed by

$$\frac{D(\dot{x}, \dot{y}, \dot{z})}{Dt} = (\ddot{x} - \omega^2 x - 2\omega\dot{y}, \ddot{y} - \omega^2 y + 2\omega\dot{x}, \ddot{z}).$$

Notice that $(-x\omega^2, -y\omega^2, 0)$ represents centripetal acceleration, whereas $(-2\omega\dot{y}, 2\omega\dot{x}, 0)$ represents Coriolis acceleration due to the rotation of the reference frame. *Hint*: start by using (6.4) to show that the nonzero Christoffel symbols for the rotating frame are $\tilde{\Gamma}^x_{tt} = -\omega^2 x$, $\tilde{\Gamma}^y_{tt} = -\omega^2 y$, $\tilde{\Gamma}^x_{ty} = \tilde{\Gamma}^x_{yt} = -\omega$, and $\tilde{\Gamma}^y_{tx} = \tilde{\Gamma}^y_{xt} = \omega$.

6.8 Application: The Foucault Pendulum

As a simple mechanical illustration of the theory presented in this chapter, we will consider the Foucault pendulum. The physical apparatus is easily described: a large, spherical pendulum that is free to swing back and forth in any vertical plane passing through the pivot point. The pendulum is carefully initialized so that it swings back and forth in a particular plane. If the pendulum is situated on the North Pole, it will swing back and forth in the same plane in inertial space. Relative to an observer sitting on the ground watching the pendulum, the pendulum's plane of swing will precess about the Earth's axis of rotation, completing one full precession each day. On the other hand, if the pendulum is situated on the Earth's equator, the plane of swing will remain the same. A much more interesting effect, however, is realized for Foucault pendula located elsewhere on the planet's surface. Not only does the swing plane of such a pendulum precess about the vertical axis, but it does not return to its initial plane after a 24-hour period! This phenomenon may be precisely understood in terms of the parallel transport of the pendulum's angular velocity as the pendulum circumscribes a complete revolution about the Earth, as was shown in [124].

To analyze this phenomenon, we will compute the result of parallel transport of a tangent vector to the sphere S^2 as it traces a circular path of constant latitude about the polar axis. We will use the local spherical coordinates (θ, φ) to describe locations on the sphere, where θ is the polar angle and φ the azimuthal angle.

Exercise 6.21. Show that under the change of coordinates

$$x = \sin\theta\cos\varphi, \quad y = \sin\theta\sin\varphi, \quad z = \cos\theta$$

the Euclidean metric $ds^2 = dx^2 + dy^2 + dz^2$ in $\mathbb{R}^3$ restricted to the sphere S^2 may be expressed in local spherical coordinates (θ, φ), $0 < \theta < \pi$ and $0 \le \varphi < 2\pi$ by the relation

$$ds^2 = d\theta^2 + \sin^2\theta d\varphi^2. \tag{6.11}$$

Exercise 6.22. Show that the nonzero components of the Christoffel symbols of the Levi-Civita connection of metric (6.11) are given by

$$\Gamma^1_{22} = -\sin\theta\cos\theta \quad \text{and} \quad \Gamma^2_{12} = \Gamma^2_{21} = \cot\theta.$$

Show that the geodesics on S^2 are great circles.

Now let us consider the smooth curve $c : [0, 2\pi] \to S^2$ that may be described in local coordinates (θ, φ) by $\theta(t) = \alpha$, $\varphi(t) = t$, for constant α and $0 \le t \le 2\pi$, and an initial tangent vector

$$\xi_0 = \xi_0^1 \frac{\partial}{\partial\theta} + \xi_0^2 \frac{\partial}{\partial\varphi} \in T_{(\alpha,0)}S^2.$$

Note that the vectors $\frac{\partial}{\partial\theta}$ and $\frac{\partial}{\partial\varphi}$ represent southward- and eastward-pointing unit vectors, respectively. We now define $\xi(t) \in T_{c(t)}S^2$ to be the result of parallel transport of the vector ξ_0 along the curve $c : [0, 2\pi] \to S^2$. From (6.5) and the parallel transport requirement that $\nabla_{\dot{c}}\xi = 0$, we see that the coordinates $\xi(t) = (\xi^1(t), \xi^2(t))$ satisfy the initial value problem

$$\dot{\xi} = \begin{bmatrix} 0 & \sin\alpha\cos\alpha \\ -\cot\alpha & 0 \end{bmatrix} \cdot \xi, \quad \xi(0) = (\xi_0^1, \xi_0^2)^T.$$

The solution of this equation is easily shown to be

$$\xi(t) = \xi_0^1 \begin{bmatrix} \cos(\omega t) \\ -\sin(\omega t) \end{bmatrix} + \xi_0^2 \begin{bmatrix} \sin(\omega t) \\ \cos(\omega t) \end{bmatrix},$$

where $\omega = \cos(\alpha)$. Note that the polar angle α of the pendulum's location is related to the latitude ℓ by $\alpha = \pi/2 - \ell$ for $\ell \in [-\pi/2, \pi/2]$. It follows that the precession rate of the plane of the pendulum's swing is given in terms of the pendulum's latitude by the relation

$$\omega = \sin(\ell).$$

Thus, at the North Pole ($\ell = \pi/2$), as we expected, the precession rate is precisely $\omega = 1$, indicating that one complete precession of the pendulum's swing plane is encountered each day, whereas at the equator, the precession rate is $\omega = 0$. At other latitudes, the precession rate lies somewhere in between. For example, at a latitude of $\ell = \pi/4$, the precession rate is $\omega = 1/\sqrt{2}$, and after a full revolution of the Earth, the pendulum's swing plane will only have rotated by an angle of $2\pi/\sqrt{2}$ (approximately 105° short of a full revolution).

Additional insight into the precession of Foucault pendula may be obtained by recalling the result of Exercise 6.19, which states that the angle between two parallel vector fields remains constant along the curves along which they are parallel. Since the Earth's equator is a geodesic, it transports its own tangent vector $\frac{\partial}{\partial\varphi}$ in a parallel fashion. Therefore, along the equator, the angle between the parallel vector field $\xi(t)$ and the equator's tangent vector $\frac{\partial}{\partial\varphi}$ must remain unchanged, yielding no precession. At all other latitudes, the circular paths $c(t)$ are not geodesics, and therefore the tangent vector field $\dot{c}(t)$ of the path is not a parallel vector field; hence the angle between $\xi(t)$ and $\dot{c}(t)$ will not remain fixed, the plane of the pendulum's swing does not remain fixed, and precession results.

Another astonishing application of parallel transport may be found within the field of general relativity. If a gyroscope is placed on board a satellite in circular orbit about the Earth, its angular momentum vector should point in the same direction once the satellite has made a complete revolution. However, due to the curvature of space-time, this is not the case, and the gyroscope undergoes a small precession known as a *Thomas precession*, which was first noted and applied to the spin of relativistic electrons in [282]; see [212] and [290] for additional details of this effect.

6.9 Application: General Relativity

Einstein's general theory of relativity is largely based on Riemannian geometry. In this context, our manifold is *space-time* itself. Moreover, space and time have different signatures in the metric, and hence it is called a *pseudo-Riemannian* or *Lorentzian* metric. For a flat space-time (e. g., in the case of special relativity), the metric used is the *Lorentz metric*

$$ds^2 = c^2dt^2 - dx^2 - dy^2 - dz^2,$$

where c is the speed of light. This metric measures distance in a flat space-time. (Notice it is no longer positive definite.)

In general relativity, one considers *Einstein's field equations*, which connect the stress-energy tensor from physics with the Einstein tensor, which is derived from the metric and its associated connection. Particles travel along geodesics of this metric. Thus, matter tells space-time how to curve, and space-time tells matter how to move.

If you would like to see a geodesic on a space-time manifold, simply toss a coin in the air or drop a ball and watch its trajectory. Thus, you can directly observe the curvature of space-time. If space-time were flat, releasing a ball would result in the ball staying put. It will continue its trajectory *through time*. However, space and time are curved, and the best it can do is follow a geodesic. This geodesic curves the ball away from a path parallel to our local time axis towards the earth.

The metric responsible for such behavior is the *Schwarzschild metric*,

$$ds^2 = \left[1 - \frac{2GM}{rc^2}\right] c^2 dt^2 - \left[1 - \frac{2GM}{rc^2}\right]^{-1} dr^2 - r^2(d\theta^2 + \sin^2(\theta) d\phi^2),$$

where (r, θ, ϕ) are standard spherical coordinates, M is a parameter (the mass of the object), and G is Newton's gravitational constant. This is the general spherically symmetric solution to Einstein's field equations in vacuum (i. e., obtained by setting the stress-energy tensor to zero). Notice that something very peculiar happens as $r \to 2GM/c^2$—space and time reverse roles! The sign of the dt^2 component becomes negative (which we call *space-like*) and the sign of the dr^2 component becomes positive (or *time-like*)! This is exciting stuff. As a result of this, the radius $r_s = 2GM/c^2$ is given the special name *Schwarzschild radius*, and any spherically symmetric mass (i. e., star) that collapses so that its radius is *less than* its Schwarzschild radius becomes a *black hole*. Once a particle enters a black hole (the boundary of which is defined by the *horizon*—$r = r_s$), it may never return. On the inside of a black hole, the radial coordinate itself becomes time-like, and what the rest of the universe experiences as time becomes *space-like*. Thus, the particle will start "aging" towards the center of the black hole, and the horizon will be in that particle's past. A particle can no sooner move away from the center of the black hole as you or I can travel backwards in time, for the simple reason that is precisely what it would have to do, travel back in time. The spherical horizon of the black hole will no longer be perceived as a sphere—or as anything at all! It will become a part of that particle's past and, with the rest of the universe, will be gone forever.

Another strange consequence of the Schwarzschild metric is that it takes an infinite amount of (coordinate) time to fall into a black hole. *Coordinate* time is time "at infinity," so it is the time Δt that passes for a distant observer. This is distinct from *proper time*, which is the time that passes from the viewpoint of the particle. A particle will fall into a black hole in finite proper time, but infinite coordinate time. So, as far as the rest of the universe is concerned, that particle will never actually reach the horizon; rather it will become slower and slower until it appears frozen at the surface. From the particle's point of view, the rest of the universe will age infinitely fast and, once within the horizon, time will proceed in the inward radial direction. It is easiest to see this by considering a photon, which travels on a *null geodesic*, where $ds = 0$. Assuming that a photon (traveling at the speed of light) starts outside the black hole and travels along a radial trajectory, we obtain:

$$cdt = \frac{dr}{1 - 2GM/(rc^2)}.$$

By integrating from, say, r_s to $2r_s$, we obtain an improper integral on the right-hand side that diverges:

$$c\Delta t = \int_{r_s}^{2r_s} \frac{dr}{1 - r_s/r} = \infty.$$

This shows that, even for a photon, an infinite amount of coordinate time elapses as it falls into the black hole.

What happens at the center of a black hole? Nobody knows. There are theories that you will pass into different universes; see *Kruskal coordinates*. And there are wild conjectures surrounding quantum gravity. Birkhoff's theorem states that any spherically symmetric mass distribution that has a radius less than 9/8ths of its Schwarzschild radius *must* collapse, i. e., there is no stable equilibrium. For this reason physicists assume that the star must collapse all the way to a single point—*the singularity*. I, however, will take the liberty to present here my own conjecture.

Black holes do not exist. At least, not within the framework of the Schwarzschild metric and at least not *now*. There are two problems which I foresee.

The first problem is that every star rotates. This immediately excludes the viability of the Schwarzschild metric from practical discussion. If a star were truly stationary, the collision with the smallest asteroid would result in a small, perhaps unnoticeable rotation. As that start collapsed, however, its rotation would speed up due to the conservation of angular momentum. By the time its radius got anywhere close to the Schwarzschild radius, it would be rotating furiously, and the rest of space and time along with it. Thus, the star would firmly be in the realm of a different metric known as the *Kerr metric*, which describes rotating black holes.

The second problem is that of time. It takes an infinite amount of *coordinate time* for a particle to fall into a black hole. Coordinate time is the dt^2, the time as experienced by the rest of the universe. From the particle's perspective, it only takes a *finite time* to enter the black hole. That is wild stuff. The particle will just pass through the boundary, but the rest of the universe will observe the particle becoming slower and slower and slower and *never* actually entering the black hole. It would literally take forever. Why does this prevent the formation of a black hole in the first place? Because the star itself must first collapse into a black hole. And the metric that applies to a particle in the vacuum surrounding the star must also apply to any particle at the outmost edge of the star—the atoms farthest from the center of the black hole. (A different space-time metric takes over in the interior of the black hole.) Thus, if it takes a particle *outside* of a black hole an infinite amount of time (as observed by the rest of the universe) to enter into a black hole, then so too must it take the particles on the surface of a collapsing star an infinite amount of time to enter the horizon. Though Birkhoff's theorem states that there is no equilibrium solution for such a star, we are not talking about one. We are talking about a start that asymptotically approaches a black hole as $t \to \infty$. Thus, a black hole can never fully form, at least not in finite time as far as the rest of the universe is concerned.

7 Lagrangian Mechanics

At the time of its publication, Newtonian mechanics [224] represented the greatest leap in our understanding of the laws of nature that the world had ever seen. For the first time in human history, scholars realized the power of the newly discovered theory of calculus and its ability to formulate a concise set of equations that would describe the motion of particles and bodies. However, the theory had its limitations. The laws of motion were explicitly formulated for particles and bodies that could be described using Euclidean coordinates as state variables. The imposition of simple constraints—such as the constraint that a bob of a pendulum can trace only a circular path—requires additional equations that later turned out to be unnecessary. This issue was resolved with the subtle and elegant generalization of Newtonian mechanics known as Lagrangian mechanics. In [172] a set of equations of motion for mechanical systems is derived, expressed in terms of what Lagrange called *generalized coordinates*. For example, instead of describing a pendulum bob with two Euclidean coordinates (x, y) and a *holonomic* constraint $x^2 + y^2 = 1$, one could instead describe the system with a single generalized coordinate $\theta \in S^1$ representing the angle between the pendulum bob and the vertical. The astute reader might have already made the connection between these so-called *generalized coordinates* and our discussion of differentiable manifolds from Chapter 6. What Lagrange actually accomplished was the first mathematical description of mechanical systems whose configuration space was that of a differentiable manifold; Lagrange's generalized coordinates were the objects mathematicians would later call *local coordinates*. Thus, geometric mechanics was officially born. In this chapter, we will present the theory of Lagrangian mechanics from the modern point of view—as equations of motion for mechanical systems evolving on differentiable manifolds. For a classical treatment of the subject, see texts such as [8], [104], [102], or [293].

7.1 Hamilton's Principle

Lagrangian mechanics is *variational* in nature and is derivable from a variational principle known as *Hamilton's principle*. Variational problems of the calculus of variations are similar to optimization problems of differential calculus, which seek to find *paths* that are optimal with respect to some sort of performance measure. For example, there are many paths that you may take to go from your home to your local grocery store; one of those paths, however, minimizes the distance that you would need to travel. As it turns out, all laws of nature are variational in nature, which means that nature is always trying to find optimal paths relative to some cost function or performance measure. Before we discuss this further, we will introduce the notion of Lagrangian function.

https://doi.org/10.1515/9783110597806-007

Let Q be an n-dimensional configuration manifold for a mechanical system. A *Lagrangian* is a function $L : TQ \to \mathbb{R}$. For most mechanical systems, the Lagrangian is simply given by the *kinetic minus potential energy*.

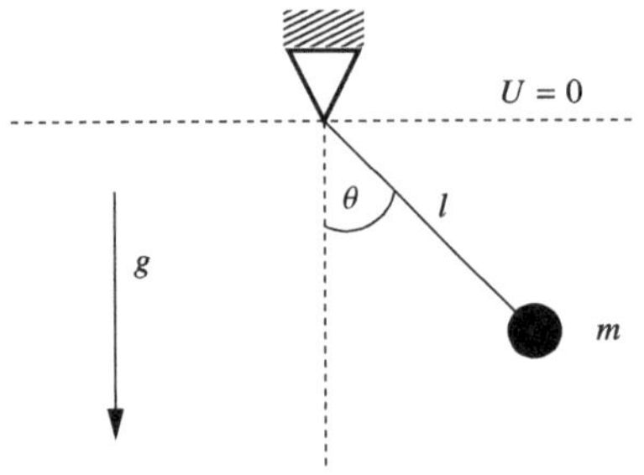

Figure 7.1: A simple pendulum.

Example 7.1. Let us consider the example of a simple pendulum that was introduced in Example 2.4 and that is depicted in Figure 7.1. The configuration manifold for our system is $Q = S^1$, which may be described by the local coordinates (θ), $\theta \in (-\pi, \pi)$, where θ is the angle between the pendulum bob and the vertical, as measured from its "down" position. The speed of the bob is $l|\dot{\theta}|$, so that its kinetic energy is given by

$$T = \frac{1}{2} m l^2 \dot{\theta}^2.$$

The gravitational potential energy U, as measured from the horizontal dashed line, is given by $U = -mgl\cos\theta$. Therefore, the Lagrangian for the system, $L : TQ \to \mathbb{R}$, is given by $L = K - U$, or

$$L = \frac{1}{2} m l^2 \dot{\theta}^2 + mgl\cos\theta.$$

We will return to this example later on, during our discussion of the Euler–Lagrange equation. □

Example 7.2. Consider a dumbbell on an inclined plane, as depicted in Figure 7.2. The configuration manifold for the system is $Q = \mathbb{R}^2 \times S^1$, which may be described in terms

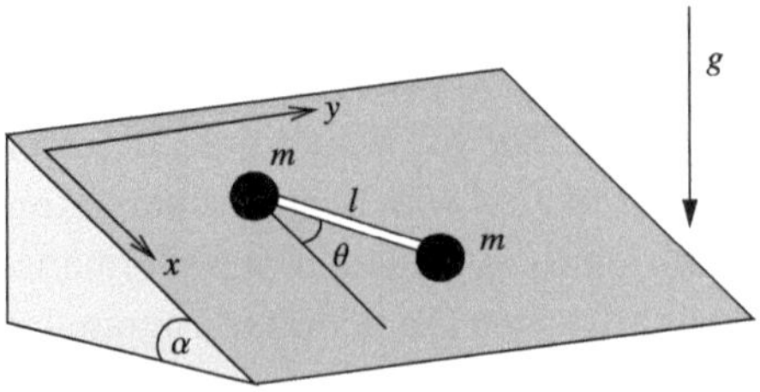

Figure 7.2: Dumbbell on an inclined plane.

of local coordinates (x, y, θ), where (x, y) is the position of the first mass and θ is the angle between the bar and the x-axis. Let us suppose that the two weights, each of mass m, are connected by a massless rod of length l, that the plane makes an angle of α with the horizontal, and that gravity acts with constant downward acceleration g. In terms of local coordinates, the kinetic energy for the system is given by

$$T = m\left(\dot{x}^2 + \dot{y}^2 + \frac{1}{2}l\dot{\theta}^2 - l\dot{x}\dot{\theta}\sin\theta + l\dot{y}\dot{\theta}\cos\theta\right).$$

See, for example, [106] for a review of kinematics and a discussion on the calculation of kinetic energy for systems of particles and rigid bodies. Meanwhile, the gravitational potential energy for our system may be expressed in terms of our local coordinates as

$$\begin{aligned} U &= -mgx\sin\alpha - mg(x + l\cos\theta)\sin\alpha \\ &= -mg(2x + l\cos\theta)\sin\alpha. \end{aligned}$$

Therefore, the Lagrangian for the system $L : TQ \to \mathbb{R}$, given by the *kinetic minus potential energy*, may be expressed as

$$L = m\left(\dot{x}^2 + \dot{y}^2 + \frac{1}{2}l\dot{\theta}^2 - l\dot{x}\dot{\theta}\sin\theta + l\dot{y}\dot{\theta}\cos\theta\right) + mg(2x + l\cos\theta)\sin\alpha.$$

The kinetic energy for the system may be expressed as a bilinear, symmetric, quadratic form acting on the system's velocity. Let us define the matrix-valued function on Q

$$[g_{ij}] = \begin{bmatrix} 2m & 0 & -l\sin\theta \\ 0 & 2m & l\cos\theta \\ -l\sin\theta & l\cos\theta & l \end{bmatrix}.$$

The symmetric quadratic form $[g_{ij}]$ is an example of a *metric* on manifold Q, and hence our manifold is a *Riemannian* manifold. Riemannian metrics, such as the one presented here, commonly arise in geometric mechanics and are known as *kinetic-energy metrics*. □

We are now able to state Hamilton's principle. Hamilton's principle reveals the variational nature of mechanics since the actual motions realized by mechanical systems consist precisely of the paths that obey this principle.

Definition 7.1 (Hamilton's Principle). Let Q be the configuration manifold for a mechanical system and $L : TQ \to \mathbb{R}$ its Lagrangian function. If a smooth curve $\gamma : [a, b] \to Q$, connecting the points $q(a) = q_1$ and $q(b) = q_2$, yields an extreme value for the integral

$$\int_a^b L(\gamma(t), \dot{\gamma}(t))\, dt,$$

as compared to the value of this integral computed using nearby smooth paths, then the curve γ is said to satisfy *Hamilton's principle*.

As we shall soon see, the equations of motion may be derived directly from Hamilton's principle. Hamilton's principle may be used to describe any mechanical system for which all forces (with the exception of constraint forces) may be derived from a scalar potential, which is a function of position, velocity, and time.

In §7.3, we will use Hamilton's principle to derive the Euler–Lagrange equation, the central equations of motion for Lagrangian mechanics. However, we will first require a precise formulation for the idea of a variation of a path.

7.2 Variations of Curves and Virtual Displacements

In this paragraph, we introduce the notion of a *variation* to a curve, which can be thought of as a smooth, one-parameter family of smooth curves with fixed endpoints that passes through a given fiducial curve. This concept gives rise to the notion of *virtual displacements* and is fundamental to properly framing variational principles and subsequently deriving equations of motion for their solutions. The mathematical theory of variations is laid out in [98] and explained relative to the variational principles of mechanics in [173]. However, this history dates back to [85, 86], [119, 120], [135], [97], [172], and [209]. The earliest presentation of the theory may actually even go as far back as [133], in which it is shown that the law of reflection [84] follows from the assumption that light travels between two points along the shortest possible path. This was expanded to include the law of refraction in [151] and later generalized to the principle of least time by Fermat [64]. Recently, the analogy between optics and nonholonomic systems was discussed in [26]. For a more detailed history of the principle of least action, see [251].

We first introduce the notion of a *path space*, which will be taken as our domain of consideration when formulating variational principles.

Definition 7.2. Let Q be a differentiable manifold and fix $q_1, q_2 \in Q$ and $[a,b] \subset \mathbb{R}$. The *path space* from q_1 to q_2 is the set

$$\Omega(q_1, q_2, [a,b]) = \{\gamma : [a,b] \to Q : \gamma \text{ is a } C^2 \text{ curve}, \gamma(a) = q_1, \gamma(b) = q_2\}.$$

A given point $\gamma \in \Omega$ is actually a C^2 curve $\gamma : [a,b] \to Q$ with endpoints $\gamma(a) = q_1$ and $\gamma(b) = q_2$. We will sometimes abbreviate $\Omega(q_1, q_2, [a,b])$ with Ω when q_1, q_2, and $[a,b]$ are understood.

Definition 7.3. A *functional* on a manifold Q is a mapping of the form

$$\mathcal{I} : \Omega(q_1, q_2, [a,b]) \to \mathbb{R},$$

i. e., it is a real-valued function of curves.

The problem of the calculus of variations is to seek extreme values of functionals of the form

$$\mathcal{I}[y] = \int_a^b L(y(t), \dot{y}(t))\, dt,$$

where $y \in \Omega(q_1, q_2, [a, b])$ and $L : TQ \to \mathbb{R}$. The integral $\mathcal{I}[y]$ is called the *action*. According to Hamilton's principle, the laws of mechanics are variational in nature, as the actual equations of motion yield extreme values of the action $\mathcal{I}[y]$, where $L : TQ \to \mathbb{R}$ is the mechanical Lagrangian for the system.

Definition 7.4. Consider a smooth curve $c : [a, b] \to Q$. A *variation* of c is a C^2 mapping $\vartheta : [-\varepsilon, \varepsilon] \times [a, b] \to Q$ such that
1. $\vartheta(0, t) = c(t)$ for all $t \in [a, b]$ and
2. $\vartheta(s, a) = c(a)$ and $\vartheta(s, b) = c(b)$ for all $s \in [-\varepsilon, \varepsilon]$.

Given a variation of a nominal curve c, one defines a vector field along c in the direction of the variation that is known as a *virtual displacement* or, in some works, an *infinitesimal variation*. The precise definition follows.

Definition 7.5. Given a variation $\vartheta : [-\varepsilon, \varepsilon] \times [a, b] \to Q$ of a smooth curve $c : [a, b] \to Q$, the *virtual displacement* is the vector field $\delta c : [a, b] \to TQ$, where $\delta c(t) \in T_{c(t)}Q$ and $\delta c(a) = \delta c(b) = 0$, defined by

$$\delta c(t) = \left. \frac{\partial \vartheta(s, t)}{\partial s} \right|_{s=0}.$$

The path space $\Omega(q_1, q_2, [a, b])$ itself may be regarded as an *infinite dimensional differential manifold*. As noted above, a point $y \in \Omega$ in this space corresponds to a smooth curve on the manifold, and it can be further shown that tangent vectors to Ω at y correspond to the possible virtual displacements of the curves y; see [196] for the details of this viewpoint. An illustration of a nominal curve along with a given variation and virtual displacement is given in Figure 7.3.

To define the variation for a function on the tangent bundle, we will first require the following definition.

Definition 7.6. Let $c : [a, b] \to Q$ be a smooth curve. Then the *canonical lift* of c to the tangent bundle is the smooth path $\hat{c} : [a, b] \to TQ$ defined by $\hat{c}(t) = (c(t), \dot{c}(t))$.

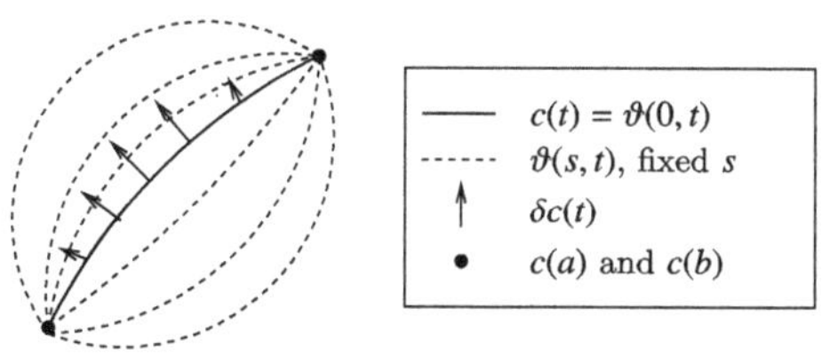

Figure 7.3: A variation of $c \in \Omega(q_1, q_2, [a, b])$ and the virtual displacement $\delta c \in T_cQ$.

Similarly, if $\vartheta : [-\varepsilon, \varepsilon] \times [a, b] \to Q$ is a variation of c, then the *canonical lift* of ϑ is the variation $\hat{\vartheta} : [-\varepsilon, \varepsilon] \times [a, b] \to Q$ of $\hat{c}$ defined by $\hat{\vartheta}(s, t) = (\vartheta(s, t), \vartheta_t(s, t))$.

One may now apply Definition 7.5 to the lift $\hat{c}$ and its variation $\hat{\vartheta}$ to define a vector field $\delta\hat{c} : [a, b] \to T^2Q$, which, in local coordinates, may be described by

$$\delta\hat{c} := \left.\frac{\partial\hat{\vartheta}(s, t)}{\partial s}\right|_{s=0} = \delta q^i \frac{\partial}{\partial q^i} + \delta\dot{q}^i \frac{\partial}{\partial \dot{q}^i}.$$

Definition 7.7. Let $c : [a, b] \to Q$ be a smooth curve, $\vartheta : [-\varepsilon, \varepsilon] \times [a, b] \to Q$ a variation, and $f : TQ \to \mathbb{R}$ a function on the tangent bundle. Then the *variation of* f *with respect to* ϑ is defined as the directional derivative

$$\delta f := \delta\hat{c}[f].$$

This is given in local coordinates $(q^1, \ldots, q^n)$ by the relation

$$\delta\hat{c}[f] = \frac{\partial f}{\partial q^i}\delta q^i + \frac{\partial f}{\partial \dot{q}^i}\delta\dot{q}^i.$$

7.3 Euler–Lagrange Equation

In this paragraph, we present the equations of motion for Lagrangian systems. We first discuss the fundamental lemma of the calculus of variations, which is then applied during the derivation of the Euler–Lagrange equation. We then examine the link between the Euler–Lagrange equation and Riemannian geometry.

Fundamental Lemma of Calculus of Variations

To obtain a set of differential equations for the configuration variables of a mechanical system, we will make use of the theory of calculus of variations on manifolds. Before we begin this discussion, we will require the establishment of a fundamental lemma upon which the theory hinges.

Lemma 7.1. *If a continuous function* $f : [0, 1] \to \mathbb{R}$ *satisfies the equation*

$$\int_0^1 f(t)h(t)\, dt = 0$$

for all continuous functions $h : [0, 1] \to \mathbb{R}$ *with the property* $h(0) = h(1) = 0$, *then* $f(t) \equiv 0$.

Proof. We will proceed by contradiction. Suppose the continuous function f is not equivalently equal to zero. Then there must exist a certain $t^* \in (0,1)$ such that $|f(t^*)| > 0$. Without loss of generality, let us suppose that $f(t^*) > 0$. Since f is continuous, there exist an $\varepsilon > 0$ and a $c > 0$ such that $f(t) > c$ for all t in the open neighborhood $t \in B_\varepsilon(t^*) \subset (0,1)$, where $B_\varepsilon(t^*) := (t^* - \varepsilon, t^* + \varepsilon)$ is the open ε-ball about x^*.

Now, let us choose a continuous function $h : [0,1] \to \mathbb{R}$ such that
- $h(t) = 0$ for $t \in [0,1] \setminus B_\varepsilon(t^*)$;
- $h(t) \geq 0$ for $t \in B_\varepsilon(t^*)$; and
- $h(t) = 1$ for $t \in B_{\varepsilon/2}(t^*)$.

This kind of continuous function obviously exists. Figure 7.4 presents an example of such a function $h(t)$. Then, clearly, we have

$$\int_0^1 f(t)h(t)\,dt \geq \varepsilon c > 0,$$

contradicting our hypothesis that $\int_0^1 f(t)h(t)dt = 0$ for *all* continuous test functions $h : [0,1] \to \mathbb{R}$. Therefore, $f(t)$ must vanish identically on the interval $[0,1]$. □

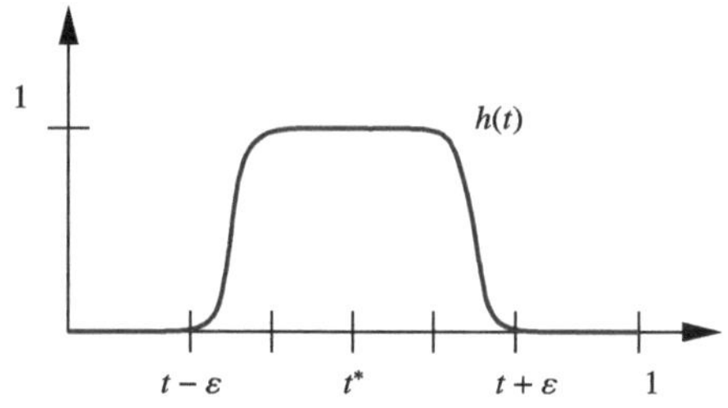

Figure 7.4: Continuous function $h(t)$ satisfying prescribed conditions.

Euler–Lagrange Equation

Now that we have covered some basic background on variations and virtual displacements, we are ready to state the central theorem of Lagrangian mechanics and, consequently, of the calculus of variations.

Theorem 7.1 (Euler–Lagrange Equation). *Let Q be the configuration manifold of a mechanical system with Lagrangian $L : TQ \to \mathbb{R}$. Given two fixed points $q_1, q_2 \in Q$ and interval $[a,b] \subset \mathbb{R}$, the curve $c : [a,b] \to Q$ is an extremal curve of the action $\mathcal{I} : \Omega(q_1, q_2, [a,b]) \to \mathbb{R}$, defined by*

$$\mathcal{I}[c] = \int_a^b L(c(t), \dot{c}(t))\,dt,$$

if and only if it satisfies the Euler–Lagrange equation

$$\frac{d}{dt}\frac{\partial L}{\partial \dot{q}} - \frac{\partial L}{\partial q} = 0. \tag{7.1}$$

Proof. Let $\vartheta : [-\varepsilon, \varepsilon] \times [a, b] \to Q$ be an arbitrary variation of curve c. For each $s \in [-\varepsilon, \varepsilon]$, we define $c_s \in \Omega(q_1, q_2, [a, b])$ by the relation $c_s(t) = \vartheta(s, t)$. Evaluating $\mathcal{I} : \Omega \to \mathbb{R}$ at c_s, we obtain

$$\mathcal{I}[c_s] = \int_a^b L(c_s(t), \dot{c}_s(t))dt.$$

Differentiating this expression with respect to s yields what we call the *variation of* $\mathcal{I}$ *with respect to* ϑ:

$$\delta\mathcal{I}[c; \vartheta] := \left.\frac{d\mathcal{I}[c_s]}{ds}\right|_{s=0} = \int_a^b \delta\hat{c}[L]\, dt = \int_a^b \left(\frac{\partial L}{\partial q^i}\delta q^i + \frac{\partial L}{\partial \dot{q}^i}\delta \dot{q}^i\right) dt,$$

which is expressed in terms of the local coordinates (q^i). Now, by definition of the virtual displacement $\delta c(t)$, we have

$$\delta c(t) = (\delta q^1(t), \ldots, \delta q^n(t)) = \left.\frac{\partial \vartheta(s, t)}{\partial s}\right|_{s=0}.$$

Since the variation ϑ is C^2, we may use integration by parts to obtain

$$\begin{aligned}\delta\mathcal{I}[c; \vartheta] &= \int_a^b \left(\frac{\partial L}{\partial q^i}\delta q^i + \frac{\partial L}{\partial \dot{q}^i}\frac{d\delta q^i}{dt}\right) dt \\ &= \left.\frac{\partial L}{\partial \dot{q}^i}\delta q^i\right|_{t=a}^{t=b} + \int_a^b \left(\frac{\partial L}{\partial q^i}\delta q^i - \frac{d}{dt}\frac{\partial L}{\partial \dot{q}^i}\delta q^i\right) dt.\end{aligned}$$

Since $\delta c \in T_c\Omega$, it follows that the variations vanish at the endpoints of the curve, i. e., $\delta c(a) = \delta c(b) = 0$. Thus we are left with

$$\delta\mathcal{I}[c; \vartheta] = \int_a^b \left(\frac{\partial L}{\partial q^i} - \frac{d}{dt}\frac{\partial L}{\partial \dot{q}^i}\right)\delta q^i\, dt = 0.$$

For an extremal curve c, this integral must vanish relative to an *arbitrary* variation ϑ. Thus, by the fundamental lemma of the calculus of variations (Lemma 7.1), we obtain the Euler–Lagrange equation. □

Example 7.3. Let us consider again the simple pendulum of Example 7.1. Recall that the configuration manifold for this system is $Q = S^1$ and that the mechanical Lagrangian is given by

$$L = \frac{1}{2} m l^2 \dot{\theta}^2 + mgl \cos \theta.$$

To write out the Euler–Lagrange equation, we will require the following ingredients:

$$\begin{aligned} \frac{\partial L}{\partial \dot{\theta}} &= m l^2 \dot{\theta}, \\ \frac{d}{dt} \frac{\partial L}{\partial \dot{\theta}} &= m l^2 \ddot{\theta}, \\ \frac{\partial L}{\partial \theta} &= -mgl \sin \theta. \end{aligned}$$

Combining the preceding equations into the Euler–Lagrange equation, we obtain

$$m l^2 \ddot{\theta} + mgl \sin \theta = 0,$$

which is equivalent to the pendulum equation (2.29). See Example 3.6 for a discussion of the solutions to this equation. □

Exercise 7.1. Compute the Euler–Lagrange equation for the dumbbell on the inclined plane (Example 7.2).

Exercise 7.2. Compute the Euler–Lagrange equation for the Lagrangian

$$L = \frac{1}{2}(\dot{x}^2 + \dot{y}^2) - y\dot{x} + x\dot{y} + U(x, y), \tag{7.2}$$

where $U : \mathbb{R}^2 \to \mathbb{R}$ is a smooth function on $Q = \mathbb{R}^2$. Have we discussed any systems that have this form?

Lagrangian Mechanics on Riemannian Manifolds

It is almost a universal practice in mechanical systems to give the Lagrangian $L : TQ \to \mathbb{R}$ as the kinetic energy $T(q, \dot{q})$ minus the potential energy $U(q)$, where the kinetic energy may usually be expressed locally in terms of a symmetric, bilinear form g_{ij} as

$$T = \frac{1}{2} g_{ij} \dot{q}^i \dot{q}^j.$$

The kinetic energy of such mechanical systems therefore defines a Riemannian structure $\langle \cdot, \cdot \rangle_q$ on Q known as the *kinetic-energy metric*. Whenever such a kinetic-energy metric is given, we obtain the following relation between the Euler–Lagrange equation and covariant differentiation on Riemannian manifolds.

Proposition 7.1. *Let Q be a differentiable manifold with Riemannian structure given by the kinetic energy and let $L : TQ \to \mathbb{R}$ be a mechanical Lagrangian of the form*

$$L(q,\dot{q}) = \frac{1}{2}\langle \dot{q},\dot{q}\rangle_q - U(q).$$

Then the Euler–Lagrange equation (7.1) *is equivalent to the equation*

$$\nabla_{\dot{q}}\dot{q} = -dU^{\sharp}, \tag{7.3}$$

where $\nabla : TQ \times TQ \to TQ$ is the Levi-Civita connection associated with the kinetic-energy metric.

Proof. A simple calculation shows that the Euler–Lagrange equation for Lagrangians of this form yields

$$g_{rj}\ddot{q}^j + \left(\frac{\partial g_{rj}}{\partial q^i} - \frac{1}{2}\frac{\partial g_{ij}}{\partial q^r}\right)\dot{q}^i\dot{q}^j = -\frac{\partial U}{\partial q^r}.$$

However, the left-hand side is equal to $g_{kr}(\ddot{q}^k + \Gamma^k_{ij}\dot{q}^i\dot{q}^j)$ by (6.9), and the result follows. □

Note that for mechanical systems with no potential energy, the resulting solution flow on manifold Q is therefore precisely the geodesic flow (6.6) given by the Levi-Civita connection associated with the kinetic-energy metric. However, the much more striking thing to realize about (7.3) is that it is completely analogous to Newton's equations of motion for particles in a potential-energy field, i. e., $\ddot{x} = -\nabla U$. On the left-hand side of (7.3), we find exactly the covariant acceleration of the particle, defined using the Levi-Civita connection associated with the kinetic-energy metric, whereas on the right-hand side, we find the negative potential-energy gradient, as mapped from the cotangent bundle to the tangent bundle using the kinetic-energy metric. Thus, the Euler–Lagrange equation, represented in the form of (7.3), is literally a covariant, geometric form of the classical Newtonian equations.

7.4 Distributions and Frobenius' Theorem

In this section, we lay out some basic definitions regarding the theory of distributions. The notions of distributions and integrability will play a fundamental role in our subsequent discussion of constrained systems.

Definition 7.8. A k-dimensional *distribution* on a differentiable manifold Q is the smooth assignment of a k-dimensional subspace to the tangent space at each point.

A distribution is also referred to as a *vector subbundle* of Q. If we let $\Delta \subset TQ$ be a distribution, then we refer to the selected subspace at $q \in Q$ by $\Delta_q \subset T_qQ$. By a "*smooth* assignment of a k-dimensional subspace to the tangent space at each point" we mean the following: for each open set $U \subset Q$, we can define a set of k smooth vector fields $X_1, \dots, X_k \in TU$ such that $\Delta_q \subset T_q Q = \operatorname{span}\{X_1(q), \dots, X_k(q)\}$, for all $q \in U$.

Definition 7.9. A distribution is *involutive* if, whenever $X, Y \in \Delta$, it follows that $[X, Y] \in \Delta$.

Definition 7.10. A k-dimensional distribution Δ is *integrable* if, for each $q \in Q$, there exists a local, k-dimensional submanifold N of Q containing the point q whose tangent bundle equals the distribution Δ restricted to N, i. e.,

$$TN = \Delta|_N.$$

The local manifold N is called an *integral manifold* of Δ. If a distribution Δ is integrable, we say that Q can be *foliated* with integral manifolds of Δ. A collection of such integral manifolds that covers Q is called a *foliation* of Q by integral manifolds of Δ.

See Figure 7.5 for an illustration of integral manifolds of an integrable distribution Δ forming a foliation of manifold Q. Our next theorem connects the notions of involutivity and integrability of a distribution.

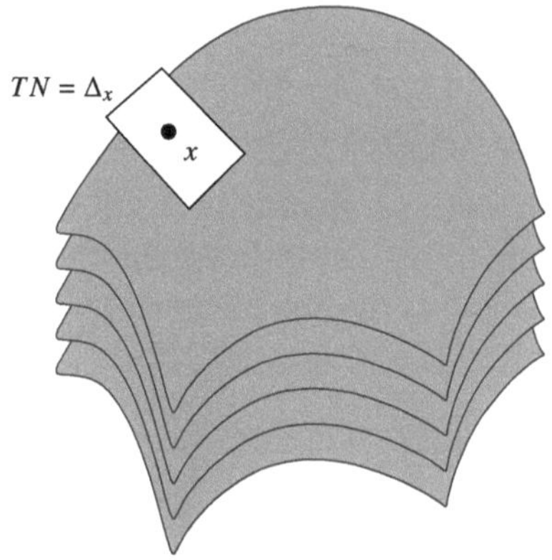

Figure 7.5: Integral manifolds of a distribution Δ, forming a foliation of manifold Q.

Theorem 7.2 (Frobenius' Theorem). *A smooth, k-dimensional distribution Δ of a differentiable manifold Q is integrable if and only if it is involutive.*

Frobenius' theorem was first proved in [66], in which sufficient conditions for the theorem are established, and [55], in which the necessary conditions are described. The theorem was first applied to Pfaffian systems by his namesake in [96]. We refer the reader to these sources for the proof and to [2] and [178] for expositions.

The statement of Frobenius' theorem is directly related to the geometric interpretation of the Jacobi–Lie bracket, discussed in Proposition 6.5: given a noninvolutive distribution, one may concatenate flow along two basis vector fields to form a third, independent direction.

Example 7.4. Consider the manifold $Q = \mathbb{R}^3$, with local coordinates (x, y, z), and the vector fields

$$X_1 = \frac{\partial}{\partial x}, \quad X_2 = \frac{\partial}{\partial y}.$$

Since these vector fields are smooth, they define a 2-D distribution throughout $\mathbb{R}^3$, defined by the subspace assignment

$$\Delta_q = \text{span}\left\{\frac{\partial}{\partial x}, \frac{\partial}{\partial y}\right\} \subset T_q\mathbb{R}^3,$$

for each $q \in \mathbb{R}^3$. To show that Δ is involutive, consider the arbitrary smooth vector fields $X = \alpha_1 X_1 + \alpha_2 X_2$ and $Y = \beta_1 X_1 + \beta_2 X_2$, where $\alpha_1, \alpha_2, \beta_1, \beta_2 \in C^\infty(\mathbb{R}^3)$ are smooth functions on Q. A simple calculation shows

$$\begin{aligned}
[X, Y] &= (\alpha_1 X_1 + \alpha_2 X_2)(\beta_1 X_1 + \beta_2 X_2) - (\beta_1 X_1 + \beta_2 X_2)(\alpha_1 X_1 + \alpha_2 X_2) \\
&= \left(\alpha_1 \frac{\partial \beta_1}{\partial x} + \alpha_2 \frac{\partial \beta_1}{\partial y} - \beta_1 \frac{\partial \alpha_1}{\partial x} - \beta_2 \frac{\partial \alpha_1}{\partial y}\right) X_1 \\
&\quad + \left(\alpha_1 \frac{\partial \beta_2}{\partial x} + \alpha_2 \frac{\partial \beta_2}{\partial y} - \beta_1 \frac{\partial \alpha_2}{\partial x} - \beta_2 \frac{\partial \alpha_2}{\partial y}\right) X_2;
\end{aligned}$$

therefore, $[X, Y] \in \Delta$ for all smooth vector fields $X, Y : \mathbb{R}^3 \to \Delta$. Hence, the distribution is involutive. By Frobenius' theorem, the distribution Δ must also be integrable. Moreover, for this simple example, one immediately recognizes the one-parameter family of integral submanifolds

$$N_\lambda := \{(x, y, z) \in \mathbb{R}^3 : z = \lambda\},$$

which foliate the manifold $\mathbb{R}^3$. □

Exercise 7.3. Consider the 2-D distribution $\Delta \subset T\mathbb{R}^3$ that is spanned by the vector fields

$$X_1 = y\frac{\partial}{\partial x} - x\frac{\partial}{\partial y} \quad \text{and} \quad X_2 = xz\frac{\partial}{\partial x} + yz\frac{\partial}{\partial y} - (x^2 + y^2)\frac{\partial}{\partial z}.$$

Is the distribution involutive? Can you determine a simple description for its integral submanifolds?

Definition 7.11. Given a k-dimensional distribution Δ, the *accessibility distribution* $\mathcal{C}$ of Δ is the smallest Lie algebra of vector fields containing Δ. Moreover, if $\dim \mathcal{C}_q = n$, for all $q \in Q$, the distribution Δ is called *completely nonintegrable*.

In other words, $\mathcal{C}$ is the span of all possible Lie brackets of the vector fields in Δ. The distribution $\mathcal{C}$ is involutive, since $[X, Y] \in \mathcal{C}$ whenever $X, Y \in \mathcal{C}$ (by definition), and is therefore integrable. If $\dim \mathcal{C}_q = n$ for all $q \in Q$, then its integral manifold must be Q itself. A completely nonintegrable distribution is therefore a distribution for which all points in Q may be accessed by following the flow of the vector fields and their brackets.

7.5 Mechanical Systems with Holonomic Constraints

Let Q be the n-dimensional configuration manifold of a Lagrangian system that is subject to m *holonomic* constraints $f_\alpha : Q \to \mathbb{R}$, where $m < n$. The descriptor *holonomic* refers to the fact that the constraints may be rendered as a function of the position variables; one may contrast this with *nonholonomic* constraints, or nonintegrable velocity constraints, which we will take up in Section 7.6. Hamilton's principle can be modified by replacing the action integral with

$$\int_a^b \left(L(\gamma(t), \dot{\gamma}(t)) + \sum_{\alpha=1}^{m} \lambda_\alpha f_\alpha(\gamma(t)) \right) dt,$$

for Lagrange multipliers λ_α, with $\alpha = 1, \ldots, m$. Following a similar procedure as our derivation of the Euler–Lagrange equations, one may show that the resulting equations of motion are

$$\frac{d}{dt}\frac{\partial L}{\partial \dot{q}} - \frac{\partial L}{\partial q} + \sum_{\alpha=1}^{m} \lambda_\alpha \frac{\partial f_\alpha}{\partial q} = 0.$$

By defining the generalized constraint forces

$$Q = -\sum_{\alpha=1}^{m} \lambda_\alpha \frac{\partial f_\alpha}{\partial q},$$

we can recast this as

$$\frac{d}{dt}\frac{\partial L}{\partial \dot{q}} - \frac{\partial L}{\partial q} = Q.$$

It is important to note that since the sign of the Lagrange multipliers is arbitrary, the generalized constraint force Q gives the *magnitude*, and not the direction of the resulting constraint forces. We must appeal to our physics intuition to determine the appropriate direction in which the constraint forces operate.

Example 7.5. Consider a particle of mass m sliding down a hemisphere with radius a, under the influence of a constant gravitational acceleration g. Let θ be the angle from the top of the sphere to the particle. Let us further select the initial conditions so that the particle's motion is constrained to lie within the x-z plane. (This is true if, for example, we select coordinates such that the initial velocity does not have a y-component.) The Lagrangian is given by the kinetic minus the potential energy, or

$$L = \frac{1}{2}m(\dot{x}^2 + \dot{z}^2) - mgz.$$

The constraint is $a - \sqrt{x^2 + z^2} = 0$. Transforming to polar coordinates, so that $r^2 = x^2 + z^2$ and $\cos(\theta) = x/z$, the constraint equation becomes $a - r = 0$, and the Lagrangian becomes

$$L = \frac{1}{2}mr^2\dot{\theta}^2 - mgr\cos\theta,$$

with holonomic constraint $f(r) = a - r = 0$. The Euler–Lagrange equations yield

$$\begin{aligned} \frac{d}{dt}\frac{\partial L}{\partial \dot{r}} - \frac{\partial L}{\partial r} + \lambda\frac{\partial f}{\partial r} &= -mr\dot{\theta}^2 + mg\cos\theta - \lambda = 0, \\ \frac{d}{dt}\frac{\partial L}{\partial \dot{\theta}} - \frac{\partial L}{\partial \theta} &= mr^2\ddot{\theta} - mgr\sin\theta = 0. \end{aligned}$$

Enforcing the constraint $r = a$ and solving, we obtain

$$\begin{aligned} \dot{\theta}^2 &= -\frac{2g}{a}\cos\theta + \frac{2g}{a}, \\ \lambda &= mg(3\cos\theta - 2). \end{aligned}$$

The constraint force $Q = -\lambda$ equals zero precisely when $\theta = \cos^{-1}(2/3)$; hence we deduce that the particle loses contact with the sphere at this angle. □

7.6 Nonholonomic Mechanics

In this section, we introduce the notion of *nonholonomic* systems. Nonholonomic systems can be thought of as systems with nonintegrable *velocity* constraints; thus, it is the type of motion, and not the configuration space, that is constrained.

As it turns out, mechanical systems with nonholonomic constraints no longer satisfy (a generalized, nonholonomic form of) Hamilton's principle. When taking variations of a curve in the process of carrying out the extremization of the action integral, one must apply the nonholonomic constraints either to the *varied curves* or to the *virtual displacements*; one *cannot* demand that both be kinematically admissible. Thus the theory bifurcates into two different sets of equations of motion, one named *nonholonomic mechanics* and the other *variational nonholonomic dynamics* (sometimes called *vakonomics*). See, for example, [41], [58], [164–169], [65], [103], and [232] for details on this distinction. Recent work has also been done on the notion of *equivalence*, for which systems may exhibit certain paths that simultaneously satisfy *both* nonholonomic and variational equations of motion; see [89] and [91] for further details.

Nonholonomic Constraints

We begin our discussion with an overview of nonholonomic constraints and how their mathematical representation may be described in terms of distributions.

Definition 7.12. A *nonholonomic mechanical system* is a triple $(Q, L, \mathcal{D})$, where Q is an n-dimensional configuration manifold, $L : TQ \to \mathbb{R}$ is a mechanical Lagrangian, and $\mathcal{D} \subset TQ$ is an $(n-m)$-dimensional, completely nonintegrable distribution called the *constraint distribution.*

Definition 7.13. A curve $c : [a,b] \to Q$ on a manifold in a nonholonomic system is said to be *kinematically admissible* if $\dot{c}(t) \in \mathcal{D}_{c(t)}$ for all $t \in [a,b]$. Similarly, a vector field $X : Q \to TQ$ is said to be *kinematically admissible* if $X(q) \in \mathcal{D}_q$ for all $q \in Q$.

A nonholonomic system with an $(n-m)$-dimensional completely nonintegrable distribution is said to have m *velocity constraints*. These are given by a set of m nonexact differential forms ω^a, for $a = 1, \ldots, m$, whose kernel spans the distribution $\mathcal{D}$. In other words, a vector is kinematically admissible at q, i. e., $v \in \mathcal{D}_q$, if and only if $\omega^a(v) = 0$ at q for $a = 1, \ldots, m$.

Example 7.6 (Vertical rolling disc). We now consider the example of a vertical rolling disc. The configuration manifold is given by $Q = S^1 \times S^1 \times \mathbb{R}^2$, which may be described by the set of local coordinates (φ, θ, x, y), where (x, y) is the point of contact of the disc and the x-y plane, φ is the angle between the vertical plane of the disc and the x-axis, and θ is the angle of a contact point P on the rim of the disc with the vertical; see Figure 7.6.

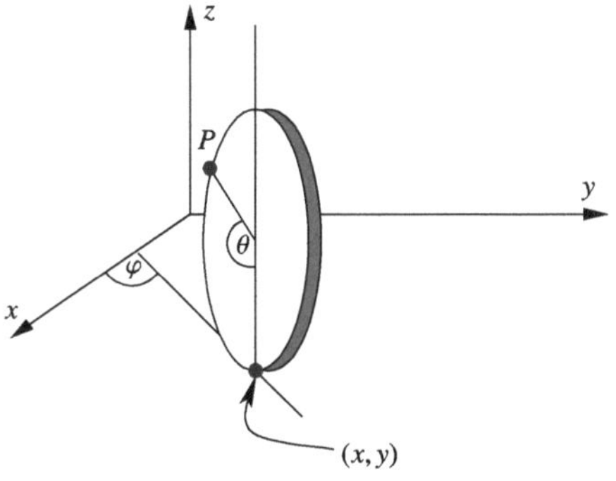

Figure 7.6: Schematic of a vertical rolling disc.

The Lagrangian expressed in terms of the local coordinates (φ, θ, x, y) is given by

$$L = \frac{1}{2}m(\dot{x}^2 + \dot{y}^2) + \frac{1}{2}I\dot{\theta}^2 + \frac{1}{2}J\dot{\varphi}^2,$$

where I and J are the moments of inertia of the disc about an axis perpendicular to the plane of the disc and a vertical axis passing through the disc's center, respectively.

Kinematic (or velocity) constraints for our system are given by the physical restriction that the disc must roll without slipping. This condition yields two constraints, which may be expressed as

$$-R\cos\varphi\dot{\theta} + \dot{x} = 0, \tag{7.4}$$

$$-R\sin\varphi\dot{\theta} + \dot{y} = 0, \tag{7.5}$$

where R is the radius of the disc. Let us introduce the differential forms, which, for reasons of convention, we name

$$\omega^3 = -R\cos\varphi d\theta + dx,$$
$$\omega^4 = -R\sin\varphi d\theta + dy.$$

These are known as the *constraint forms*, as the nonholonomic constraints may be expressed via the relations $\omega^3(\dot{q}) = \omega^4(\dot{q}) = 0$, where

$$\dot{q} = \dot{\varphi}\frac{\partial}{\partial\varphi} + \dot{\theta}\frac{\partial}{\partial\theta} + \dot{x}\frac{\partial}{\partial x} + \dot{y}\frac{\partial}{\partial y}.$$

The kernel of the constraint forms is the distribution spanned by the vector fields

$$X_1 = \begin{bmatrix} 1 \\ 0 \\ 0 \\ 0 \end{bmatrix} \quad \text{and} \quad X_2 = \begin{bmatrix} 0 \\ 1 \\ R\cos\varphi \\ R\sin\varphi \end{bmatrix},$$

or, in alternative notation,

$$X_1 = \frac{\partial}{\partial\varphi} \quad \text{and} \quad X_2 = \frac{\partial}{\partial\theta} + R\cos\varphi\frac{\partial}{\partial x} + R\sin\varphi\frac{\partial}{\partial y}.$$

Exercise 7.4. Check that $\omega^3(X_1) = \omega^3(X_2) = \omega^4(X_1) = \omega^4(X_2) = 0$ and that $[X_1, X_2] \neq 0$.

This verifies the claim that vector fields X_1 and X_2 span the kernel of the constraint forms ω^3 and ω^4 and that the constraints are nonholonomic (since the distribution $\mathcal{D} = \operatorname{span}\{X_1, X_2\}$ is nonintegrable, by Frobenius' theorem).

The vertical rolling disc is an important and standard example to which we will return after deriving the equations of motion for systems with nonholonomic constraints. □

Exercise 7.5 (Ice-skating on a cylinder). Consider the configuration manifold $Q = \mathbb{R}^3 \times S^1$, with local coordinates (x, y, z, ψ), and constraint distribution $\mathcal{D}$ that is spanned by the vector fields

$$X_1 = -y\cos\psi\frac{\partial}{\partial x} + x\cos\psi\frac{\partial}{\partial y},$$

$$X_2 = \sin\psi \frac{\partial}{\partial z},$$
$$X_3 = \frac{\partial}{\partial \psi}.$$

(a) Show that the accessibility distribution of $\mathcal{D}$ is 3-D and conclude that $\mathcal{D}$ is *not* completely nonintegrable.
(b) Show that the system is equivalent to one with a 3-D configuration manifold $Q' = Z^1 \times S^1$, where $Z^1 = \{(x, y, z) \in \mathbb{R}^3 : x^2 + y^2 = 1\}$, and a certain 2-D completely nonintegrable constraint distribution.

This system may be thought of as a knife edge constraint on a cylinder.

Variational Principles in Nonholonomic Systems

A remarkable feature of systems with nonintegrable velocity constraints (i. e., nonholonomic systems) is that two distinct kinds of motion can arise depending on the precise variational principle used in determining the equations of motion. We will name these motions *nonholonomic mechanics* and *variational nonholonomic dynamics*. It is important to recognize that they literally correspond to different motions for the system and not just different representations of the underlying equations of motion. We will now discuss the geometrical reasons for the distinction of two separate types of motions.

Let us consider a smooth, kinematically admissible curve $c : [a, b] \to Q$ connecting the endpoints $q_1, q_2 \in Q$. Let $\vartheta : [-\varepsilon, \varepsilon] \times [a, b] \to Q$ be an arbitrary C^2 variation of the curve c and let us consider the vector fields $V, \delta c : [-\varepsilon, \varepsilon] \times [a, b] \to TQ$ defined by

$$V(s, t) = \frac{\partial \vartheta(s, t)}{\partial t}, \tag{7.6}$$
$$\delta q(s, t) = \frac{\partial \vartheta(s, t)}{\partial s}. \tag{7.7}$$

In local coordinates $(q^1, \dots, q^n)$, these vector fields may be expressed by the relations

$$V = \frac{\partial q^i}{\partial t}\frac{\partial}{\partial q^i} \quad \text{and} \quad \delta q = \frac{\partial q^i}{\partial s}\frac{\partial}{\partial q^i}.$$

These vector fields have the following interpretation. For a fixed value of $s = s^*$, the vector field $V(s^*, t)$ is the velocity of the varied curve $\vartheta(s^*, t) \in \Omega(q_1, q_2, [a, b])$. On the other hand, the vector field δq may be given two interpretations: first, for a fixed value $s = s^*$, the vector field $\delta q(s^*, t)$ may be thought of as a virtual displacement for the varied curve $\vartheta(s^*, t)$; second, for a fixed value $t = t^*$, the vector field $\delta q(s, t^*)$ may be thought of as the velocity of the virtual curve $\vartheta(s, t^*)$. Of particular importance, the vector field $\delta q(0,t) = \delta c(t)$, defined along fiducial curve c, is the virtual displacement of fiducial curve c.

The question we now pose, and the question that is fundamental to the understanding of the distinct types of motion mentioned previously, is as follows. In the presence of nonholonomic constraints, is it possible for both the virtual displacements $\delta c(t) \in T_{c(t)}Q$ and the velocity of the varied curves $V(s,t)$ to be simultaneously kinematically admissible? In other words, is it possible to construct a closed, differential quadrilateral in the $[-\varepsilon,\varepsilon] \times [a,b]$ space, which parameterizes the variation ϑ, using only kinematically admissible segments? The answer to this question turns out to be negative, as we will see in the following proposition.

Proposition 7.2. *Let Q be a differentiable manifold and $\mathcal{D}$ a nonintegrable distribution on TQ. If $c : [a,b] \to Q$ is a smooth curve and $\vartheta : [-\varepsilon,\varepsilon] \times [a,b] \to Q$ is an arbitrary C^2 variation of c, then it is impossible for both vector fields $V(s,t)$ and $\delta q(s,t)$, as defined by* (7.6) *and* (7.7)*, to be kinematically admissible.*

Proof. We will proceed by contradiction. Suppose that both vector fields (7.6) and (7.7) are kinematically admissible for an arbitrary C^2 variation $\vartheta : [-\varepsilon,\varepsilon] \times [a,b] \to Q$ of curve c, i. e., suppose that $\delta q(s,t) \in \mathcal{D}_{\vartheta(s,t)}$ and $V(s,t) \in \mathcal{D}_{\vartheta(s,t)}$ for all $(s,t) \in [-\varepsilon,\varepsilon] \times [a,b]$. Since the variation ϑ was arbitrary, it follows from Frobenius' theorem that the bracket $[V,\delta q]|_{s=0} \notin \mathcal{D}$, as otherwise the distribution would be integrable. However, since the variation $\vartheta(s,t) = (q^1(s,t),\dots,q^n(s,t))$ is C^2 and due to the chain rule, we have

$$[V,\delta q] = V\delta q - \delta q V = \left(\frac{\partial^2 q^i}{\partial t \partial s} - \frac{\partial^2 q^i}{\partial s \partial t}\right)\frac{\partial}{\partial q^i} = 0,$$

thereby contradicting the fact that $[V,\delta q]|_{s=0} \notin \mathcal{D}_{c(t)}$. We conclude that it is *not* possible for variations of curves in systems that are subject to nonholonomic constraints to possess *both* kinematically admissible virtual displacements *and* kinematically admissible varied curves. □

As a result of Proposition 7.2, one must, when deriving the equations of motion, select variations that *either* possess the property that the varied curves are kinematically admissible *or* the property that the virtual displacements are kinematically admissible. One cannot select both.

If one selects variations for which the virtual displacements are kinematically admissible, then the varied curves $\vartheta(s^*,t)$, for fixed $s^* \in [-\varepsilon,\varepsilon]$, will not be kinematically admissible. Such a selection is known as the *Lagrange–d'Alembert principle*; this situation is depicted in Figure 7.7.

To formalize this notion, we introduce the following definition.

Definition 7.14 (Lagrange–d'Alembert Principle). Let $(Q,L,\mathcal{D})$ be a nonholonomic system. Then the smooth curve $c : [a,b] \to Q$ is said to satisfy the *Lagrange–d'Alembert*

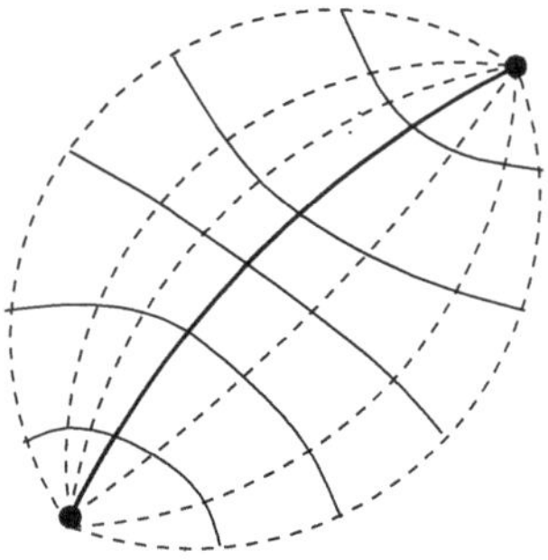

Figure 7.7: A variation of $c \in \Omega(q_1, q_2, [a, b])$ that satisfies the Lagrange–d'Alembert principle. Kinematically admissible curves are represented by solid lines.

principle if it is an extremal curve of the action integral

$$\mathcal{I}[\gamma; \vartheta] = \int_a^b L(\gamma(t), \dot{\gamma}(t))\, dt = 0, \quad \text{for } \vartheta : \delta c(t) \in \mathcal{D}_{c(t)}, \tag{7.8}$$

where variations of c are chosen such that the virtual displacement is kinematically admissible, i. e., $\delta c(t) \in \mathcal{D}_{c(t)}$ for all $t \in [a, b]$.

On the other hand, if one selects variations for which the varied curves $\vartheta(s^*, t)$, for fixed $s = s^* \in [-\varepsilon, \varepsilon]$, are kinematically admissible, then the virtual displacements cannot be kinematically admissible. This is known as the *nonholonomic form of Hamilton's principle*; such a situation is depicted in Figure 7.8.

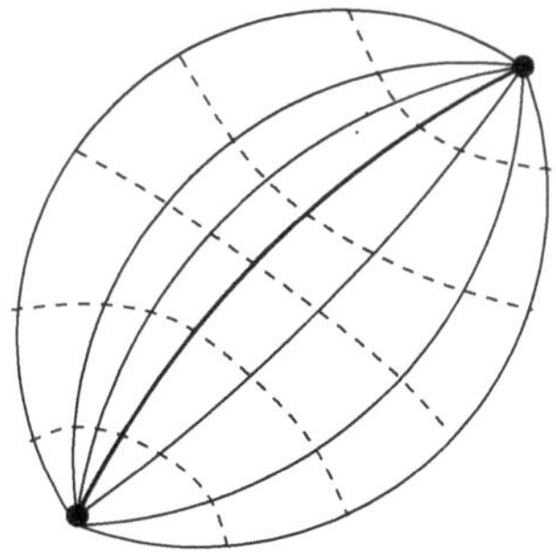

Figure 7.8: A variation of $c \in \Omega(q_1, q_2, [a, b])$ that satisfies the *nonholonomic form of Hamilton's principle*. Kinematically admissible curves are represented by solid lines.

To formalize this notion, we first define the *path space of kinematically admissible curves*,

$$\Omega_{\mathcal{D}}(q_1, q_2, [a, b]) := \left\{\gamma \in \Omega(q_1, q_2, [a, b]) : \dot{\gamma}(t) \in \mathcal{D}_{\gamma(t)}\right\},$$

which is comprised of kinematically admissible curves connecting q_1 and q_2.

Definition 7.15 (Nonholonomic Form of Hamilton's Principle). Let $(Q, L, \mathcal{D})$ be a nonholonomic system. Then the smooth curve $c : [a, b] \to Q$ is said to satisfy the *nonholo-*

nomic form of Hamilton's principle if it is an extremal curve of the action integral

$$\mathcal{I}[y] = \int_a^b L(y(t), \dot{y}(t))\, dt, \quad \text{for } y \in \Omega_{\mathcal{D}}, \tag{7.9}$$

relative to all *kinematically admissible curves* $y : [a, b] \to Q$.

The nonholonomic form of Hamilton's principle is a *true* variation principle: a curve $c : [a, b] \to Q$ satisfies the nonholonomic form of Hamilton's principle if and only if it is a critical value of the functional (7.9) on the space $\Omega_{\mathcal{D}}$.

Due to Proposition 7.2, a curve subject to nonholonomic constraints cannot satisfy *both* Hamilton's principle *and* the Lagrange–d'Alembert principle. To derive equations of motion for a nonholonomic system, one must select one principle or the other. Depending on which variational principle one selects when formulating the equations of motion, one obtains types of motion for the resulting system that, in general, do not coincide. A natural reaction is to ask how we may determine which variational principle corresponds to the one that was selected by nature itself. In other words, for real-life mechanical systems subject to nonholonomic constraints, which of the above two variational principles yields the correct equations that govern the dynamics of the ensuing motion? The answer to this question is simple. One uses the calculus of variations to derive the distinct sets of equations of motion for each principle. Then one takes an example mechanical system, such as a ball rolling on a turntable, and simulates the dynamics that arise out of each set of the underlying equations of motion; see [187] for the details of this computation. Then one goes to a thrift store, buys a turntable and a bag of marbles, and sees whether the motion of a marble on a rotating record follows paths similar to the ones predicted by Hamilton's principle or the ones predicted by the Lagrange–d'Alembert principle.

As it turns out, nature selected the *Lagrange–d'Alembert principle*. For any number of real-life mechanical systems, the motions are ones that minimize the action among variations whose *virtual displacements* are kinematically admissible. Thus, nature does *not* minimize the action given by (7.9) among all possible kinematically admissible paths connecting q_1 to q_2! Since Hamilton's principle is a true variational principle and since the resulting motion is dynamical in nature, we name the two sets of motion as follows.

Definition 7.16. Let $(Q, L, \mathcal{D})$ be a mechanical system. Smooth curves that satisfy the nonholonomic form of Hamilton's principle are said to satisfy the equations of motion for *variational nonholonomic dynamics*, whereas smooth curves that satisfy the Lagrange–d'Alembert principle are said to satisfy the equations of motion for *nonholonomic mechanics*.

Nonholonomic Mechanics

We are now poised to state the fundamental set of equations of motion for nonholonomic mechanics.

Theorem 7.3 (Fundamental Nonholonomic Form of the Euler–Lagrange Equation). *Let $(Q, L, \mathcal{D})$ be a nonholonomic system, where $\mathcal{D}$ may be described as the kernel of a set of m constraint one-forms $\omega^a = A_i^a dq^i$, for $a = (n-m+1), \ldots, n$. Then a curve $c : [a, b] \to Q$ satisfies the Lagrange–d'Alembert principle if and only if it satisfies the* fundamental nonholonomic form of the Euler–Lagrange equation,

$$\frac{d}{dt}\frac{\partial L}{\partial \dot{q}} - \frac{\partial L}{\partial q} = \lambda_a \omega^a, \quad \omega^a(\dot{q}) = 0, \tag{7.10}$$

for some choice of m Lagrange multipliers $\lambda_a : [a, b] \to \mathbb{R}$.

Proof. Let $c : [a, b] \to Q$ be a smooth curve and $\vartheta : [-\varepsilon, \varepsilon] \times [a, b] \to Q$ a variation. Upon taking variations of the action (7.8) and proceeding in a similar fashion as in the proof of Theorem 7.1, we obtain

$$\delta\mathcal{I}[c; \vartheta] = \int_a^b \left(\frac{\partial L}{\partial q^i} - \frac{d}{dt}\frac{\partial L}{\partial \dot{q}^i} \right) \delta q^i dt = 0.$$

However, the individual components of the virtual displacement δc are no longer independent but constrained by the requirement $\delta c(t) \in \mathcal{D}_{c(t)}$ for all $t \in [a, b]$, i. e., $\omega^a(\delta c(t)) = 0$. We can enforce this condition by introducing a set of Lagrange multipliers, $\lambda_a : [a, b] \to \mathbb{R}$, and writing

$$\delta\mathcal{I}[c; \vartheta] = \int_a^b \left[\left(\frac{\partial L}{\partial q^i} - \frac{d}{dt}\frac{\partial L}{\partial \dot{q}^i} \right) \delta q^i + \lambda_a \omega_i^a \delta q^i \right] dt = 0.$$

The components of the virtual displacement δq^i may now be taken as independent. Since this variation must vanish for an arbitrary variation ϑ and since we require curve c to be kinematically admissible, i. e., $\dot{c}(t) \in \mathcal{D}_{c(t)}$, so that $\omega^a(\dot{q}) = 0$, we obtain our result. □

Example 7.7 (Knife Edge on Inclined Plane). Consider a particle of mass m on an inclined plane with a *knife edge* constraint, as depicted in Figure 7.9. A *knife edge* constraint means that the particle is allowed to move in the direction of the blade, which makes an angle of φ with the x-axis, but not perpendicularly to the blade. This is what happens when you go ice skating.

The configuration manifold is $Q = S^1 \times \mathbb{R}^2$, and we take as local coordinates the variables (φ, x, y), where (x, y) is the contact point of the particle and the inclined plane and where φ is the angle between the blade of the knife and the x-axis. The knife

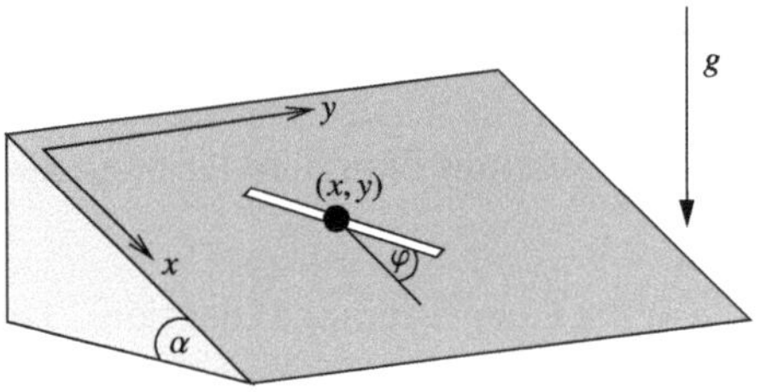

Figure 7.9: Particle with a knife edge constraint on an inclined plane.

edge constraint restricts the types of motion of the system, but by properly *steering* the particle, one may still move the system into any configuration; thus the knife edge constraint is nonholonomic.

The Lagrangian $L : TQ \to \mathbb{R}$ for our system is given by

$$L = \frac{1}{2}m(\dot{x}^2 + \dot{y}^2) + \frac{1}{2}I\dot{\varphi}^2 + mgx\sin\alpha,$$

where I is the moment of inertia of the blade and g is acceleration due to gravity. The knife edge constraint is given by the condition

$$-\sin\varphi\dot{x} + \cos\varphi\dot{y} = 0.$$

The differential constraint form $\omega^3 = -\sin\varphi dx + \cos\varphi dy$ thus allows us to write our constraint as $\omega^3(\dot{q}) = 0$.

The nonholonomic form of the Euler–Lagrange equation (7.10) yields the following mechanical equations of motion for our system:

$$\begin{aligned} I\ddot{\varphi} &= 0, \\ m\ddot{x} - mg\sin\alpha &= -\lambda\sin\varphi, \\ m\ddot{y} &= \lambda\cos\varphi, \end{aligned}$$

as can be readily checked. □

Exercise 7.6. Write out the nonholonomic form of the Euler–Lagrange equation for the vertical rolling disc of Example 7.6.

Exercise 7.7. Consider a hoop with radius a rolling down an inclined plane at angle ϕ. Let x be the distance rolled and θ the angle between a point fixed on the hoop and the contact point of the plane. Let us consider the constraint that the hoop rolls without slipping, so that $\dot{x} = r\dot{\theta}$.

1. Is this a holonomic or nonholonomic constraint? Why?
2. The Lagrangian for the system is given by

$$L = \frac{m\dot{x}^2}{2} + \frac{ma^2\dot{\theta}^2}{2} - mg(l - x)\sin\phi,$$

where l is the total length of the inclined plane. Show that the equations of motion are given by

$$\begin{aligned} m\ddot{x} - mg\sin\phi + \lambda &= 0, \\ ma^2\ddot{\theta} - \lambda r &= 0, \\ r\dot{\theta} &= \dot{x}. \end{aligned}$$

3. Show that

$$\ddot{\theta} = \frac{g\sin\phi}{2a}.$$

4. Conclude that the hoop rolls down the plane with exactly half the acceleration it would have slipping down a frictionless plane.

Variational Nonholonomic Dynamics

Next, we will actuate the nonholonomic form of Hamilton's principle to obtain a set of equations that describes the *variational motions* of the system. Again, these are not the actual, physical motions that a *mechanical* system with nonholonomic constraints would follow, but rather they are a set of equations of motion that solve a *mathematical* problem in the *constrained calculus of variations*. These equations, though mathematically (as opposed to mechanically) inspired, nevertheless have their place in real-world applications, as these motions are related to the solution motions of certain optimal control problems; see [25], [162], and [201] for details.

Theorem 7.4 (Equations for Variational Nonholonomic Dynamics). *Let $(Q, L, \mathcal{D})$ be a nonholonomic system, where $\mathcal{D}$ may be described as the kernel of a set of m constraint one-forms $\omega^a = A_i^a dq^i$ for $a = (n-m+1), \dots, n$. Then a curve $c : [a,b] \to Q$ satisfies the nonholonomic form of Hamilton's principle if and only if it satisfies the equations of motion for variational nonholonomic dynamics,*

$$\frac{d}{dt}\frac{\partial L}{\partial \dot{q}} - \frac{\partial L}{\partial q} = -\mu_a \mathbf{i}_{\dot{q}} d\omega^a - \dot{\mu}_a \omega^a, \quad \omega^a(\dot{q}) = 0, \tag{7.11}$$

for some choice of m Lagrange multipliers $\mu_a : [a,b] \to \mathbb{R}$. Here, $\mathbf{i}_{\dot{q}} d\omega^a$ is the interior product, or contraction, of $\dot{q}$ with $d\omega^a$, and $d\omega^a$ is the exterior derivative of the one-form ω^a. In local coordinates, this expression is equivalent to

$$\frac{d}{dt}\frac{\partial L}{\partial \dot{q}^i} - \frac{\partial L}{\partial q^i} = \mu_a\left(\frac{\partial A_j^a}{\partial q^i} - \frac{\partial A_i^a}{\partial q^j}\right)\dot{q}^j - \dot{\mu}_a A_i^a, \quad A_i^a \dot{q}^i = 0 \tag{7.12}$$

for $a = (n-m+1), \dots, n$ and $i, j = 1, \dots, n$.

Proof. Let $c : [a, b] \to Q$ be a smooth curve and $\vartheta : [-\varepsilon, \varepsilon] \times [a, b] \to Q$ a variation. Upon taking variations of the action (7.8) and proceeding in a similar fashion as in the proof of Theorem 7.1, we obtain

$$\delta\mathcal{I}[c;\vartheta] = \int_a^b \left(\frac{\partial L}{\partial q^i} - \frac{d}{dt}\frac{\partial L}{\partial \dot{q}^i}\right)\delta q^i dt = 0.$$

Again, the variations δq^i are not independent but must be selected to enforce the condition that $\omega^a(\dot{q}) = 0$ along the varied curves of the variation ϑ. To enforce this condition, we require the variation

$$\frac{\partial}{\partial s}(A_i^a \dot{q}^i) = 0$$

to vanish along the fiducial trajectory, i. e., this is literally the condition that $\delta\omega^a(\dot{q}) = 0$. Differentiating, one obtains

$$\frac{\partial}{\partial s}(A_i^a \dot{q}^i) = \frac{\partial A_j^a}{\partial q^i}\dot{q}^j \delta q^i + A_i^a \delta\dot{q}^i.$$

We may enforce this condition by appending the foregoing constraints to the integrand of the varied action using the Lagrange multipliers $\mu_a : [a, b] \to Q$, for $a = (n - m + 1), \ldots, n$, which yields

$$\delta\mathcal{I}[c;\vartheta] = \int_a^b \left[\left(\frac{\partial L}{\partial q^i} - \frac{d}{dt}\frac{\partial L}{\partial \dot{q}^i}\right)\delta q^i + \mu_a \frac{\partial A_j^a}{\partial q^i}\dot{q}^j \delta q^i + \mu_a A_i^a \frac{d\delta q^i}{dt}\right] dt.$$

Integrating the final term by parts and recalling the condition $\delta c(a) = \delta c(b) = 0$, we obtain

$$\delta\mathcal{I}[c;\vartheta] = \int_a^b \left[\frac{\partial L}{\partial q^i} - \frac{d}{dt}\frac{\partial L}{\partial \dot{q}^i} + \mu_a\left(\frac{\partial A_j^a}{\partial q^i} - \frac{\partial A_i^a}{\partial q^j}\right)\dot{q}^j - \dot{\mu}_a A_i^a\right]\delta q^i\, dt = 0.$$

Since this variation must vanish for an arbitrary variation ϑ, we require the coefficients of δq^i to vanish, thereby yielding our result (7.12) as expressed in terms of local coordinates.

To show this expression is equivalent to its invariant form (7.11), let us compute the exterior derivative of $\omega^a = A_i^a dq^i$, which is

$$d\omega^a = \frac{\partial A_i^a}{\partial q^j} dq^j \wedge dq^i.$$

Now, the interior product has the property that, when acting on a vector X,

$$\mathbf{i}_{\dot{q}} d\omega^a(X) = d\omega^a(\dot{q}, X) = \frac{\partial A_i^a}{\partial q^j}(\dot{q}^j X^i - \dot{q}^i X^j).$$

It follows that

$$\mathbf{i}_{\dot{q}} d\omega^a = \left(\frac{\partial A_i^a}{\partial q^j} - \frac{\partial A_j^a}{\partial \dot{q}^i} \right) \dot{q}^j dq^i,$$

which demonstrates the equivalence. □

The first term on the right-hand side of the equations of motion for variational nonholonomic dynamics (7.11) is a measure of the failure of the differential constraint forms ω^a to be exact. Were these forms exact, their exterior derivative would vanish and (7.11) would then coincide with the nonholonomic form of the Euler–Lagrange equation (7.10), with the identification $\lambda_a = -\dot{\mu}_a$. From this observation we may conclude that for *holonomic* systems, i. e., systems with integrable constraints, the mechanical and variational motions coincide, as their corresponding (exact) constraint forms possess vanishing exterior derivatives.

Exercise 7.8. Show that the equations of motion for variational nonholonomic dynamics (7.12) for the knife edge on an inclined plane (discussed in Example 7.7) work out to be

$$\begin{aligned} m\ddot{x} - mg \sin\alpha &= \mu \cos\varphi \dot{\varphi} + \dot{\mu} \sin\varphi, \\ m\ddot{y} &= \mu \sin\varphi \dot{\varphi} - \dot{\mu} \cos\varphi, \\ I\ddot{\varphi} &= -\mu \cos\varphi \dot{x} - \mu \sin\varphi \dot{y}. \end{aligned}$$

How does this motion differ from the mechanical motion of the system?

Exercise 7.9. Write out the equations of motion for variational nonholonomic dynamics (7.12) for the vertical rolling disc of Example 7.6.

Proposition 7.3. *Let $(Q, L, \mathcal{D})$ be a nonholonomic system, where $\mathcal{D}$ may be described as the kernel of a set of m constraint one-forms $\omega^a = A_i^a dq^i$ for $a = (n - m + 1), \ldots, n$. If we define the augmented Lagrangian $\Lambda : TQ \times \mathbb{R}^m \to \mathbb{R}$ as*

$$\Lambda(q, \dot{q}, \mu) = L(q, \dot{q}) + \mu_a \omega^a(\dot{q}),$$

then the equations of motion for variational nonholonomic dynamics are equivalent to the Euler–Lagrange equation for the augmented Lagrangian, *i. e.*,

$$\frac{d}{dt}\frac{\partial \Lambda}{\partial \dot{q}} - \frac{\partial \Lambda}{\partial q} = 0, \quad \omega^a(\dot{q}) = 0.$$

Exercise 7.10. Prove Proposition 7.3.

Proposition 7.3 further reinforces the notion that the constraints are applied before taking variations when formulating the equations for variational nonholonomic dynamics.

7.7 Application: Nöther's Theorem

In this section, we discuss a famous result, due to [226], that establishes a correspondence between symmetries and conservation laws. Nöther's theorem is discussed from a classical viewpoint in [8]. It is laid out in a modern geometric context in terms of momentum mappings in [24] and [196], the latter of which discusses symmetries and reduction in mechanical systems in depth.

Definition 7.17. A Lagrangian system (Q, L) *admits* the smooth mapping $h : Q \to Q$ if the Lagrangian $L : TQ \to \mathbb{R}$ is invariant under the mapping $Dh : T_qQ \to T_{h(q)}Q$ for all $q \in Q$, i. e., for any $q \in Q$ and $v \in T_qQ$,

$$L(q, v) = L(h(q), Dh \cdot v).$$

Theorem 7.5 (Nöther). *If the Lagrangian system (Q, L) admits the one-parameter group of diffeomorphisms $h^s : Q \to Q$ for all $s \in \mathbb{R}$, then there exists an integral of motion $J : TQ \to \mathbb{R}$. Moreover, this integral may be expressed as the inner product*

$$J(q, \dot{q}) = \left\langle \frac{\partial L}{\partial \dot{q}}, \left. \frac{dh^s(q)}{ds} \right|_{s=0} \right\rangle .$$

Proof. Let $c : [a, b] \to Q$ be a solution of the Euler–Lagrange equation (7.1). Since the Lagrangian is invariant under the mapping $Dh^s : T_{c(t)}Q \to T_{h^s(c(t))}Q$ for all $t \in [a, b]$, it follows that $h^s \circ c : [a, b] \to Q$ is also a solution of the Euler–Lagrange equation for each fixed $s \in [-\varepsilon, \varepsilon]$. In fact, this defines a *variation* $\vartheta : [-\varepsilon, \varepsilon] \times [a, b] \to Q$ of curve c by the relation

$$\vartheta(s, t) = h^s(c(t)).$$

This particular variation holds the special property that each of the varied curves $c_s : [a, b] \to Q$, defined by $c_s(t) = \vartheta(s, t)$, for each fixed $s \in [-\varepsilon, \varepsilon]$, is itself a solution to the Euler–Lagrange equation. Since the Lagrangian admits the one-parameter group of diffeomorphisms h^s, it follows that its variation must vanish along the fiducial curve, i. e.,

$$\delta L = \delta\hat{c}[L] = \left. \frac{\partial L(\vartheta, \dot{\vartheta})}{\partial s} \right|_{s=0} = 0.$$

Note that this condition holds only for our special variation ϑ and not for arbitrary variations of curve c. Using local coordinates, so that $\vartheta(s, t) = (q^1(s, t), \dots, q^n(s, t))$, we compute

$$\begin{aligned} \frac{\partial L}{\partial s} &= \frac{\partial L}{\partial q^i} \frac{\partial q^i}{\partial s} + \frac{\partial L}{\partial \dot{q}^i} \frac{\partial \dot{q}^i}{\partial s} \\ &= \frac{d}{dt} \frac{\partial L}{\partial \dot{q}^i} \frac{\partial q^i}{\partial s} + \frac{\partial L}{\partial \dot{q}^i} \frac{\partial^2 q^i}{\partial t \partial s} \end{aligned}$$

$$= \frac{d}{dt}\left[\frac{\partial L}{\partial \dot{q}^i}\frac{\partial q^i}{\partial s}\right].$$

The first term of the second line is obtained since curve c satisfies the Euler–Lagrange equation (7.1). The result follows, noting the relation

$$\frac{\partial q^i}{\partial s}\frac{\partial}{\partial q^i} = \delta q^i \frac{\partial}{\partial q^i} = \frac{dh^s}{ds}$$

and evaluating at $s = 0$. □

Remark. A one-parameter group of diffeomorphisms that is admitted by a Lagrangian is also known as a *symmetry* of the system. Nöther's theorem states that every symmetry yields an integral of motion.

Example 7.8 (Translational Symmetry). In this example, we will consider a free particle with mass m. The configuration manifold $Q = \mathbb{R}^3$ may be described in local Cartesian coordinates (x, y, z). The Lagrangian for a free particle consists solely of kinetic energy, i. e.,

$$L = \frac{m}{2}(\dot{x}^2 + \dot{y}^2 + \dot{z}^2) = \frac{m}{2}\|\dot{q}\|^2.$$

Now let us introduce the one-parameter symmetry group

$$h^s(x, y, z) = (x + s, y, z),$$

corresponding to translations in the x-direction. The differential of this mapping, $Dh^s : T_qQ \to T_{h^s(q)}Q$, is given by $Dh^s = I$, so that $Dh^s \cdot v = v$ for all $v \in T_qQ$. It follows that $L(q, v) = L(h^s(q), Dh^s(v))$, and thus L admits the one-parameter group of diffeomorphisms h^s.

Nöther's theorem therefore yields the following integral of motion:

$$J(q, \dot{q}) = \left\langle \frac{\partial L}{\partial \dot{q}}, \left.\frac{dh^s}{ds}\right|_{s=0} \right\rangle.$$

Since $dh^s/ds|_{s=0} = (1, 0, 0)$, it follows that

$$J(q, \dot{q}) = \frac{\partial L}{\partial \dot{x}} = m\dot{x}$$

is a conserved quantity for our system. This quantity corresponds to the x-component of the particle's linear momentum. Similarly, one may show that $m\dot{y}$ and $m\dot{z}$ are also conserved quantities of the system. We thus obtain the *conservation of linear momentum* for a free particle. □

Example 7.9 (Rotational Symmetry). Consider again a particle with mass m and configuration manifold $Q = \mathbb{R}^3$. Let $r = (x, y, z)^T$ be a set of local coordinates for the particle and suppose that the particle is found in a central force field, i. e., a potential-energy field that depends only on the distance to the origin $\|r\|$. The Lagrangian for such a system is given by

$$L = \frac{m}{2}\dot{r}^T\dot{r} - U(\|r\|),$$

where $U(\|r\|)$ is the potential-energy function.

Consider the *special orthogonal group* SO(3), i. e., the group of 3×3 orthogonal matrices with unit determinant

$$\mathrm{SO}(3) = \{A \in \mathbb{R}^{3\times 3} : A^T A = I \text{ and } \det(A) = 1\}.$$

The matrices in SO(3) correspond to rigid rotations of $\mathbb{R}^3$. For all matrices $A \in \mathrm{SO}(3)$, the smooth mapping $h : Q \to Q$ defined by $h(r) = Ar$ is admitted by the Lagrangian L. To see this, note that $Dh = A$; therefore, $Dh \cdot \dot{r} = A \cdot \dot{r}$. Since A is orthogonal, it preserves $\|r\|$, so that $\|h(r)\| = \|Ar\| = \|r\|$. Thus the Lagrangian is invariant under the mapping h, i. e.,

$$L(h(r), Dh \cdot \dot{r}) = \frac{m}{2}\dot{r}^T A^T A\dot{r} - U(\|Ar\|) = \frac{m}{2}\dot{r}^T\dot{r} - U(\|r\|) = L(r, \dot{r}).$$

Now, let us consider the particular one-parameter group of diffeomorphisms $h^s : Q \to Q$ defined by

$$h^s(r) = \begin{bmatrix} 1 & 0 & 0 \\ 0 & \cos s & -\sin s \\ 0 & \sin s & \cos s \end{bmatrix} r.$$

This is a symmetry for our system, as the coefficient matrix $A(s) \in \mathrm{SO}(3)$ for all s. Therefore, we obtain an integral of motion through the relation

$$J(r, \dot{r}) = \left\langle \frac{\partial L}{\partial \dot{r}}, \left.\frac{dh^s}{ds}\right|_{s=0} \right\rangle.$$

To compute this, consider

$$\begin{aligned} \left.\frac{dh^s}{ds}\right|_{s=0} &= \left.\begin{bmatrix} 0 & 0 & 0 \\ 0 & -\sin s & -\cos s \\ 0 & \cos s & -\sin s \end{bmatrix}\right|_{s=0} \cdot \begin{bmatrix} x \\ y \\ z \end{bmatrix} \\ &= \begin{bmatrix} 0 & 0 & 0 \\ 0 & 0 & -1 \\ 0 & 1 & 0 \end{bmatrix} \begin{bmatrix} x \\ y \\ z \end{bmatrix} = \begin{bmatrix} 0 \\ -z \\ y \end{bmatrix}. \end{aligned}$$

It follows that $J(r, \dot{r}) = -mz\dot{y} + my\dot{z}$ is a conserved quantity. We recognize this as the x-component of the *angular momentum* $H : TQ \to TQ$ of the particle about the origin, i. e., as

$$H = r \times (m\dot{r}),$$

where "×" is the usual cross product on $\mathbb{R}^3$.

Exercise 7.11. Use Nöther's theorem and two cleverly chosen paths through SO(3) to show that the y- and z-components of the angular momentum H are also conserved quantities for a particle in a central-force field.

This example and exercise demonstrate the *conservation of angular momentum* for classical particles in central force fields. Since gravitational fields constitute a particular instance of central-force fields, this result yields conservation of angular momentum for satellites, say, that are in orbit about the Earth. □

Exercise 7.12. The *spherical pendulum* is a Lagrangian system with configuration manifold $Q = S^2$ and Lagrangian L, which, in terms of local coordinates (θ, φ), may be expressed as

$$L = \frac{1}{2}mr^2(\dot{\theta}^2 + \sin^2\theta\dot{\varphi}^2) - mgr\cos\theta,$$

where r is the length of the pendulum rod and g is acceleration due to gravity. The local coordinates θ and φ represent the polar and azimuthal angles for spherical coordinates, respectively. Write out Lagrange's equations for the spherical pendulum. Show that the Lagrangian admits the one-parameter group of diffeomorphisms

$$h^s(\theta, \varphi) = (\theta, \varphi + s).$$

Determine the corresponding integral of motion yielded by Nöther's theorem. What does this integral represent physically?

Example 7.10 (Time Symmetry). In this example, we relax our condition that our Lagrangian must be *time-independent*. To accomplish this, we will consider our Lagrangian to be defined on the *extended phase space* $L : TQ \times \mathbb{R} \to \mathbb{R}$. We will proceed a bit differently in this example. To begin, let us consider the function $H : TQ \times \mathbb{R} \to \mathbb{R}$ defined by

$$H = \left\langle \frac{\partial L}{\partial \dot{q}}, \dot{q} \right\rangle - L(q, \dot{q}, t).$$

Taking the derivative of H, we obtain

$$\frac{dH}{dt} = \frac{\partial L}{\partial \dot{q}^i}\ddot{q}^i + \frac{d}{dt}\frac{\partial L}{\partial \dot{q}^i}\dot{q}^i - \frac{\partial L}{\partial q^i}\dot{q}^i - \frac{\partial L}{\partial \dot{q}^i}\ddot{q}^i - \frac{\partial L}{\partial t}.$$

Along solution trajectories, we may enforce the Euler–Lagrange equation, thus simplifying this expression to

$$\frac{dH}{dt} = -\frac{\partial L}{\partial t}.$$

Now suppose we introduce the one-parameter group of diffeomorphisms on the extended phase space given by the mapping $t \mapsto t + s$. Note that a Lagrangian admits the group of diffeomorphisms

$$L(q, \dot{q}, t) = L(q, \dot{q}, t + s)$$

if and only if L is independent of time. This is called *time symmetry*. From our foregoing discussion, we see that all systems with time-independent Lagrangians possess a certain integral of motion,

$$H = \left\langle \frac{\partial L}{\partial \dot{q}}, \dot{q} \right\rangle - L(q, \dot{q}),$$

known as the *total mechanical energy* of the system. We will discuss this topic further in the next chapter. □

8 Hamiltonian Mechanics

During our discussion of nonlinear dynamics in Euclidean space, we saw how one might convert the second-order systems of differential equations arising from Newton's equations of motion for mechanics into a larger first-order system defined on the phase space. This technique generalizes to geometric mechanics and forms the connection between *Lagrangian* and *Hamiltonian mechanics*. As discussed in Chapter 7, the Euler–Lagrange equation for a mechanical system with an n-dimensional configuration manifold yields (in local coordinates) a system of n second-order ordinary differential equations that govern the dynamics of the mechanical system. In this chapter, we address the subject of transforming the Euler–Lagrange equation for mechanics into an equivalent system of $2n$ first-order ordinary differential equations. The appropriate geometric setting for these new equations is the *cotangent bundle*; the phase space variables for our system of $2n$ first-order ordinary differential equations may therefore be interpreted as *generalized coordinates* and their corresponding *conjugate momenta*. Such systems possess a rich geometric structure and have recently been generalized, leading to the birth of the study of *symplectic manifolds*.

8.1 Legendre Transform

Our first goal of the chapter will be to introduce *Hamilton's equations of motion*. As mentioned in the chapter introduction, Hamilton's equations constitute a set of first-order differential equations on the cotangent bundle of the configuration manifold and further govern the mechanical evolution of the system. Before we introduce the equations, we will need to define a certain function $H : T^*Q \to \mathbb{R}$, known as the *Hamiltonian*, for our system. This function will be defined with the aid of a certain vector-space isomorphism that exists between the tangent bundle and the cotangent bundle.

Definition 8.1. Let (Q, L) be a Lagrangian system. The *fiber derivative*, $\mathbb{F}L$, of the Lagrangian L is a fiber-preserving mapping $\mathbb{F}L : TQ \to T^*Q$ that is defined by

$$\mathbb{F}L(v)[w] = \frac{d}{ds}L(q, v + sw)\bigg|_{s=0}, \tag{8.1}$$

where $v, w \in T_qQ$.

Note that the quantity $\mathbb{F}L(v)$ is a one-form and that (8.1) shows how this one-form maps vectors w in the tangent bundle to $\mathbb{R}$. The fiber derivative $\mathbb{F}L : TQ \to T^*Q$ is literally a derivative in the *fiber direction*. For a more general exposition on fiber derivatives, see [2].

Of particular importance is the *generalized momentum*, a one-form defined by

$$p = \mathbb{F}L(\dot{q}).$$

https://doi.org/10.1515/9783110597806-008

The foregoing definition manifests itself in a useful way when examined using a set of local coordinates $q = (q^1, \ldots, q^n)$. In particular, we have

$$\mathbb{F}L(v)[w] = \frac{\partial L(q, v)}{\partial \dot{q}^i} w^i.$$

Hence, the components of $\mathbb{F}L(\dot{q})$ relative to the basis $\{dq^1, \ldots, dq^n\}$ of T^*Q may be expressed by the relations

$$p_i = \frac{\partial L}{\partial \dot{q}^i}, \tag{8.2}$$

so that $p = p_i dq^i$. The components of the generalized momentum $p \in T^*Q$ relative to the basis $\{dq^1, \ldots, dq^n\}$ are referred to as the *conjugate momenta* of the generalized coordinates q^i. The coordinate–momentum pairs (q^i, p_i) are referred to as *symplectic pairs*. A set of coordinates on T^*Q is said to be *canonical* if its constituents can be arranged in symplectic pairs.

Exercise 8.1. Compute the conjugate momenta to the local coordinates x and y for the Lagrangian (7.2).

Definition 8.2. A Lagrangian $L : TQ \to \mathbb{R}$ is *hyperregular* if its fiber derivative $\mathbb{F}L : TQ \to T^*Q$ is a diffeomorphism.

For a hyperregular Lagrangian L, one can invert the mapping $\mathbb{F}L : TQ \to T^*Q$ and solve for

$$\dot{q}^i = \dot{q}^i(q^1, \ldots, q^n, p_1, \ldots, p_n) = \dot{q}^i(q, p).$$

As it turns out, mechanical systems in general possess hyperregular Lagrangians. In order for a Lagrangian to fail to be hyperregular, it would have to be independent of one or more components of a system's velocity. This is not possible for a mechanical system, since the kinetic energy is obtained by a nondegenerate, symmetric, bilinear quadratic form (the kinetic-energy metric) acting on the velocity. However, nonhyperregular Lagrangians do arise in the study of kinematic sub-Riemannian optimal control problems; see, for example, [16], [27], and [35].

Definition 8.3. Let (Q, L) be a Lagrangian system. If L is hyperregular, then we define its corresponding *Hamiltonian*, $H : T^*Q \to \mathbb{R}$, by the relation

$$H(q, p) = \langle p, \dot{q} \rangle - L(q, \dot{q}), \tag{8.3}$$

where $\dot{q} = \dot{q}(q, p)$. The change of L on TQ to H on T^*Q is known as the *Legendre transform*.

Example 8.1 (Simple Pendulum). Let us consider the case of the simple pendulum discussed in Example 7.1. Recall that the configuration manifold was given by $Q = S^1$, with local coordinates $q^1 = \theta$ and the Lagrangian

$$L = \frac{1}{2}ml^2\dot{\theta}^2 + mgl\cos\theta.$$

The conjugate momentum of θ is therefore

$$p_\theta = \frac{\partial L}{\partial \dot{\theta}} = ml^2\dot{\theta}.$$

Thus, the Hamiltonian for the simple pendulum is given by

$$H(q,p) = \langle p, \dot{q}\rangle - L(q,\dot{q}) = \frac{p_\theta^2}{ml^2} - \frac{1}{2}\frac{ml^2 p_\theta^2}{m^2 l^4} - mgl\cos\theta,$$

which simplifies to

$$H(q,p) = \frac{1}{2}\frac{p_\theta^2}{ml^2} - mgl\cos\theta.$$

This expression equals the total mechanical energy for the system since the first term is the kinetic energy and the second term is the potential energy. □

Example 8.2. Let us consider a general mechanical Lagrangian of the form

$$L(q,\dot{q}) = \frac{1}{2}g_{ij}\dot{q}^i\dot{q}^j - U(q),$$

where $U(q)$ is the potential energy and g_{ij} is the kinetic-energy metric for the system, so that the double sum

$$K = \frac{1}{2}g_{ij}\dot{q}^i\dot{q}^j$$

represents the kinetic energy for the system. Note that g_{ij} is a positive definite, symmetric, quadratic form. Let us further denote the inverse of the matrix $[g_{ij}]$ as $[g^{ij}]$, so that $[g^{ij}] = [g_{ij}]^{-1}$, i. e., $g^{ik}g_{kj} = \delta^i_j$.

To compute the conjugate momentum, we take the derivative

$$p_i = \frac{\partial L}{\partial \dot{q}^i} = g_{ij}\dot{q}^j,$$

which turns out to be precisely equivalent to $p = \dot{q}^\flat$, using the notation from §6.7. We used the fact that the kinetic-energy metric is symmetric, i. e., $g_{ij} = g_{ji}$. Inverting this relation, we obtain

$$\dot{q}^i = g^{ij}p_j.$$

It follows that the kinetic energy may be expressed as

$$K = \frac{1}{2} g^{ij} p_i p_j.$$

Next, observe that

$$p_i \dot{q}^i = p_i g^{ij} p_j = g^{ij} p_i p_j.$$

Therefore, it follows that the Hamiltonian is equivalent to

$$H(q,p) = \dot{q}^i p_i - L(q,\dot{q}) = \frac{1}{2} g^{ij} p_i p_j + U(q).$$

Whereas the Lagrangian for the system is typically given by the *kinetic minus potential energies*, the Hamiltonian is given by the *kinetic plus potential energies* and therefore represents the *total mechanical energy* of the system. □

8.2 Hamilton's Equations of Motion

In this paragraph, we will prove that Hamilton's equations govern the motion of mechanical systems on their cotangent bundle. First, we will introduce an important lemma.

Lemma 8.1. *Let Q be a configuration manifold for a mechanical system, $L : TQ \to \mathbb{R}$ a hyperregular Lagrangian, and $H : T^*Q \to \mathbb{R}$ its associated Hamiltonian. Then the system's velocity and Hamiltonian are related by the equation*

$$\dot{q} = \frac{\partial H(q,p)}{\partial p}.$$

Proof. By definition of the Legendre transformation, we have

$$H(q,p) = \langle p, \dot{q} \rangle - L(q,\dot{q}),$$

where $\dot{q} = \dot{q}(q,p)$ is given by the inverse fiber derivative $\mathbb{F}L^{-1} : T^*Q \to TQ$. Taking the derivative with respect to p_i, the conjugate momentum of q^i, and using the chain rule, we obtain

$$\frac{\partial H}{\partial p_i} = \dot{q}^i + p_j \frac{\partial \dot{q}^j}{\partial p_i} - \frac{\partial L}{\partial \dot{q}^j} \frac{\partial \dot{q}^j}{\partial p_i}.$$

However, we recognize

$$p_j = \frac{\partial L}{\partial \dot{q}^j},$$

by definition of p_j, thereby proving our result. □

Theorem 8.1 (Hamilton's Equations). *Let (Q, L) be a Lagrangian system with hyperregular Lagrangian and $H : T^*Q \to \mathbb{R}$ its associated Hamiltonian. Then the solution to* Hamilton's equations,

$$\dot{q} = \frac{\partial H}{\partial p} \quad \textit{and} \quad \dot{p} = -\frac{\partial H}{\partial q}, \tag{8.4}$$

on the cotangent bundle T^*Q, trace out the same curve, when projected to the configuration manifold Q, as the solution to the Euler–Lagrange equation on TQ.

Proof. By Theorem 7.1, the solution of the Euler–Lagrange equation is the curve that satisfies Hamilton's principle (Definition 7.1). By (8.3), Hamilton's principle is equivalent to selecting the curve $\gamma : [a, b] \to Q$, which extremizes the action

$$\mathcal{I}[\gamma] = \int_a^b (\langle p, \dot{q} \rangle - H(q, p))\, dt.$$

Taking variations, we obtain

$$\delta\mathcal{I}[\gamma; \vartheta] = \int_a^b \left(\dot{q}^i \delta p_i + p_i \delta \dot{q}^i - \frac{\partial H}{\partial q^i} \delta q^i - \frac{\partial H}{\partial p_i} \delta p_i \right) dt.$$

Integrating the second term of the integrand by parts and recalling the boundary conditions $\delta q(a) = \delta q(b) = 0$, we obtain

$$\delta\mathcal{I}[\gamma; \vartheta] = \int_a^b \left[\left(\dot{q}^i - \frac{\partial H}{\partial p_i} \right) \delta p_i - \left(\dot{p}_i + \frac{\partial H}{\partial q^i} \right) \delta q^i \right] dt.$$

By Lemma 8.1, the coefficients of δp vanish identically, and we are left with

$$\delta\mathcal{I}[\gamma; \vartheta] = -\int_a^b \left(\dot{p}_i + \frac{\partial H}{\partial q^i} \right) \delta q^i dt.$$

Now, the individual components δq^i are independent. Hence, in order for $\delta\mathcal{I}[\gamma; \vartheta]$ to vanish for an arbitrary C^2 variation of γ, we require

$$\dot{p}_i = -\frac{\partial H}{\partial q^i}.$$

Combining this result with Lemma 8.1 proves the theorem. □

Care must be exercised when performing numerical simulations of Hamiltonian dynamical systems. Standard numerical integration methods for ordinary differential equations do not preserve fundamental quantities and structures when applied to Hamiltonian systems. As such, special integration techniques, known as *geometric integrators* or *symplectic integrators*, have been developed to integrate Hamilton's equations of motion while preserving the symplectic structure. See, for example, [114], [181], and [197] and the references therein for a discussion on geometric integrators.

Exercise 8.2. Show that Hamilton's equations for the simple pendulum of Examples 7.1 and 8.1 yield

$$\dot{\theta} = \frac{p_\theta}{ml^2} \quad \text{and} \quad \dot{p}_\theta = -mgl \sin\theta.$$

How do these compare with the Euler–Lagrange equation?

Exercise 8.3. Compute the conjugate momenta and Hamiltonian for the spherical pendulum of Exercise 7.12. Write out Hamilton's equations for this system.

Exercise 8.4. Show that the Hénon–Heiles system (4.1) may be written as a Hamiltonian system with Hamiltonian

$$H = \frac{1}{2}(u^2 + v^2) + \frac{1}{2}(x^2 + y^2) + x^2 y - \frac{y^3}{3}. \tag{8.5}$$

Compute Hamilton's equations for the system.

Exercise 8.5. Consider the *Lotka–Volterra* equations for the dynamics of a *predator–prey ecosystem*,

$$\begin{aligned} \dot{S} &= -rSI + \mu S, \\ \dot{I} &= rSI - \gamma I. \end{aligned}$$

Here, S represents the population of the *prey* and I the population size of the *predators*; with $r > 0, \mu > 0, \gamma > 0$ being positive parameters. Show that under the transformation $p = \ln S$, $q = \ln I$, the system becomes

$$\begin{aligned} \dot{p} &= \mu - re^q, \\ \dot{q} &= re^p - \gamma. \end{aligned}$$

Show this constitutes a Hamiltonian system, with Hamiltonian

$$H(q,p) = re^p - \gamma p + re^q - \mu q.$$

Verify that the Hamiltonian is conserved along solution trajectories.

8.3 Hamiltonian Vector Fields and Conserved Quantities

In this paragraph, we introduce some of the basic concepts that led to the development of *symplectic structures* and *symplectic manifolds*, which are discussed later in §8.5.

Definition 8.4. Let $(q^1,\dots,q^n,p_1,\dots,p_n)$ be a set of canonical coordinates on T^*Q. Then we define the *symplectic rotation* as the mapping $\mathcal{J} : T^*T^*Q \to TT^*Q$ determined by the relation

$$\mathcal{J}(\omega) = \beta^i \frac{\partial}{\partial q^i} - \alpha_j \frac{\partial}{\partial p_j}, \tag{8.6}$$

where the one-form $\omega \in T^*T^*Q$ is given in terms of local coordinates as

$$\omega = \alpha_i dq^i + \beta^j dp_j.$$

The mapping $\mathcal{J} : T^*T^*Q \to TT^*Q$ in the previous definition may be thought of as the $2n \times 2n$ matrix

$$J = \begin{bmatrix} O_n & I_n \\ -I_n & O_n \end{bmatrix},$$

where O_n and I_n are the $n\times n$ zero matrix and $n\times n$ identity matrix, respectively. Let $x = (q^1,\dots,q^n,p_1,\dots,p_n)$, which represents a set of local coordinates for T^*Q. We will see the utility of such a definition by first considering the one-form $dH \in T^*T^*Q$, which is given by

$$dH = \frac{\partial H}{\partial q^i} dq^i + \frac{\partial H}{\partial p_i} dp_i.$$

One immediately sees that Hamilton's equations (8.4) may be expressed in the form

$$\dot{x} = \mathcal{J}(dH). \tag{8.7}$$

The term $\mathcal{J}(dH)$ is known as the *symplectic gradient*, and it may be thought of as the operation of a gradient followed by the operation of symplectic rotation.

Definition 8.5. Given a hyperregular Hamiltonian H, we denote the *Hamiltonian vector field associated with* H by the vector field $X_H \in \mathfrak{X}(T^*Q)$ given by the expression

$$X_H = \mathcal{J}(dH), \tag{8.8}$$

where $\mathcal{J} : T^*T^*Q \to TT^*Q$ is the symplectic rotation (8.6).

Exercise 8.6. Show that, in local coordinates $(q^1,\dots,q^n,p_1,\dots,p_n)$, the Hamiltonian vector field associated with the Hamiltonian H is given by

$$X_H = \frac{\partial H}{\partial p_i}\frac{\partial}{\partial q^i} - \frac{\partial H}{\partial q^i}\frac{\partial}{\partial p_i}.$$

Hamilton's equations therefore determine the flow of X_H on T^*Q.

Example 8.3. Consider the one-degree-of-freedom system with Hamiltonian

$$H = \frac{1}{2}(q^2 + p^2).$$

The symplectic gradient $\mathcal{J}(dH)$ is given by

$$X_H = \mathcal{J}(dH) = p\frac{\partial}{\partial q} - q\frac{\partial}{\partial p}.$$

This is perpendicular to the gradient dH for each (q,p), so that $\mathcal{J}(dH)$ lies tangent to each of the constant H contours; see Figure 8.1. □

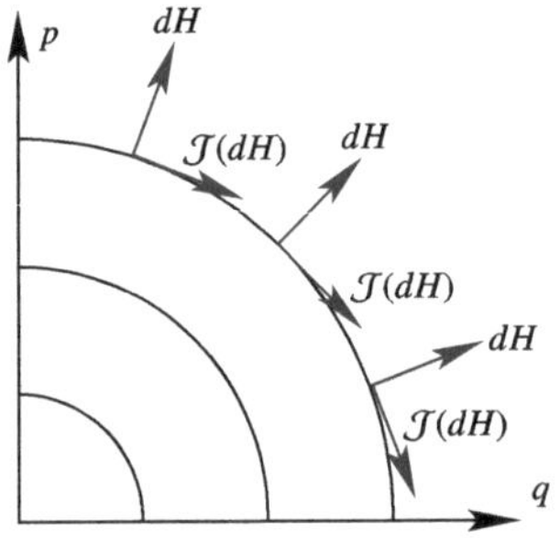

Figure 8.1: Constant H contours for Example 8.3, plotted with several gradient vectors dH and symplectic gradient vectors $\mathcal{J}(dH)$.

Theorem 8.2 (Conservation of Energy). *Let $H : T^*Q \to \mathbb{R}$ be a time-independent Hamiltonian. Then H is conserved along the Hamiltonian phase flow of the vector field X_H given by* (8.8).

Proof. Taking the material derivative of the function H along solution curves of the vector field X_H, we obtain

$$\frac{DH}{Dt} = \frac{\partial H}{\partial q^i}\frac{dq^i}{dt} + \frac{\partial H}{\partial p_i}\frac{dp_i}{dt} = \frac{\partial H}{\partial q^i}\frac{\partial H}{\partial p_i} - \frac{\partial H}{\partial p_i}\frac{\partial H}{\partial q^i} = 0.$$

This proves the theorem. □

Next, let us consider the material derivative of an arbitrary function $F : T^*Q \to \mathbb{R}$ along the Hamiltonian vector field X_H associated with the Hamiltonian $H : T^*Q \to \mathbb{R}$. We obtain

$$\frac{DF}{Dt} = \frac{\partial F}{\partial q^i}\frac{dq^i}{dt} + \frac{\partial F}{\partial p_i}\frac{dp_i}{dt} = \frac{\partial F}{\partial q^i}\frac{\partial H}{\partial p_i} - \frac{\partial F}{\partial p_i}\frac{\partial H}{\partial q^i}.$$

This motivates the following definition.

Definition 8.6. The *Poisson bracket* $\{\cdot,\cdot\} : C^\infty(T^*Q)\times C^\infty(T^*Q) \to C^\infty(T^*Q)$ of any two functions $F, H : T^*Q \to \mathbb{R}$ is defined by

$$\{F,H\} = \frac{\partial F}{\partial q^i}\frac{\partial H}{\partial p_i} - \frac{\partial F}{\partial p_i}\frac{\partial H}{\partial q^i}. \tag{8.9}$$

As an immediate consequence of this definition and the preceding calculation, we obtain the following proposition.

Proposition 8.1. *The smooth function $F : T^*Q \to \mathbb{R}$ is a conserved quantity along the flow of a Hamiltonian vector field $X_H = \mathcal{J}(dH)$, associated with a Hamiltonian $H : T^*Q \to \mathbb{R}$, if and only if*

$$\{F, H\} = 0.$$

Moreover, for any $F : T^*Q \to \mathbb{R}$, its material derivative along the flow of X_H is given by

$$\frac{DF}{Dt} = \{F, H\}.$$

Exercise 8.7. Show that for any smooth $H, F : T^*Q \to \mathbb{R}$, one has the relation

$$[X_H, X_F] = -X_{\{H,F\}},$$

where X_H, X_F, and $X_{\{H,F\}}$ are the Hamiltonian vector fields associated with H, F, and $\{H, F\}$, respectively.

Exercise 8.8. Show that the Poisson bracket satisfies the *Jacobi identity*

$$\{\{F, G\}, H\} + \{\{H, F\}, G\} + \{\{G, H\}, F\} = 0$$

for all smooth $F, G, H : T^*Q \to \mathbb{R}$.

Exercise 8.9. Show that if $F, G : T^*Q \to \mathbb{R}$ are integrals of motion for the phase flow of the Hamiltonian vector field X_H, then $\{F, G\}$ is also an integral of motion.

A second classical result of Hamiltonian mechanics, next to conservation of the Hamiltonian function, is the volume preservation of phase space under the Hamiltonian phase flow. The theorem first appeared in [99] and was derived using an identity due to [190].

Theorem 8.3 (Liouville). *Let $H : T^*\mathbb{R}^n \to \mathbb{R}$ be a C^2, hyperregular Hamiltonian on a Euclidean configuration manifold. Let $\Omega \subset \mathbb{R}^n$ be a compact, connected $2n$-dimensional set, $V(\Omega)$ its Euclidean volume, and $\varphi(t; \Omega)$ its evolution under the Hamiltonian phase flow X_H,*

$$\varphi(t; \Omega) = \{x \in \mathbb{R}^{2n} : \varphi(-t; x) \in \Omega\}.$$

Then

$$V(\Omega) = V(\varphi(t; \Omega)).$$

Proof. This result follows due to the fact that Hamiltonian vector fields are divergenceless. To see this, first note that

$$X_H = \left(\frac{\partial H}{\partial p_1}, \ldots, \frac{\partial H}{\partial p_n}, -\frac{\partial H}{\partial q^1}, \ldots, -\frac{\partial H}{\partial q^n}\right).$$

Computing the divergence of X_H, we obtain

$$\nabla \cdot X_H = \frac{\partial^2 H}{\partial q^1 \partial p_1} + \cdots + \frac{\partial^2 H}{\partial q^n \partial p_n} - \frac{\partial^2 H}{\partial p_1 \partial q^1} - \cdots - \frac{\partial^2 H}{\partial p_n \partial q^n} = 0.$$

Volume conservation of the phase flow then follows from the fact that

$$\left.\frac{dV(t;\Omega)}{dt}\right|_{t=0} = \oint_{\partial\Omega} X_H \cdot dS = \int_{\Omega} \nabla \cdot X_H dV = 0,$$

where the second equality follows from Gauss' divergence theorem. □

8.4 Routh's Equations

We will now return to our discussion of symmetry that we began in §7.7. Recall that Nöther's theorem (Theorem 7.5) tells us that every symmetry of a mechanical system corresponds to a conserved quantity. We saw explicit examples where Lagrangians that were independent of one or more of the generalized coordinates each yielded a conserved momentum of the system. Routh's approach explicitly formulates a set of reduced Euler–Lagrange equations for systems with similar conserved quantities. Routh's equations first appeared in [254, 255]. They are also discussed in classical references such as [104, 106] and [293] and, in terms of modern geometry and variational principles, in [200], [259], and [297].

Recall that the conjugate momentum, p_i, of the generalized coordinate q^i is defined by the relation

$$p_i = \frac{\partial L}{\partial \dot{q}^i}.$$

The Euler–Lagrange equation (7.1) therefore relates the time derivatives of the conjugate momenta with the Lagrangian function, i. e.,

$$\frac{dp_i}{dt} = \frac{\partial L}{\partial q^i}.$$

Hence, if a Lagrangian is independent of one or more of its generalized coordinates, then the corresponding conjugate momenta will be conserved along solution flows. This motivates the following definition.

Definition 8.7. Let (Q, L) be a Lagrangian system. If one of the local coordinates q^α does not explicitly appear in the Lagrangian $L(q, \dot{q})$, then we refer to that coordinate as an *ignorable coordinate* or a *cyclic coordinate.*

The foregoing discussion contains the observation that the conjugate momentum to each cyclic coordinate is conserved. The following theorem shows how the equations of motion may be reduced by utilizing these conserved momenta.

Theorem 8.4 (Routh). *Suppose a Lagrangian system (Q, L) has k cyclic coordinates $q^1, \ldots, q^k$, where $k < n$, so that $L = L(q^{k+1}, \ldots, q^n, \dot{q}^1, \ldots, \dot{q}^n)$. Then the Euler–Lagrange equation produces the same motions as* Routh's equation, *i. e.,*

$$\frac{d}{dt}\frac{\partial R}{\partial \dot{q}^a} - \frac{\partial R}{\partial q^a} = 0, \quad \text{for } a = (k+1), \ldots, n,$$

where $R = R(q^{k+1}, \ldots, q^n, \dot{q}^{k+1}, \ldots, \dot{q}^n, \beta_1, \ldots, \beta_k)$ is the classical Routhian

$$R = L - \sum_{\alpha=1}^{k} \beta_\alpha \dot{q}^\alpha$$

and where the velocity components $\dot{q}^\alpha$, for $\alpha = 1, \ldots, k$, may be eliminated by inverting the relations

$$\beta_\alpha = \frac{\partial L}{\partial \dot{q}^\alpha}, \quad \text{for } \alpha = 1, \ldots, k.$$

Each β_α, for $\alpha = 1, \ldots, k$, is a conserved quantity.

Proof. Taking variations of the Routhian, we obtain

$$\begin{aligned}\delta R &= \frac{\partial R}{\partial q^a}\delta q^a + \frac{\partial R}{\partial \dot{q}^a}\delta \dot{q}^a + \frac{\partial R}{\partial \beta_\alpha}\delta\beta_\alpha \\ &= \frac{\partial L}{\partial q^a}\delta q^a + \frac{\partial L}{\partial \dot{q}^a}\delta \dot{q}^a + \frac{\partial L}{\partial \dot{q}^\alpha}\delta \dot{q}^\alpha - \beta_\alpha \delta \dot{q}^\alpha - \dot{q}^\alpha \delta\beta_\alpha,\end{aligned}$$

where $\alpha = 1, \ldots, k$ and $a = (k+1), \ldots, n$. Note that, by definition, $\beta_\alpha = \frac{\partial L}{\partial \dot{q}^\alpha}$, so that the third and fourth terms of the second line cancel. Moreover, one can show that

$$\dot{q}^\alpha = \frac{\partial R}{\partial \beta_\alpha}$$

by careful modification of Lemma 8.1. Hence, by comparing the coefficients of the δq^a and $\delta \dot{a}^a$ components of each line above, we see that

$$\frac{\partial R}{\partial q^a} = \frac{\partial L}{\partial q^a} \quad \text{and} \quad \frac{\partial R}{\partial \dot{q}^a} = \frac{\partial L}{\partial \dot{q}^a}.$$

Substituting these relations into the Euler–Lagrange equation, we obtain our result. □

We observe that Routh's equation represents an intermediary between the Lagrangian and Hamiltonian formalisms: it combines the advantages of Hamilton's equations for cyclic coordinates while retaining the advantages of the Lagrangian formalism for noncyclic coordinates. To illustrate this, consider the following example.

Example 8.4. Consider a planar particle subject to an inverse-power central-force field, with resulting Lagrangian

$$L = \frac{m}{2}(\dot{r}^2 + r^2\dot{\theta}^2) + \frac{k}{r^n}.$$

We observe that θ is a cyclic coordinate. Defining the conjugate momentum

$$\beta = \frac{\partial L}{\partial \dot{\theta}} = mr^2\dot{\theta},$$

we may express the Routhian as

$$R = L - \beta\dot{\theta} = \frac{m\dot{r}^2}{2} - \frac{\beta^2}{2mr^2} + \frac{k}{r^n}.$$

Routh's equations yield

$$\frac{d}{dt}\frac{\partial R}{\partial \dot{r}} - \frac{\partial R}{\partial r} = m\ddot{r} - \frac{\beta^2}{mr^3} + \frac{nk}{r^{n+1}} = 0,$$
$$\dot{\beta} = 0, \quad \beta = mr^2\dot{\theta} = \text{const.}$$

□

Exercise 8.10. Compute Routh's equation for the spherical pendulum of Exercise 7.12.

8.5 Symplectic Manifolds

Hamiltonian dynamical systems possess interesting geometric structures whose existence has given rise to the field of *symplectic topology*. Some modern references include [15], [141], and [210]. We will touch upon this vast subject in the next few pages and hopefully provide the reader with a brief glimpse into this beautiful subject.

Definition 8.8. A *symplectic form* on an even-dimensional differentiable manifold is a skew-symmetric, nondegenerate, bilinear form. A *symplectic manifold* is a pair (M, ω) where M is an even-dimensional differentiable manifold and ω is a symplectic form on M.

Example 8.5. Let us consider the simple case of $Q = \mathbb{R}^n$, with $T^*\mathbb{R}^n \cong \mathbb{R}^{2n}$, with local coordinates given by $(q^1, \ldots, q^n, p_1, \ldots, p_n)$. We define the *standard symplectic form on* $\mathbb{R}^{2n}$, denoted by the symbol ω_0, by the relation

$$\omega_0(u, v) = \langle u, Jv \rangle,$$

for all $u, v \in \mathbb{R}^{2n}$, where J is the matrix defined by

$$J = \begin{bmatrix} 0_n & I_n \\ -I_n & 0_n \end{bmatrix}. \tag{8.10}$$

To show that ω_0 is antisymmetric, consider the computation

$$\begin{aligned} \omega_0(u, v) &= u^T J v = (J^T u)^T v = -(Ju)^T v = -\langle Ju, v\rangle \\ &= -\langle v, Ju\rangle = -\omega_0(v, u), \end{aligned}$$

which demonstrates our result. □

Exercise 8.11. Show that the standard symplectic form on $\mathbb{R}^{2n}$ may be equivalently expressed as the two-form

$$\omega_0 = dq \wedge dp,$$

where $dq \wedge dp = dq^i \wedge dp_i = dq^1 \wedge dp_1 + \cdots + dq^n \wedge dp_n$.

The standard symplectic form ω_0 on $\mathbb{R}^{2n}$ plays a fundamental role in the study of symplectic manifolds, due to the following theorem.

Theorem 8.5 (Darboux). *Every symplectic manifold (M, ω) is locally diffeomorphic to $(\mathbb{R}^{2n}, \omega_0)$.*

Darboux's theorem was first proven in [215] and [291]. A more geometric proof is given in [8]. See also [10], [141], and [210].

Definition 8.9. A *Hamiltonian dynamical system* is a triple, (M, ω, H), such that (M, ω) is a symplectic manifold and $H \in C^\infty(M)$ is a smooth function, called the *Hamiltonian*, on M.

To each Hamiltonian dynamical system there is an associated *Hamiltonian vector field* which defines a dynamical system on M. We will now introduce a method for determining the associated Hamiltonian vector field given a Hamiltonian function.

Proposition 8.2. *Given a symplectic manifold (M, ω), the mapping from TM into T^*M, which sends $X \mapsto \mathbf{i}_X\omega$, defines an isomorphism between TM and T^*M.*

Exercise 8.12. Prove Proposition 8.2.

Definition 8.10. Let (M, ω) be a symplectic manifold. The *symplectic rotation* is a mapping $\mathcal{J} : T^*M \to TM$ that sends $\alpha \mapsto X$, where $X \in TM$ is the unique vector, given by Proposition 8.2, that satisfies $\mathbf{i}_X\omega = \alpha$.

Definition 8.11. Given a symplectic manifold (M, ω) and Hamiltonian $H : M \to \mathbb{R}$, we define the *Hamiltonian vector field* $X_H \in \mathfrak{X}(M)$ associated with the Hamiltonian H as the vector field

$$X_H = \mathcal{J}(dH),$$

i. e., X_H is the unique vector field, given by Proposition 8.2, for which

$$\mathbf{i}_{X_H}\omega = dH.$$

Example 8.6. Let us consider again the example of $M = \mathbb{R}^{2n}$ with the standard symplectic form ω_0. The gradient of the Hamiltonian H is given by

$$dH = \frac{\partial H}{\partial q^1}dq^1 + \cdots + \frac{\partial H}{\partial q^n}dq^n + \frac{\partial H}{\partial p_1}dp_1 + \cdots \frac{\partial H}{\partial p_n}dp_n.$$

If we let dH act on an arbitrary vector field Y, then we obtain

$$dH(Y) = \langle dH, Y\rangle.$$

Similarly, if we let the one-form $\mathbf{i}_{X_H}\omega_0$ act on the same arbitrary vector field Y, we obtain

$$\mathbf{i}_{X_H}\omega_0(Y) = \omega_0(X_H, Y) = \langle X_H, JY\rangle = -\langle JX_H, Y\rangle.$$

Hence, the condition that $\mathbf{i}_{X_H}\omega_0(Y) = dH(Y)$, for the arbitrary vector field $Y \in T(\mathbb{R}^{2n})$, coupled with the fact that $J^2 = -I_{2n}$, implies that

$$X_H = JdH,$$

which is equivalent to the vector field of Definition 8.5. □

Definition 8.12. Let (M, ω) be a symplectic manifold. A *symplectomorphism*, or *canonical transformation*, is a diffeomorphism $f : M \to M$ that preserves the symplectic form, i. e.,

$$\omega(u, v) = \omega(Df(u), Df(v)),$$

for all $u, v \in TM$.

Let us again consider $(\mathbb{R}^{2n}, \omega_0)$. Then a diffeomorphism $f : \mathbb{R}^{2n} \to \mathbb{R}^{2n}$ is a symplectomorphism if and only if

$$\begin{aligned}\langle u, Jv\rangle &= \langle Df(u), JDf(v)\rangle \\ &= u^T(Df)^T J(Df)v \\ &= \langle u, (Df)^T J(Df)(v)\rangle,\end{aligned}$$

i. e., if and only if $(Df)^T J(Df) = J$. This example motivates the following definition.

Definition 8.13. A matrix $A \in \mathbb{R}^{2n}$ is called *symplectic* if and only if

$$A^T JA = J,$$

where J is defined in (8.10).

Theorem 8.6. *Let (M, ω) be a symplectic manifold. Then the diffeomorphism $f : M \to M$ is a symplectomorphism if and only if its differential Df is a symplectic matrix.*

Definition 8.14. The *symplectic group* $\mathrm{Sp}(2n)$ is the group of $2n \times 2n$ symplectic matrices, i. e.,

$$\mathrm{Sp}(2n) = \{A \in \mathrm{GL}(2n; \mathbb{R}) : A^T J A = J\}.$$

Exercise 8.13. Show that if A, B are symplectic matrices, then so are AB, A^{-1}, and A^T.

For a given matrix $A \in \mathrm{Sp}(2n)$, it follows from the condition $A^T J A = J$ that $\det(A)^2 = 1$. In fact, one can also show that $\det(A) = 1$. This result is equivalent to Liouville's theorem, which states that all symplectomorphisms are volume preserving.

Theorem 8.7. *Let (M, ω) be a symplectic manifold and H a Hamiltonian. Then the Hamiltonian phase flow (the flow along the vector field X_H) is a one-parameter group of symplectomorphisms, i. e., the Hamiltonian phase flow preserves the symplectic structure.*

Proof. Let $\varphi : \mathbb{R} \times T^*Q \to T^*Q$ be the Hamiltonian phase flow. Recall that Hamilton's equations may be expressed by the relation $\dot{x} = J \cdot DH$, where $x = (q, p)$. The state-transition matrix therefore satisfies the matrix differential equation

$$\dot{\Phi} = J \cdot D^2 H \cdot \Phi, \tag{8.11}$$

with initial condition $\Phi(0) = I_{2n}$, along solutions to Hamilton's equations. The matrix $D^2 H$ is the symmetric, *Hessian* matrix of H, defined by

$$[D^2 H]_{ij} = \frac{\partial^2 H}{\partial x^i \partial x^j}.$$

Premultiplying each side of (8.11) by J, we obtain the relation

$$J \cdot \dot{\Phi} = -D^2 H \cdot \Phi.$$

Taking the transpose, while noting that $J^T = -J$ and $D^2 H^T = D^2 H$, we obtain

$$\dot{\Phi}^T \cdot J = \Phi^T \cdot D^2 H.$$

Now, φ is a one-parameter group of diffeomorphisms if and only if $\Phi(t) \in \mathrm{Sp}(2n)$ for all t. Initially, we have $\Phi(0) = I_{2n} \in \mathrm{Sp}(2n)$ since $I^T J I = J$. Now,

$$\begin{aligned} \frac{d}{dt}\left(\Phi^T J \Phi\right) &= \dot{\Phi}^T J \Phi + \Phi^T J \dot{\Phi} \\ &= \Phi^T D^2 H \Phi - \Phi^T D^2 H \Phi = 0. \end{aligned}$$

It follows that

$$\Phi^T(t) J \Phi(t) = \Phi^T(0) J \Phi(0) = J$$

for all time; hence φ is a one-parameter group of symplectomorphisms. □

The symplectic group $\mathrm{Sp}(2n)$ is one example of a *matrix Lie group*, which is a matrix group that possesses a manifold structure. We will discuss matrix Lie groups in more depth in Chapter 9.

Next, we will introduce the idea of an *infinitesimal symplectomorphism*. Let $A : \mathbb{R} \to \mathrm{Sp}(2n)$ be a smooth curve through $\mathrm{Sp}(2n)$ such that $A(0) = I_{2n}$. Differentiating the identity $A(t)^T JA(t) = J$ along curve A at $t = 0$, we obtain

$$\frac{d}{dt}(A^T JA)\bigg|_{t=0} = A^T J\dot{A} + \dot{A}^T JA|_{t=0} = J\dot{A} + \dot{A}^T J = 0.$$

This motivates the following definition.

Definition 8.15. The *symplectic Lie algebra* is the vector space

$$\mathfrak{sp}(2n) = \left\{A \in \mathbb{R}^{2n\times 2n} : JA + A^T J = 0\right\}.$$

Since the symplectic Lie algebra literally consists of the tangent vectors to all smooth curves through $\mathrm{Sp}(2n)$ passing through the identity I_{2n} at time $t = 0$, we may make the identification that

$$\mathfrak{sp}(2n) = T_I \mathrm{Sp}(2n),$$

i. e., the symplectic Lie algebra $\mathfrak{sp}(2n)$ is the tangent space to the symplectic group at the identity element. We will see more of this in Chapter 9.

Exercise 8.14. Write $X \in \mathrm{GL}(2n; \mathbb{R})$ as

$$X = \begin{bmatrix} A & B \\ C & D \end{bmatrix},$$

where $A, B, C, D \in \mathbb{R}^{n\times n}$. What are the conditions on the $n \times n$ matrices A, B, C, D such that the matrix $X \in \mathrm{Sp}(2n)$? Use these conditions to determine the dimension of the manifold $\mathrm{Sp}(2n)$.

Exercise 8.15. Prove that if $A \in \mathrm{Sp}(2n)$ and $\lambda \in \mathbb{C}$ is an eigenvalue of A, then $\frac{1}{\lambda}$, $\bar{\lambda}$, and $\frac{1}{\bar{\lambda}}$ are also eigenvalues of A.

Exercise 8.16. Prove that if $A \in \mathfrak{sp}(2n)$ and $\lambda \in \mathbb{C}$ is an eigenvalue of A, then $-\lambda$, $\bar{\lambda}$, and $-\bar{\lambda}$ are also eigenvalues of A.

Exercise 8.17. Prove that if $\lambda \in \mathbb{C}$ and $\xi \in \mathbb{C}^{2n}$ are an eigenvalue–eigenvector pair for the matrix $\Psi = \Phi^T \Phi$, where $\Phi \in \mathrm{Sp}(2n)$, then so are λ^{-1} and $J \cdot \xi$.

Exercise 8.18. Prove that if χ is a Lyapunov exponent of the Hamiltonian phase flow, then so is $-\chi$.

8.6 Symplectic Invariants

In this final paragraph on Hamiltonian mechanics, we will discuss the concept of *symplectic invariants*, which may be thought of as invariant quantities that exist on various subsets of the phase space. We will begin our discussion with a bit of nomenclature regarding several types of submanifolds found in symplectic manifolds.

Definition 8.16. Let (V, ω) be a symplectic vector space. The *symplectic complement* of a linear subspace $W \subset V$ is defined as the subset

$$W^{\omega} = \{v \in V : \omega(v, w) = 0, \text{ for all } w \in W\}.$$

As a consequence of this definition, we have the following proposition.

Proposition 8.3. *Let W be a subspace of the symplectic vector space (V, ω). Then*

$$\dim W + \dim W^{\omega} = \dim V, \quad W^{\omega\omega} = W.$$

Proof. Define a mapping $\iota_{\omega} : V \to V^*$ by the relation

$$\iota_{\omega}(v) \cdot w = \omega(v, w).$$

Since ω is nondegenerate, ι_{ω} is an isomorphism that identifies W^{ω} with the annihilator $W^{\perp}$ of W in V^*. Since $\dim W + \dim W^{\perp} = \dim V$, the result follows. □

Definition 8.17. Let (V, ω) be a symplectic vector space. Then the subspace $W \subset V$ is
i. *isotropic* if $W \subset W^{\omega}$;
ii. *coisotropic* if $W^{\omega} \subset W$;
iii. *symplectic* if $W \cap W^{\omega} = \{0\}$; and
iv. *Lagrangian* if $W = W^{\omega}$.

Exercise 8.19. Let $(\mathbb{R}^6, \omega_0)$ be a symplectic Euclidean space, with local coordinates $(q^1, q^2, q^3, p_1, p_2, p_3)$, with the standard symplectic form ω_0. Let e_i be the ith basis vector for $i = 1, \ldots, 6$. Identify the type of each of the subspaces $\text{span}\{e_1, e_2\}$, $\text{span}\{e_1, e_4\}$, $\text{span}\{e_1, e_2, e_3\}$, and $\text{span}\{e_1, e_2, e_3, e_5\}$.

Note that W is isotropic if and only if ω vanishes on W and W is symplectic if and only if ω, when restricted to the subspace W, is nondegenerate.

These definitions play a fundamental role in identifying types of submanifolds of symplectic manifolds. First, note that if (M, ω) is a symplectic manifold, then $(T_qM, \omega|_q)$ is a symplectic vector space. This allows for the following definition.

Definition 8.18. Let (M, ω) be a symplectic manifold and $N \subset M$ a smooth submanifold. Then N is said to be
i. *isotropic* if T_qN is an isotropic subspace of $(T_qM, \omega|_q)$;
ii. *coisotropic* if T_qN is a coisotropic subspace of $(T_qM, \omega|_q)$;

iii. *symplectic* if T_qN is a symplectic subspace of $(T_qM, \omega|_q)$; and
iv. *Lagrangian* if T_qN is a Lagrangian subspace of $(T_qM, \omega|_q)$

for all $q \in N$.

Notice that N is a symplectic submanifold of (M, ω), pursuant to the above definition, if and only if $(N, \omega|_N)$ is *itself* a symplectic manifold.

Remark. A submanifold N of a symplectic manifold (M, ω) must not necessarily fall into one of the classifications of Definition 8.18. For example, a submanifold of $(\mathbb{R}^{2n}, \omega_0)$ might have a region that is parallel to the q_1-p_2 plane and a region that is parallel to the q_1-q_2 plane.

Integral Invariants

In the classical study of mechanics, a certain form of conservation law known as *integral invariants* constituted the earliest predecessor to modern symplectic invariants.

Definition 8.19. A differential k-form α is called an *integral invariant* of the map $f : M \to M$ if

$$\int_{f\Sigma} \alpha = \int_{\Sigma} \alpha,$$

for all k-chains Σ.

Theorem 8.8. *Let (M, ω) be a symplectic manifold. Then ω is an integral invariant of any Hamiltonian phase flow on M.*

Proof. This statement follows immediately from Theorem 8.7 and the foregoing definition. □

Consider the exterior powers of the symplectic form ω on the symplectic manifold (M, ω):

$$\omega^2 = \omega \wedge \omega, \quad \omega^3 = \omega \wedge \omega \wedge \omega,$$

and so on. The preceding theorem yields an immediate corollary.

Corollary 8.1. *Let (M, ω) be a symplectic manifold. Then each of the forms $\omega, \omega^2, \ldots, \omega^n$ is an integral invariant of any Hamiltonian phase flow on M.*

The integral invariants $\omega, \omega^2, \ldots, \omega^n$ of the Hamiltonian phase flow are related to the *integral invariants of Poincaré–Cartan*, introduced in [43] and [238–240], by simple application of Stokes' theorem. The integral invariants of Poincaré–Cartan are literally the invariance of the *closed* path integral

$$\oint pdq - Hdt$$

along smooth families of closed curves contained in any *tube* of trajectories (i. e., the flow of all points of a given closed path) in the extended phase space $(M \times \mathbb{R})$. For more details regarding these integral invariants, see, for example, [8], [11], [104], and [293]. In variational calculus and optimal control, this integral invariant is also known as the *Hilbert integral invariant.*

The most efficacious way to visualize the integral invariant ω (Theorem 8.8) is in Euclidean space $(\mathbb{R}^{2n}, \omega_0)$ with the standard symplectic form $\omega_0 = dq^i \wedge dp_i$. Let $\Sigma \subset \mathbb{R}^{2n}$ be an oriented, 2-D surface (a submanifold with boundary). Then the integral

$$\int_\Sigma \omega$$

represents the *sum of the signed area projections* of Σ onto the n symplectic planes $(q_1, p_1), \ldots, (q_n, p_n)$. The *sign* of the area projection is obtained by the value of the orientation of the projected surface on the given symplectic plane. For example, consider the area element $d\Sigma \in (\mathbb{R}^4, \omega_0)$ and its *signed* symplectic area projections, as depicted in Figure 8.2. The projection of $d\Sigma$ onto the q_1-p_1 plane, which we will denote by $\pi_1(d\Sigma)$, has a positive orientation, whereas the projection of $d\Sigma$ onto the q_2-p_2 plane, denoted by $\pi_2(d\Sigma)$, has a *negative* orientation. Denoting the (unsigned, positive) areas of $\pi_1(d\Sigma)$ and $\pi_2(d\Sigma)$ by A_1 and A_2, respectively, we conclude

$$\int_{d\Sigma} \omega = A_1 - A_2.$$

The integral invariants of Poincaré–Cartan may therefore be interpreted in the following fundamental way. For any 2-D surface Σ, the sum of the signed area projections

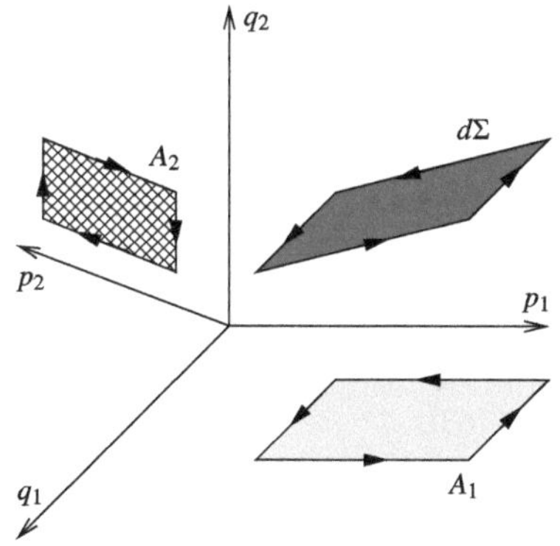

Figure 8.2: Illustration of an area element $d\Sigma$ in $(\mathbb{R}^4, \omega_0)$ and its signed area projections onto the q_1-p_1 and q_2-p_2 planes. Positive oriented projections are represented with light shading and negatively oriented projections are represented with a crosshatch lattice shading.

onto the n symplectic planes is preserved under any Hamiltonian phase flow, i. e.,

$$\int_{\varphi_t(\Sigma)} \omega = \int_{\Sigma} \omega,$$

where $\varphi_t : \mathbb{R}^{2n} \to \mathbb{R}^{2n}$ is a Hamiltonian phase flow. Now let Ω be a $2k$-dimensional submanifold of (M, ω). The integral invariants of Corollary 8.1 may be similarly interpreted as

$$\frac{1}{k!} \int_{\varphi_t(\Omega)} \omega^k = \frac{1}{k!} \int_{\Omega} \omega^k.$$

This integral represents the sum of the oriented $2k$-volume projections of the $2k$-dimensional subvolume Ω onto each symplectic "$2k$ plane." As a special case, if we select $k = n$, then we simply obtain *Liouville's theorem*, or volume conservation of the Hamiltonian phase flow.

As an added consequence of Theorem 8.8 and Corollary 8.1, the absolute values $|\omega|$ and $|\omega^k|$ are also integral invariants. These were named *Wirtinger-type* integral invariants in [205], which also introduced the following physical interpretation. We first introduce the following inequality due to [295].

Proposition 8.4 (Wirtinger). *Let $X_1, \ldots, X_{2k}$ in $\mathbb{R}^{2n}$ be a set of vectors in $\mathbb{R}^{2k}$. Then the $2k$-volume of the parallelepiped spanned by these vectors is bounded below by* Wirtinger's inequality, *i. e.,*

$$\frac{1}{k!} |\omega_0^k(X_1, \ldots, X_{2k})| \leq \mathrm{Vol}_{2k}(X_1, \ldots, X_{2k}),$$

where ω_0 is the standard symplectic form on $\mathbb{R}^{2n}$.

As an immediate consequence of Wirtinger's inequality, we have the following lower bound on the attainable $2k$-volume of Ω:

$$\frac{1}{k!} \int_{\Omega} |\omega^k| \leq \mathrm{Vol}_{2k}(\varphi_t(\Omega)),$$

where φ_t is again a Hamiltonian phase flow.

Gromov's Nonsqueezing Theorem

In our final paragraph on symplectic invariants, we will introduce the famed *Gromov's nonsqueezing theorem*, which acts as a classical analogy to Heisenberg's uncertainty principle in quantum mechanics. Gromov's nonsqueezing theorem has been applied to spacecraft uncertainty analysis in [150], [264], and [263].

We will now introduce the notion of a *symplectic capacity*, first introduced for subsets of $\mathbb{R}^{2n}$ in [82, 83] and later generalized to symplectic manifolds in [140].

Definition 8.20. Let (M,ω) be a symplectic manifold. A *symplectic capacity* is a map $(M,\omega) \mapsto c(M,\omega)$ that associates a nonnegative number or ∞ to every symplectic manifold such that

i. **[Monotonicity]:**

$$c(M,\omega) \le c(N,\tau),$$

if there exists a symplectic embedding $(M,\omega) \to (N,\tau)$;

ii. **[Conformality]:**

$$c(M,a\omega) = |a|c(M,\omega),$$

for all $a \in \mathbb{R}, a \neq 0$; and

iii. **[Nontriviality]:**

$$c(B(1),\omega_0) = \pi = c(Z(1),\omega_0),$$

where $B(1)$ and $Z(1)$ are the open unit ball and the open symplectic cylinder in the standard symplectic space $(\mathbb{R}^{2n},\omega_0)$, respectively.

It is important to note that $Z(r)$ is considered a *symplectic* cylinder, i. e., it is the direct product of a circle on a given symplectic plane (q_1,p_1) with $\mathbb{R}^{2n-2}$, i. e.,

$$Z(r) = \{(q,p) \in \mathbb{R}^{2n} : q_1^2 + p_1^2 < r\}.$$

Exercise 8.20. If c is a symplectic capacity, show that $c(B(r)) = \pi r^2$ and $c(Z(r)) = \pi r^2$.

Given the existence of a capacity (a difficult thing to prove), one immediately obtains the following result, due to [109].

Theorem 8.9 (Gromov's Nonsqueezing Theorem). *If there exists a symplectic embedding $\varphi : B(r) \to Z(R)$, then $r \le R$.*

Proof. The existence of a symplectic capacity is proven in [140]. Let c be a symplectic capacity. Using the result of Exercise 8.20 and the fact that φ is a symplectic embedding, we obtain the inequality

$$\pi r^2 = c(B(r)) \le c(Z(R)) = \pi R^2,$$

thereby proving the result. □

The statement of Gromov's nonsqueezing theorem is illustrated in Figure 8.3.

The hallmark symplectic capacity of symplectic topology is the *Gromov width*, which appeared in [109].

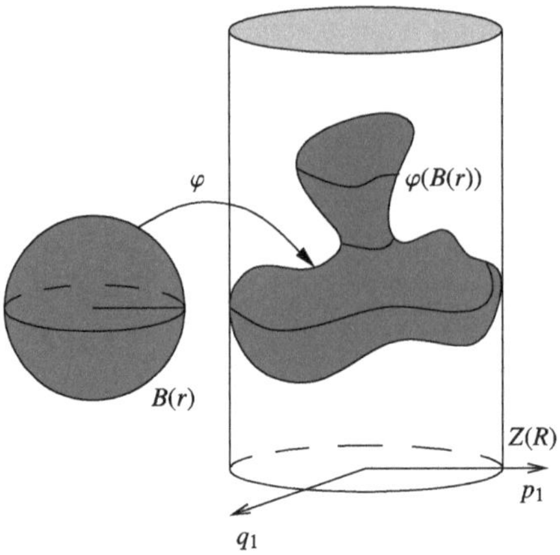

Figure 8.3: Gromov's nonsqueezing theorem. Illustration depicts a symplectic embedding $\varphi : B(r) \to Z(R)$, $r \leq R$, in the standard symplectic space $(\mathbb{R}^{2n}, \omega_0)$.

Definition 8.21. The *Gromov width*, $D(M,\omega)$, of the symplectic manifold (M,ω) is the number

$$D(M,\omega) = \sup\{\pi r^2 : \text{there exists a } \varphi \in \mathcal{S}((B(r),\omega_0);(M,\omega))\},$$

where $\mathcal{S}((M,\omega);(N,\tau))$ is the set of symplectic embeddings from (M,ω) into (N,τ).

The Gromov width is always well defined. Due to Darboux's theorem, there always exists a symplectic embedding $\varphi(B(\varepsilon),\omega_0) \to (M,\omega)$, for ε small enough, so that the set of such embeddings that appears in the preceding definition is nonempty. Our next theorem shows that the Gromov width is a symplectic capacity and that it is also the *smallest* symplectic capacity.

Theorem 8.10. *The Gromov width $D(M,\omega)$ is a symplectic capacity. Moreover,*

$$D(M,\omega) \leq c(M,\omega)$$

for every symplectic capacity c.

Gromov's nonsqueezing theorem leads to unintuitive consequences: every $2n$-dimensional set has a certain *symplectic width*, so that it cannot be squeezed in such a way that the area of its shadow, as projected onto any of the various symplectic planes, should ever be less than that width. This is analogous to the situation in quantum mechanics, where Heisenberg's uncertainty principle (see [127]) places a lower bound on the projection of a particle's uncertainty distribution onto any of the symplectic (coordinate-momentum) planes.

8.7 Application: Optimal Control and Pontryagin's Principle

In this section, we apply the results of Hamiltonian dynamical systems theory and the results from the calculus of variations, discussed in Chapter 7, to the optimal control problem, which we now state. For additional background on optimal control, see, for example, [4], [32], [37], [158], [177], and [225].

Definition 8.22. Given an n-dimensional manifold Q, a k-dimensional, simple connected subset $\Omega \subset \mathbb{R}^k$, two functions $f, g : Q \times \Omega \to \mathbb{R}$, and two endpoints $q_0, q_T \in Q$, the associated *optimal control problem* is to determine the path $u : [0,T] \to \Omega$ that minimizes the functional

$$\mathcal{I} = \int_0^T g(q,u)dt \tag{8.12}$$

subject to the control feedback law $\dot{q} = f(q,u)$. The set Ω is referred to as the *set of admissible controls.*

We now state a fundamental theorem in the field of optimal control theory, known as *Pontryagin's maximum principle*, which provides a set of necessary conditions for an optimal control in terms of a certain Hamiltonian function, as introduced by Pontryagin [244], Pontryagin et al. [245]. See also [24] and [277, 278] for a geometric discussion.

Theorem 8.11 (Pontryagin's Maximum Principle). *Given an optimal control problem, define the* associated Hamiltonian $H : T^*Q \times \Omega \to \mathbb{R}$ *by the relation*

$$H(q,p,u) = \langle p, f(q,u)\rangle - g(q,u). \tag{8.13}$$

Then a necessary condition for a particular control $u^* : [0,T] \to \Omega$ *to be optimal is that*

$$H(q(t),p(t),u^*(t)) = \max_{u\in\Omega} H(q(t),p(t),u) \tag{8.14}$$

for all $t \in [0,T]$.

Proof. To enforce the constraint $\dot{q} = f(q,u)$, we may append Lagrange multiplier functions $p : [0,T] \to T^*Q$ to the integrand of (8.12), thereby obtaining the action

$$\mathcal{I}' = \int_0^T \left(g(x,u) + \langle p, \dot{q} - f(q,u)\rangle\right) dt,$$

where the multipliers $p(t)$ are selected so that the variations are independent. However, given definition (8.13), we may rewrite this equation as

$$\mathcal{I}' = \int_0^T \left(\langle p,\dot{q}\rangle - H(q,p,u)\right) dt. \tag{8.15}$$

Taking variations and integrating by parts, we obtain

$$\delta\mathcal{I}' = \int_0^T \left(\left\langle \delta p, \dot{q} - \frac{\partial H}{\partial p}\right\rangle - \left\langle \dot{p} + \frac{\partial H}{\partial q}, \delta q\right\rangle - \left\langle \frac{\partial H}{\partial u}, \delta u\right\rangle\right) dt = 0.$$

However, since the variations δq, δp, and δu are independent, we arrive at the following necessary conditions for an optimal control:

$$\dot{q} = \frac{\partial H}{\partial p}, \quad \dot{p} = -\frac{\partial H}{\partial q}, \quad \text{and} \quad \left\langle \frac{\partial H}{\partial u}, \delta u \right\rangle = 0. \tag{8.16}$$

The first two conditions imply that solutions to the optimal control problem constitute a Hamiltonian dynamical system. The third condition must be considered separately for two possible cases, one in which, at a given time $t \in [0, T]$, the optimal control $u(t) \in \Omega^o = \Omega \setminus \partial\Omega$ lies in the *interior* of the set Ω and the second in which the optimal control lies on the boundary $\partial\Omega$ of the set of admissible controls.

1. If, for a given $t \in [0, T]$, the optimal control satisfies $u(t) \in \Omega^o$, then δu is unconstrained, and therefore one requires the condition

$$\frac{\partial H}{\partial u} = 0. \tag{8.17}$$

2. If, on the other hand, for a given $t \in [0, T]$, the optimal control satisfies $u(t) \in \partial\Omega$, then one enforces the condition $u \in \Omega$ by requiring that $\delta u \in T_{u(t)}\Omega$. In this case, let us suppose that we may represent the boundary $\partial\Omega$ as the level set of a function $h(u)$, so that $u \in \partial\Omega$ if and only if $h(u) = 0$. Then the condition $\langle H_u, \delta u\rangle = 0$ implies that H_u must be perpendicular to the tangent space $T_{u(t)}\Omega$, which is equivalent to requiring that

$$\frac{\partial H}{\partial u} = \lambda \frac{\partial h}{\partial u}$$

for some Lagrange multiplier λ.

But conditions 1 and 2 precisely constitute the procedure for locating extreme values inside a bounded domain (check for critical points of the function in the interior and then extreme values, via Lagrange multipliers, on the boundary). Thus, the optimal control $u^*(t)$ extremizes the Hamiltonian function. Since our goal is to minimize the functional (8.15), we therefore seek a *maximum* value of H, thereby obtaining the necessary condition (8.14) and completing the proof. □

The solutions to optimal control problems therefore constitute a Hamiltonian dynamical system due to the equations of motion (8.16). Also, requirement (8.17) is not as general as the requirement in the maximum principle (8.14) as it fails to take into account the optimal control condition when the control lies on the boundary of the admissible control set.

Several notes are in order. First and foremost, the maximum principle provides necessary conditions for the optimal control, but those conditions are not necessarily sufficient; for counterexamples, see, for example, [268]. *Singular optimal controls* are ones that cannot be determined by the maximum principle alone. In these cases, one requires higher-order conditions; see, for example, [170]. There are also cases where the optimal control is determined by the constraints alone. These cases are referred to as *abnormal*; for a discussion and examples on abnormal controls, see, for example, [24]. Even in the nonsingular cases for which the maximum principle determines the optimal control, one must still solve a boundary value problem to determine the optimal controls. The solution to such boundary value problems is outside the scope of this work; we instead refer the interested reader to, for example, [14] and [22].

Example 8.7 (Minimum-Fuel Orbital Rendezvous). As an example, we consider a minimum-fuel orbital rendezvous problem, for which one wishes to navigate a spacecraft from its current orbit to a target orbit in a fixed amount of time. For boundary conditions, we specify the initial position and velocity of the spacecraft as well as the final position and velocity to be reached at the final time. Such a problem is referred to as a *rendezvous* problem as one is literally trying to match a target spacecraft's position and velocity at a given moment in time. Related optimal control problems in the field of spacecraft navigation include optimal orbit transfer and intercept problems. The optimal orbit transfer problem entails reaching a target orbit in an unspecified amount of time, i. e., the precise point on the target orbit no longer matters, as long as you reach the target orbit. The intercept problem specifies a fixed transfer time and a final position, but the final velocity is left unspecified: the spacecraft merely needs to intercept a target spacecraft with no regard for the difference between their velocities at the point of intercept. Intercept problems would be relevant to, for example, missile defense. Seminal work in the orbital transfer problem was carried out in [142], in which it was conjectured that the minimum-fuel transfer orbit between two coplanar circular orbits was an ellipse inscribed between the two circular orbits, with periapsis and apoapsis radii equal to the radii of the initial and final circular orbit, as shown in Figure 8.4. The maneuver, known today as the *Hohmann transfer*, is performed using two impulsive engine burns: the first burn increases the velocity of the spacecraft so that it is at the periapsis of the elliptical transfer orbit; the second burn, which is performed when the spacecraft reaches the apoapsis of the transfer orbit, again

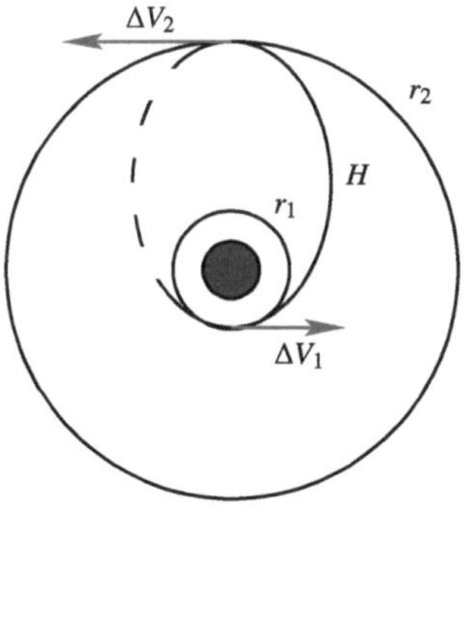

Figure 8.4: Illustration of a Hohmann transfer: elliptic transfer orbit H connects coplanar circular orbits with radii r_1 and r_2, using two impulsive engine burns ΔV_1 and ΔV_2.

increases the spacecraft's velocity so that it is on a circular orbit at the larger radius. Hohmann's conjecture was later proven in [18] and [247]. The theory of minimum-fuel transfer orbits for variable-thrust rockets was further developed by Lawden [175, 176]. For more recent developments, see also [57].

In this example, however, we will focus on the related rendezvous problem between two coplanar circular orbits, in which one further specifies the final time at which the point in the target orbit is reached. This problem was analyzed using Pontryagin's principle in [246].

The control system for a variable-thrust rocket problem is given by

$$\dot{r} = v, \tag{8.18}$$

$$\dot{v} = g(r) + \frac{v_e u}{m}\hat{v}, \tag{8.19}$$

$$\dot{m} = -u, \tag{8.20}$$

where $r \in \mathbb{R}^3$ is the position of the spacecraft, $v \in T_r\mathbb{R}^3 \cong \mathbb{R}^3$ its velocity, $m \in \mathbb{R}^+$ its mass, $g : \mathbb{R}^3 \to T_r\mathbb{R}^3$ the gravitational acceleration, $v_e \in \mathbb{R}$ the (constant) exhaust speed, and $u \in \mathbb{R}^+$ the control. In this problem, we assume that the direction of the rocket nozzle is antiparallel to the spacecraft's velocity; otherwise, one must replace the unit vector $\hat{v}$ with a control direction. The state space for the problem is given by $Q = T\mathbb{R}^3 \times \mathbb{R}^+$, with local coordinates $(r, v, m) = (x, y, z, \dot{x}, \dot{y}, \dot{z}, m)$, and the control vector field on Q may be written as

$$X = \frac{v_e}{\sqrt{\dot{x}^2 + \dot{y}^2 + \dot{z}^2}}\left(\dot{x}\frac{\partial}{\partial \dot{x}} + \dot{y}\frac{\partial}{\partial \dot{y}} + \dot{z}\frac{\partial}{\partial \dot{z}}\right) - \frac{\partial}{\partial m}.$$

Note that (8.19) at first pass seems inconsistent with Newton's second law of motion, given by equation (2.27). One would expect to encounter $-v$ in the place of $+v_e\hat{v}$ on the right-hand side of (8.19), a vector in the opposite direction with magnitude $|v|$ instead of magnitude v_e. The reason for the discrepancy, however, is that Newton's second law in the form (2.27) does not take into account momentum flux across the boundary of the rocket. Equation (8.19), on the other hand, is derived from Newton's second law for fluid mechanics,

$$\frac{d}{dt}\left(\iiint_{\Omega} \mathbf{V}\rho\, dV\right) + \iint_{\partial\Omega} \mathbf{V}\rho(\mathbf{V}_{\text{rel}} \cdot \hat{\mathbf{n}})\, dS = \mathbf{F},$$

where Ω is a control volume, $\mathbf{V}$ and ρ respectively the velocity and density of the fluid, and $\mathbf{V}_{\text{rel}}$ the *relative velocity* of a fluid particle passing through the surface of the body, i. e., the velocity of the fluid particle relative to the velocity of the body. See [5] for the derivation of this equation as well as applications to supersonic and hypersonic compressible flow. The result, as applied to a rocket with exhaust velocity v_e, is known as the *rocket equation*, which may be expressed in the form (8.19). This equation is attributed to [284], though it was recently discovered that the equation was developed independently in the early work by Moore [213].

Exercise 8.21. Show that, in the absence of gravity, the total change of velocity achieved by an engine burn is related to the initial and final mass of the rocket, m_0 and m_f, respectively, by the equation

$$\Delta v = v_e \ln \frac{m_0}{m_f}.$$

You may assume motion in one dimension.

To minimize the fuel spent during the orbit transfer, we minimize the cost function

$$\mathcal{I} = \int_0^T u(t)\, dt.$$

The corresponding Hamiltonian (8.13) is therefore given by

$$H = \mu \cdot v + \nu \cdot \left(g + \frac{v_e u}{m}\hat{v}\right) - (\lambda + 1)u(t),$$

where the multipliers (costates) are given by $p = (\mu, \nu, \lambda) \in T^*Q$. The costate equations (the $\dot{p}$ equation in (8.16)) are given by

$$\begin{aligned}
\dot{\mu} &= -\nu \cdot \frac{\partial g}{\partial r},\\
\dot{\nu} &= -\mu - \frac{v_e u}{m}\nu \cdot \frac{\partial \hat{v}}{\partial v},\\
\dot{\lambda} &= \frac{v_e u}{m^2}\nu \cdot \hat{v}.
\end{aligned}$$

Despite the complexity of the costate equations (whose initial conditions are unknown and must be selected so that the boundary conditions on the state equations are met), the fact that the Hamiltonian is linear in the control variable immediately leads to an interesting conclusion. Since

$$\frac{\partial H}{\partial u} = -(\lambda + 1),$$

the optimal control must be given by toggling the throttle between maximum throttle, when $\lambda < -1$, and zero throttle, when $\lambda > -1$. Such optimal controls commonly arise in various contexts, such as minimum-time problems and problems for which the Hamiltonian is linear in the controls, and are known as *bang–bang controls.* □

Exercise 8.22. The following example is due to [272]. Consider the control system

$$\begin{aligned}
\dot{x}_1 &= u_1,\\
\dot{x}_2 &= u_2,\\
\dot{x}_3 &= x_1 u_2
\end{aligned}$$

and the cost function

$$\mathcal{I} = \int_0^T \frac{1}{2}(u_1^2 + u_2^2)dt.$$

Use (8.16) to show that a necessary condition for a control to be optimal is that it must take the form

$$u_1(t) = a\cos(p_3 t) - b\sin(p_3 t),$$
$$u_2(t) = a\sin(p_3 t) + b\cos(p_3 t),$$

where p_3 is the costate of x_3 and is a constant.

8.8 Application: Symplectic Probability Propagation

In this final application of the chapter, we will consider probability propagation in Hamiltonian dynamical systems. This topic arises naturally in the field of orbit uncertainty and spacecraft navigation; see, for example, [264]. The general definition of a probability density function is as follows.

Definition 8.23. A *probability density function* on an n-dimensional manifold Q is a smooth function $\rho : Q \to \mathbb{R}$ that satisfies the properties

$$\rho(q) \geq 0 \text{ for all } q \in Q \quad \text{and} \quad \int_Q \rho(q)dq^1 \wedge \cdots \wedge dq^n = 1.$$

We will denote the class of probability distributions by $\mathcal{P}(Q)$.

However, we will restrict our attention to the special case of time-dependent probability density functions on the cotangent bundle, i. e., for each $t \in \mathbb{R}$, we will consider $\rho_t \in \mathcal{P}(T^*Q)$ such that $\rho_t(q,p) \geq 0$ and $\int_{T^*Q} \rho_t(q,p)\omega^n = 1$, where $\omega = dq \wedge dp$ is the standard symplectic form on T^*Q and ω^n the associated volume form.

In practice, let us suppose that we make an initial measurement of our system at time $t = 0$, which yields an initial probability distribution $\rho_0 \in \mathcal{P}(T^*Q)$ on the cotangent bundle. For concreteness, this measurement might be a measurement of the position and momentum of a spacecraft. Now, the time-dependent probability distribution must evolve so that individual trajectories of the Hamiltonian dynamical system form the characteristic curves of the evolution partial differential equation of the probability distribution. This motivates the following definition.

Definition 8.24. Given a Hamiltonian $H : T^*Q \to Q$ on a manifold Q and an initial probability density function $\rho_0 \in \mathcal{A}^{2n}(T^*Q)$ on T^*Q, the *symplectically propagated probability distribution* is the solution to the *collisionless Boltzmann equation* ([33])

$$\frac{D\rho}{Dt} = \frac{\partial \rho}{\partial t} + \{\rho, H\} = 0, \quad \rho(0) = \rho_0, \tag{8.21}$$

where $\{\rho, H\}$ is the *Poisson bracket* of the functions ρ and H.

The implication of the collisionless Boltzmann equation on galactic dynamics is discussed in [23]. Due to the definition of the Poisson bracket (8.9), the characteristic curves of the partial differential equation (8.21) are precisely the individual curves in the phase flow of the Hamiltonian system. Since Hamiltonian dynamical systems also preserve the volume form ω^n, it follows that the probability of finding the system state in $\varphi_t(\Omega)$, given some initial $2n$-dimensional subset $\Omega \subset T^*Q$, i. e.,

$$P(\Omega) = \int\limits_{\varphi_t(\Omega)} \rho(q,p)\omega^n,$$

is a constant.

Bayesian Update

Now, suppose we make an initial measurement of our system, which results in a probability distribution $\rho_A \in \mathcal{P}(T^*Q)$. This probability distribution is evolved according to the partial differential equation (8.21), so that it may be thought of as a function of time. At a subsequent time, a second measurement of the system is made, resulting in a second probability distribution $\rho_B \in \mathcal{P}(T^*Q)$, which may also be evolved. Each probability distribution thus contains separate information as to the system state. This information may be combined following Bayes' theorem, yielding a new probability distribution,

$$\rho(q,p) = \frac{\rho_A(q,p)\rho_B(q,p)}{\int_{T^*Q} \rho_A(\overline{q},\overline{p})\rho_B(\overline{q},\overline{p})\overline{\omega}^N},$$

where $\overline{q}$ and $\overline{p}$ are the variables of integration and $\overline{\omega} = d\overline{q} \wedge d\overline{p}$. The densities ρ_A and ρ_B, in the preceding formula, must be evolved to a common time, typically the time at measurement B.

9 Lie Groups and Rigid-Body Mechanics

Many mechanical systems may be described by configuration manifolds that themselves have an additional structure—that of a *group*. Primary among such systems is the free rigid body. As we saw in our studies on Lagrangian and Hamiltonian mechanics, it is important not only to understand the *classical* equations of motion but to simultaneously foster an appreciation for the underlying geometric structure on which these systems evolve. Such an appreciation is a prerequisite for the construction of highly effective numerical integrators that preserve these underlying structures, a topic of much recent interest. These advantages are only compounded in the case of mechanics on Lie groups. We begin this chapter with a review of the basic formalism and results for Lie groups. In our discussion on rigid-body mechanics, we introduce *Euler angles* as local coordinates on the Lie group SO(3), derive *Euler's equation* for rigid-body mechanics, and demonstrate the advantages of kinematic Lie group integrators over their classical counterparts.

9.1 Lie Groups and Their Lie Algebras

We begin by laying out a few of the fundamental definitions that we will use later on in our discussion of the geometry of rigid-body mechanics. For background on group theory and Lie groups, see, for example, [13], [134], and [289]. For a discussion of Lie groups and mechanics, see [24], [37], [144, 145], and [196]. For more information on symmetry and reduction theory, see also [47], [161], and [198, 199].

The topic of numerical integration of dynamical systems on Lie groups has also been a topic of recent research; see, for example, the work of [60], [61], and [217–219]. Lie group integrators have also found interesting applications to astrodynamical systems; see [179, 180].

Definition 9.1. A *Lie group* is a smooth manifold G that is also a group for which the operations of multiplication and inverse are smooth.

We recall the basic definition of a *group* as follows.

Definition 9.2. A *group* is a set G with a binary operation $\star : G \times G \to G$ that satisfies the following properties:
1. $a \star (b \star c) = (a \star b) \star c$, for all $a, b, c \in G$ (associativity);
2. there exists an element $e \in G$, known as the *identity element*, such that $a \star e = e \star a = a$ for all $a \in G$;
3. for all $a \in G$ there exists a unique element $a^{-1} \in G$, known as the *inverse element of* a, such that $a \star a^{-1} = a^{-1} \star a = e$.

https://doi.org/10.1515/9783110597806-009

Of particular interest in our studies are objects called *matrix Lie groups*.

Definition 9.3. A *matrix Lie group* is a set of invertible $n \times n$ matrices that is closed under matrix multiplication and is also a submanifold of $\mathbb{R}^{n\times n}$.

Every Lie group has an associated *Lie algebra*. We begin with the general definition of a Lie algebra.

Definition 9.4. A *Lie algebra* is a real vector space V endowed with a binary operation $[\cdot,\cdot] : V \times V \to V$, called the *bracket*, that satisfies the following axioms:
i. $[\xi,\eta] = -[\eta,\xi]$, for all $\xi,\eta \in V$ (anticommutativity); and
ii. $[[\xi,\eta],\zeta] + [[\zeta,\xi],\eta] + [[\eta,\zeta],\xi] = 0$, for all $\xi,\eta,\zeta \in V$ (Jacobi identity).

Definition 9.5. For any pair of matrices $A, B \in \mathbb{R}^{n\times n}$, we define the *matrix Lie bracket* $[\cdot,\cdot] : \mathbb{R}^{n\times n} \times \mathbb{R}^{n\times n} \to \mathbb{R}^{n\times n}$ by the relation

$$[A,B] = AB - BA$$

for all $A, B \in \mathbb{R}^{n\times n}$.

Exercise 9.1. Prove that for all $n \times n$ matrices A, B
(a) $[A,B] = -[B,A]$ and
(b) $[[A,B],C] + [[C,A],B] + [[B,C],A] = 0$,

thereby demonstrating that the set $\mathbb{R}^{n\times n}$ of real matrices with the matrix Lie bracket constitutes a Lie algebra.

For an arbitrary matrix Lie group G we wish to show that the tangent space to manifold G at identity element I forms a Lie algebra, with brackets given by the matrix Lie bracket. This follows immediately as a result of the preceding exercise, as long as the tangent space to the identity is *closed* under the bracket operation. We will denote this Lie algebra by $\mathfrak{g} = T_I G$. To prove that $A, B \in \mathfrak{g}$ implies $[A,B] \in \mathfrak{g}$, we will require the following lemma.

Lemma 9.1. *Let G be a matrix Lie group, $R \in G$, and $B \in T_I G$. Then $RBR^{-1} \in T_I G$.*

Proof. Let $\gamma : \mathbb{R} \to G$ be a smooth curve in G such that $\gamma(0) = I$ and $\dot{\gamma}(0) = B$. Such a curve exists by the definition of a tangent vector to a manifold at a point. Next, let

$$S(t) = R\gamma(t)R^{-1} \in G$$

for all t. Since $S(0) = I$, it follows that $\dot{S}(0) \in T_I G$. But $\dot{S}(0) = RBR^{-1}$, thereby completing the proof. □

The next proposition demonstrates closure of the tangent space to the identity under the binary operation of the matrix Lie bracket, thereby completing the proof that this tangent space is a Lie algebra.

Proposition 9.1. *Let G be a matrix Lie group, with $A, B \in T_I G$. Then $[A, B] = AB - BA \in T_I G$.*

Proof. Let $\gamma : \mathbb{R} \to G$ be a curve in G such that $\gamma(0) = I$ and $\dot{\gamma}(0) = A$. By Lemma 9.1 the matrices

$$S(t) = \gamma(t) B \gamma(t)^{-1} \in T_I G,$$

for all t. Hence $\dot{S}(t) \in T_I G$. In particular, we have

$$\dot{S}(0) = AB - BA \in T_I G,$$

thereby completing the proof. □

The preceding proposition proves that the tangent space to the identity of a Lie group, $\mathfrak{g} = T_I G$, constitutes a Lie algebra. We will now review a number of examples.

Example 9.1. The *general linear group* $\mathrm{GL}(n; \mathbb{R})$ is the group of all $n \times n$ real, invertible matrices. Its Lie algebra is given by $\mathfrak{gl}(n; \mathbb{R}) = \mathbb{R}^{n \times n}$. □

Example 9.2. The *orthogonal group* is the set

$$O(n) = \{X \in \mathrm{GL}(n; \mathbb{R}) : XX^T = I\},$$

which consists of orthogonal matrices. □

Note that $X \in O(n)$ implies that $\det(X) = \pm 1$. Hence, the group $O(n)$ has *two* connected components. Of particular importance will be the connected component of orthogonal matrices with positive determinant.

Example 9.3. The *special orthogonal group* is the set

$$\mathrm{SO}(n) = \{X \in O(n) : \det X = +1\}.$$

Hence, $\mathrm{SO}(n)$ is the connected component of $O(n)$ with matrices of determinant $+1$. Note that $I \in \mathrm{SO}(n)$, so that $T_I O(n) = T_I SO(n)$.

We will now show that $\mathfrak{so}(n) = T_I SO(n)$, the Lie algebra to the special orthogonal group, consists of antisymmetric matrices. To show this, let $A : \mathbb{R} \to \mathrm{SO}(n)$ be a smooth curve through $\mathrm{SO}(n)$ such that $A(0) = I$. Now, $AA^T = I$ for all $t \in \mathbb{R}$; therefore

$$\frac{d}{dt}(AA^T) = A\dot{A}^T + \dot{A}A^T = 0$$

for all t. In particular, at $t = 0$, we have the condition that

$$\dot{A}^T(0) + \dot{A}(0) = 0.$$

By definition, $\dot{A}(0) \in \mathfrak{so}(n)$, so that the condition for being a tangent vector to the identity of SO(n) is precisely

$$\mathfrak{so}(n) = \{A \in \mathbb{R}^{n\times n} : A^T + A = 0\}, \tag{9.1}$$

i. e., the tangent space to the identity of SO(n) consists of $n \times n$ antisymmetric matrices. Since the dimension of $\mathfrak{so}(n)$ is $\frac{n^2-n}{2}$, we conclude that the Lie group SO(n) must be an $\frac{n^2-n}{2}$-dimensional manifold. □

Example 9.4. Consider the *symplectic group*

$$\mathrm{Sp}(2n) = \{X \in \mathrm{GL}(2n; \mathbb{R}) : XJX^T = J\},$$

where J is the matrix defined by (8.10). Its associated Lie algebra is given by

$$\mathfrak{sp}(2n) = \{X \in \mathbb{R}^{2n\times 2n} : JX^T + XJ = 0\}.$$

For a proof, see the discussion leading up to Definition 8.15. □

Example 9.5. Consider the *special Euclidean group*

$$\mathrm{SE}(3) = \left\{X \in \mathrm{GL}(4; \mathbb{R}) : X = \begin{bmatrix} R & v \\ 0 & 1 \end{bmatrix}, R \in \mathrm{SO}(3), v \in \mathbb{R}^3\right\}.$$

This group corresponds to rigid motions in $\mathbb{R}^3$, i. e., translations coupled with rigid rotations. This Lie group is very useful when one is studying the dynamics of rigid bodies in space. Its corresponding Lie algebra is 6-D and consists of matrices of the form

$$\mathfrak{se}(3) = \left\{X \in \mathbb{R}^{4\times 4} : X = \begin{bmatrix} A & v \\ 0 & 0 \end{bmatrix}, A \in \mathfrak{so}(3), v \in \mathbb{R}^3\right\}.$$

A basis for the vector space $\mathfrak{se}(3)$ is therefore given by

$$\begin{bmatrix} 0 & 0 & 0 & 0 \\ 0 & 0 & -1 & 0 \\ 0 & 1 & 0 & 0 \\ 0 & 0 & 0 & 0 \end{bmatrix}, \quad \begin{bmatrix} 0 & 0 & 1 & 0 \\ 0 & 0 & 0 & 0 \\ -1 & 0 & 0 & 0 \\ 0 & 0 & 0 & 0 \end{bmatrix}, \quad \begin{bmatrix} 0 & -1 & 0 & 0 \\ 1 & 0 & 0 & 0 \\ 0 & 0 & 0 & 0 \\ 0 & 0 & 0 & 0 \end{bmatrix},$$

$$\begin{bmatrix} 0 & 0 & 0 & 1 \\ 0 & 0 & 0 & 0 \\ 0 & 0 & 0 & 0 \\ 0 & 0 & 0 & 0 \end{bmatrix}, \quad \begin{bmatrix} 0 & 0 & 0 & 0 \\ 0 & 0 & 0 & 1 \\ 0 & 0 & 0 & 0 \\ 0 & 0 & 0 & 0 \end{bmatrix}, \quad \begin{bmatrix} 0 & 0 & 0 & 0 \\ 0 & 0 & 0 & 0 \\ 0 & 0 & 0 & 1 \\ 0 & 0 & 0 & 0 \end{bmatrix}.$$ □

We will now return our focus to the special orthogonal group, restricting our attention to three dimensions.

Example 9.6. Consider the following one-parameter family of rotation matrices in SO(3):

$$R_1(t) = \begin{bmatrix} 1 & 0 & 0 \\ 0 & \cos t & -\sin t \\ 0 & \sin t & \cos t \end{bmatrix},$$

$$R_2(t) = \begin{bmatrix} \cos t & 0 & \sin t \\ 0 & 1 & 0 \\ -\sin t & 0 & \cos t \end{bmatrix},$$

$$R_3(t) = \begin{bmatrix} \cos t & -\sin t & 0 \\ \sin t & \cos t & 0 \\ 0 & 0 & 1 \end{bmatrix}.$$

Clearly, each of these matrices represents an orthogonal matrix with positive determinant for all t. Moreover, $R_i(0) = I$ for $i = 1, 2, 3$. Hence $\dot{R}_i(0) \in \mathfrak{so}(3)$. Computing these derivatives and evaluating at $t = 0$, we obtain

$$e_1 = \begin{bmatrix} 0 & 0 & 0 \\ 0 & 0 & -1 \\ 0 & 1 & 0 \end{bmatrix}, \quad e_2 = \begin{bmatrix} 0 & 0 & 1 \\ 0 & 0 & 0 \\ -1 & 0 & 0 \end{bmatrix}, \quad e_3 = \begin{bmatrix} 0 & -1 & 0 \\ 1 & 0 & 0 \\ 0 & 0 & 0 \end{bmatrix},$$

where $e_i = \dot{R}_i(0)$ for $i = 1, 2, 3$. As it turns out, the matrices e_1, e_2, e_3 form a basis for $\mathfrak{so}(3)$, which is the set of skew-symmetric matrices. Hence, the curves $R_1(t)$, $R_2(t)$, and $R_3(t)$, which represent rotations about the x-, y-, and z-axes, respectively, are curves through SO(3) whose tangent vectors at I form a basis for the associated Lie algebra. □

Exercise 9.2. Show that

$$R_1(t) = \exp(e_1 t), \quad R_2(t) = \exp(e_2 t), \quad R_3(t) = \exp(e_3 t),$$

where exp is the matrix exponential.

The preceding exercise is an example of a more general fact, which we will state in a theorem following a simple lemma.

Lemma 9.2. *Every matrix one-parameter subgroup $\gamma : \mathbb{R} \to G$, where G is a matrix Lie group, is of the form $t \mapsto \exp(tA)$, where $A = \dot{\gamma}(0)$.*

Proof. Let $\gamma : \mathbb{R} \to G$ be a matrix one-parameter subgroup, so that

$$\gamma(t + s) = \gamma(t)\gamma(s)$$

for all $s, t \in \mathbb{R}$. It follows that

$$\frac{d}{ds}\gamma(t + s) = \gamma(t)\frac{d\gamma(s)}{ds}.$$

When we evaluate at $s = 0$, we obtain

$$\frac{d\gamma(t)}{dt} = \gamma(t)A,$$

where $A = \dot{\gamma}(0)$. Since $\gamma(0) = I$, it follows from Proposition 1.2 that $\gamma(t) = \exp(tA)$. □

Theorem 9.1. *Let G be a matrix Lie group and $\mathfrak{g} = T_I G$ its Lie algebra. If $A \in \mathfrak{g}$, then $\exp(At) \in G$ for all t. In fact, $\gamma(t) = \exp(At)$ is a curve through G with the properties that $\gamma(0) = I$ and $\dot{\gamma}(0) = A$.*

Proof. For all $A \in \mathfrak{g}$ there exists a one-parameter subgroup $\gamma : \mathbb{R} \to G$ such that $\gamma(0) = I$ and $\dot{\gamma}(0) = A$. By Lemma 9.2, γ must take the form $\gamma(t) = \exp(At)$. □

9.2 Left Translations and Adjoints

Translation maps and adjoints play a particularly important role in the mathematical description of mechanical systems with symmetry. In systems with symmetry, the tangent space at any point can be identified with the Lie algebra via a natural isomorphism, obtained by taking the Jacobian of a left-translation map. In this paragraph, we outline some of these basic definitions.

Definition 9.6. For a Lie group G and fixed element $g \in G$, the *left-translation map*, $L_g : G \to G$, is defined by $L_g(h) = gh$ for all $h \in G$.

Let $e \in G$ be the identity element of the Lie group G and let $g \in G$ be a fixed element. Since $L_g e = g$, the Jacobian $d(L_g)_e$ is a vector-space isomorphism $d(L_g)_e : T_e G \to T_g G$. See Figure 9.1 for an illustration of this. Since $L_{g^{-1}}(L_g h) = g^{-1}(gh) = h$, it follows that $(L_g)^{-1} = L_{g^{-1}}$. Hence the mapping

$$d(L_{g^{-1}})_g : T_g G \to T_e G \cong \mathfrak{g}$$

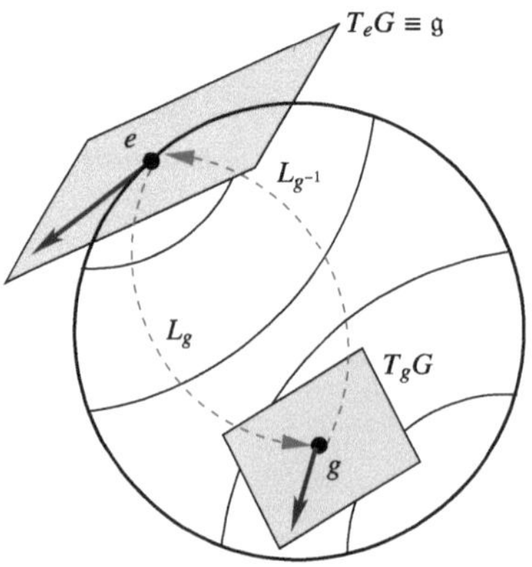

Figure 9.1: Lie group G, with tangent vectors $v \in T_e G$ and $d(L_g)_e \cdot v \in T_g G$.

is a vector-space isomorphism from the tangent space at g to the tangent space at the identity (i. e., the Lie algebra $\mathfrak{g}$).

Definition 9.7. A vector field X on G is said to be *left-invariant* if $X(gh) = d(L_g)_h(X(h))$ for all $g, h \in G$.

Any left-invariant vector field can therefore be completely identified by its value at the identity since $X(g) = d(L_g)_e(X(e))$. Hence, for any $\xi \in T_eG$, one can define an associated left-invariant vector field $\xi_L \in \mathfrak{X}(G)$ by the relation $\xi_L(g) = d(L_g)_e(\xi)$.

Next, we will use left-invariant vector fields to define a bracket on the tangent space to the identity, thereby giving it the structure of a Lie algebra. (Recall that we previously did this for the special case of *matrix* Lie groups.)

Definition 9.8. The *Lie algebra of the Lie group G* is the tangent space at the identity $\mathfrak{g} = T_eG$, with brackets given by $[\xi, \eta]_e = [\xi_L, \eta_L](e)$, for all $\xi, \eta \in \mathfrak{g}$, where $[\xi_L, \eta_L]$ is the Jacobi–Lie bracket of the left-invariant vector fields $\xi_L, \eta_L \in TG$, which are generated by $\xi, \eta \in \mathfrak{g}$.

The Jacobi–Lie bracket of vector fields satisfies the Jacobi identity, so the bracket defining the foregoing Lie algebra automatically satisfies the Jacobi identity as well, providing the tangent space T_eG with a Lie algebra structure.

One can, of course, extend the preceding concepts to *right-translation maps.*

Definition 9.9. For a Lie group G and fixed element $g \in G$, the *right-translation map*, $R_g : G \to G$, is defined by the relation $R_g(h) = hg$. For any $\xi \in \mathfrak{g} = T_eG$, we define the associated *right-invariant vector field*, $\xi_R \in \mathfrak{X}(G)$, by the relation $\xi_R(g) = d(R_g)_e(\xi)$.

The differential $d(R_{g^{-1}})_g : T_gG \to T_eG$ also constitutes a vector-space isomorphism. However, the two mappings

$$d(R_{g^{-1}})_g(X) \neq d(L_{g^{-1}})_g(X)$$

need not coincide; see Figure 9.2 for an illustration of this.

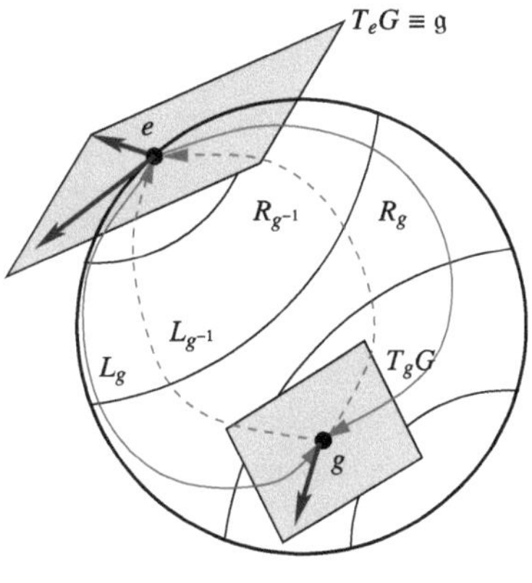

Figure 9.2: Left- and right-translation maps; a vector $v \in T_gG$ and its map to the Lie algebra under $d(R_{g^{-1}})_g : T_gG \to T_eG$ and $d(L_{g^{-1}})_g : T_gG \to T_eG$.

Though left-translation maps arise commonly, they are not ubiquitous: certain occasions merit the use of right-translations, as we will see later in this chapter.

We will now define several useful concepts of velocity for systems on Lie groups; see [196] for additional details.

Definition 9.10. Given a smooth curve $g : \mathbb{R} \to G$, we define
i. the *material velocity* of g as the vector field $\dot{g}(t) \in T_{g(t)}G$;
ii. the *body velocity* of g as $\xi^b = d(L_{g(t)^{-1}})_{g(t)}(\dot{g}(t)) \in \mathfrak{g}$; and
iii. the *spatial velocity* of g as $\xi^s = d(R_{g(t)^{-1}})_{g(t)}(\dot{g}(t)) \in \mathfrak{g}$.

For matrix Lie groups, the body and spatial velocities of a curve are given by

$$\xi^b = g^{-1}\dot{g} \quad \text{and} \quad \xi^s = \dot{g}g^{-1},$$

respectively.

For a fixed group element $g \in G$, the *adjoint map* $\mathrm{Ad}_g : \mathfrak{g} \to \mathfrak{g}$ takes the body representation of a vector into its spatial representation, i. e., it takes the pullback by the left-translation map of a vector $X \in T_gG$ to the Lie algebra into the corresponding pullback by the right-translation map. We define the adjoint formally as follows.

Definition 9.11. For a fixed element $g \in G$ and for any $\xi \in \mathfrak{g}$, the *adjoint map* is defined by

$$\mathrm{Ad}_g\xi = d(R_{g^{-1}})_g(d(L_g)_e(\xi)).$$

Note that for matrix Lie groups, the adjoint map simplifies to

$$\mathrm{Ad}_g\xi = g\xi g^{-1}.$$

Hence, we see the relation $\xi^s = \mathrm{Ad}_g\xi^b$. The adjoint map is illustrated in Figure 9.3.

Next, for vectors $\xi, \eta \in \mathfrak{g}$, one may define an operator by differentiating the adjoint map $\mathrm{Ad}_g\eta$ with respect to group element g, at $g = e$, in the direction ξ. This defines the *adjoint operator*.

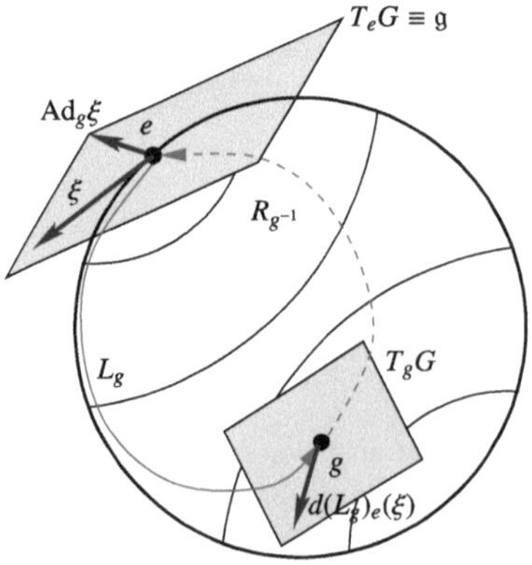

Figure 9.3: Adjoint map of vector ξ.

Definition 9.12. Let G be a Lie group and $\mathfrak{g} = T_eG$ its Lie algebra. Then, for fixed $\xi \in \mathfrak{g}$, the *adjoint operator* $\mathrm{ad}_\xi : \mathfrak{g} \to \mathfrak{g}$ is defined by

$$\mathrm{ad}_\xi \eta = \frac{d}{dt}\mathrm{Ad}_{g(t)}\eta\bigg|_{t=0},$$

where $g : [-\varepsilon, \varepsilon] \to G$ is a smooth curve such that $g(0) = e$ and $\dot{g}(0) = \xi$.

Exercise 9.3. Show that for a matrix Lie group we have

$$\mathrm{ad}_\xi \eta = [\xi, \eta],$$

where $[\xi, \eta]$ is the matrix commutator for all $\xi, \eta \in \mathfrak{g}$.

Without reference to a fixed $g \in G$ or $\xi \in \mathfrak{g}$, the maps Ad_g and ad_ξ may be thought of as the binary operations

$$\mathrm{Ad} : G \times \mathfrak{g} \to \mathfrak{g} \quad \text{and}$$
$$\mathrm{ad} : \mathfrak{g} \times \mathfrak{g} \to \mathfrak{g},$$

respectively.

Link between Exponentials and Flows on Lie Groups

Next, we define the concept of the *exponential*, which maps elements on the Lie algebra to points on the group via the flow along a certain left-invariant vector field. As we will see, this exponential map coincides with the operation of matrix exponentiation on matrix Lie groups.

Definition 9.13. Given $\xi \in G$ and its associated left-invariant vector field $\xi_L(g) = d(L_g)_e\xi$, we define the *exponential map* $\exp : \mathfrak{g} \to G$ as the flow of g along the vector field ξ_L for a unit interval of time

$$\exp(\xi) = \varphi(1; e),$$

where $\varphi : \mathbb{R} \times G \to G$ is the flow map along the vector field ξ_L.

Exercise 9.4. Show that

$$\exp(\xi t) = \varphi(t; e).$$

This demonstrates that the operation of exponentiation can be scaled to flow along the left-invariant vector field for an arbitrary duration of time.

Exercise 9.5. On a matrix Lie group G, show that

$$\mathrm{expm}(A) = \exp(A),$$

for all $A \in \mathfrak{g}$, where expm is the matrix exponential defined in (1.9).

Since left- and right-invariant vector fields are defined *globally* by their values at the single point e, one can construct a relation between the exponential map on a Lie group and the flow along left- or right-invariant vector fields emanating from an arbitrary $g \in G$.

Theorem 9.2. *Let G be a Lie group, $\xi \in \mathfrak{g}$ an element of the Lie algebra, and ξ_L the associated left-invariant vector field of ξ. Then*

$$\varphi(t;g) = L_g \circ \exp(\xi t), \tag{9.2}$$

where $\varphi : \mathbb{R} \times G \to G$ is the flow along the vector field ξ_L.

Proof. To see this, first consider the flow $h(t) = \exp(t\xi)$ for a fixed $\xi \in \mathfrak{g}$. It follows that $h(0) = e$ and $\dot{h}(t) = h(t)\xi$ since $\dot{h}(t) = d(L_h)_e\xi$.

Now consider the curve $\gamma(t) = gh(t)$ for the fixed element $g \in G$. For the curve γ, we obtain the relations $\gamma(0) = gh(0) = g$ and $\dot{\gamma}(t) = g\dot{h}(t) = gh(t)\xi = \gamma(t)\xi = \xi_L(\gamma(t))$. This demonstrates the result. □

This theorem states that the operations of left translation and flow along a left-invariant vector field commute, i. e., if one begins at the identity e, applies the left translation L_g, and then flows from g along the vector field ξ_L, then one arrives at the same terminal point as one would have obtained by first flowing from e along the vector field ξ_L and then applying the left-translation operator L_g. The corollary below states that the Jacobi–Lie bracket of two left-invariant vector fields is itself left-invariant. This will be an important observation in Chapter 10 when we connect Hamel's equation to the Euler–Poincaré equation.

Corollary 9.1. *Let G be a Lie group, $\xi, \eta \in \mathfrak{g}$ two fixed elements of the Lie algebra, and ξ_L and η_L the associated left-invariant vector fields of ξ and η, respectively. Then the Jacobi–Lie bracket of the vector fields ξ_L and η_L evaluated at a point $g \in G$ is related to its value at the identity element $e \in G$ by the relation*

$$[\xi_L, \eta_L](g) = d(L_g)_e[\xi, \eta]_e,$$

where $d(L_g)_e : \mathfrak{g} \to T_gG$ sends the tangent vector $[\xi, \eta]_e$ to a vector tangent to the Lie group at the point $g \in G$. If G is a matrix Lie group, then it further follows that

$$[\xi_L, \eta_L](g) = g[\xi, \eta] = g(\xi\eta - \eta\xi).$$

Proof. By Proposition 6.5, the Lie-group-parameterized family of curves

$$\gamma_g(t) = \varphi^{-\eta_L}_{\sqrt{t}} \circ \varphi^{-\xi_L}_{\sqrt{t}} \circ \varphi^{\eta_L}_{\sqrt{t}} \circ \varphi^{\xi_L}_{\sqrt{t}}(g)$$

has the property that $\gamma_g(0) = g$ and $\gamma_g'(0) = [\xi_L, \eta_L](g)$ for all $g \in G$. However, since ξ_L and η_L are both left-invariant vector fields, it follows from Theorem 9.2 that $\gamma_g(t) = L_g(\gamma_e(t))$. Therefore, by the chain rule, we obtain $\gamma_g'(0) = d(L_g)_e(\gamma_e'(0))$. The result follows since $\gamma_g'(0) = [\xi_L, \eta_L](g)$ and $\gamma_e'(0) = [\xi_L, \eta_L](e) = [\xi, \eta]_e$. □

This corollary implies that, for matrix Lie groups, even though the Jacobi–Lie bracket of the vector fields ξ_L and η_L evaluated at the identity e is equal to the matrix commutator $[\xi,\eta] = \xi\eta - \eta\xi$, a similar statement cannot be applied at other points g in the group, i. e., the Jacobi–Lie bracket $[\xi_L,\eta_L](g)$ does not equal the matrix commutator $[\xi_L(g),\eta_L(g)] = [g\xi, g\eta] = g\xi g\eta - g\eta g\xi$, but rather $g[\xi,\eta] = g\xi\eta - g\eta\xi$.

9.3 Euler–Poincaré Equation

In this section, we seek to generalize the Euler–Lagrange equation to an equation appropriate for mechanics on Lie groups.

Let G be a matrix Lie group and $g : \mathbb{R} \to G$ a curve through G. Recall that the *body velocity* of the curve is given by

$$\xi(t) = d(L_{g(t)^{-1}})_{g(t)}\dot{g}(t) = g(t)^{-1}\dot{g}(t), \tag{9.3}$$

where the second equality holds whenever G is a *matrix* Lie group. For example, if $G = \mathrm{SO}(3)$ represents the configuration manifold for a rigid body with a fixed point, then the components of the matrix $\xi(t)$ with respect to the standard basis for $\mathfrak{so}(3)$ are the components of the angular velocity of the rigid body expressed relative to the body-fixed frame.

Suppose now that $g : [a,b] \to G$ is a curve and $\vartheta : [-\varepsilon,\varepsilon]\times[a,b] \to G$ is a variation of g. Recall that the *virtual displacement* of curve g is the vector field

$$\delta g(t) = \left.\frac{\partial\vartheta(s,t)}{\partial s}\right|_{s=0}.$$

Definition 9.14. Let $g : [a,b] \to G$ be a curve, $\vartheta : [-\varepsilon,\varepsilon]\times[a,b] \to G$ a variation of g, and $\delta g : [a,b] \to TG|_g$ its virtual displacement. Then the *body virtual displacement* of g is the Lie-algebra-valued function $\eta : [a,b] \to \mathfrak{g}$ defined by

$$\eta(t) = d(L_{g(t)^{-1}})_{g(t)}\delta g(t) = g(t)^{-1}\delta g(t),$$

the second equality holding whenever G is a *matrix* Lie group.

Hence, the *body velocity* and *body virtual displacement* of a curve g are the pullbacks, via the left-translation map, of the material velocity and virtual displacements of the curve to the Lie algebra, respectively. To actuate the calculus of variations in a Lie group setting, we will require the following *Lie group transpositional relations*.

Lemma 9.3 (Lie Group Transpositional Relations). *Let $g : [a,b] \to G$ be a smooth curve through the matrix Lie group G, $\vartheta : [-\varepsilon,\varepsilon]\times[a,b] \to G$ a variation of g, and $\xi,\eta : [a,b] \to \mathfrak{g}$ the body velocity of g and the body virtual displacement of g, respectively. Then*

$$\delta\xi = \dot{\eta} + [\xi,\eta].$$

Proof. Let $g = g(s,t)$ represent an arbitrary position in the variation ϑ. The body velocity and body virtual displacement may be represented as

$$\xi = g^{-1}\dot{g} \quad \text{and} \quad \eta = g^{-1}\delta g,$$

respectively, so that we may think of $\xi, \eta : [-\varepsilon, \varepsilon]\times[a,b] \to \mathfrak{g}$. Differentiating, we obtain

$$\begin{aligned}\frac{\partial \xi}{\partial s} - \frac{\partial \eta}{\partial t} &= -g^{-1}\frac{\partial g}{\partial s}g^{-1}\frac{\partial g}{\partial t} + g^{-1}\frac{\partial^2 g}{\partial s \partial t} + g^{-1}\frac{\partial g}{\partial t}g^{-1}\frac{\partial g}{\partial s} - g^{-1}\frac{\partial^2 g}{\partial t \partial s}\\ &= g^{-1}\left(\frac{\partial^2 g}{\partial s \partial t} - \frac{\partial^2 g}{\partial t \partial s}\right) + [\xi, \eta]\\ &= [\xi, \eta],\end{aligned}$$

where the last equality follows since ϑ is a C^2 variation. Evaluating this equality at $s = 0$ yields our result. □

We are now in a position to derive the *Euler–Poincaré* equations for mechanics on Lie groups. These equations first appeared in the context of Lie groups in [241]. For additional discussion, see also [11], [29], [53], and [196].

Theorem 9.3 (Euler–Poincaré Equation). *Let G be a matrix Lie group, $L : TG \to \mathbb{R}$ a left-invariant Lagrangian, and $l : \mathfrak{g} \to \mathbb{R}$ its restriction to the Lie algebra. For a curve $g : [a,b] \to G$, let $\xi, \eta : [a,b] \to \mathfrak{g}$ denote its body velocity and body virtual displacement, respectively. Then the following are equivalent:*

i. *the curve $g(t)$ satisfies the Euler–Lagrange equation for L on G,*

$$\frac{d}{dt}\frac{\partial L}{\partial \dot{g}} - \frac{\partial L}{\partial g} = 0;$$

ii. *the variational principle*

$$\delta \int_a^b L(g(t), \dot{g}(t))\, dt = 0$$

holds for arbitrary, smooth variations of g (i. e., curve g satisfies Hamilton's principle);

iii. *the* Euler–Poincaré equation *holds for $l : \mathfrak{g} \to \mathbb{R}$, i. e.,*

$$\frac{d}{dt}\frac{\partial l}{\partial \xi} - \mathrm{ad}^*_\xi \frac{\partial l}{\partial \xi} = 0;$$

iv. *the variational principle*

$$\delta \int_a^b l(\xi(t))dt = 0$$

holds on $\mathfrak{g}$, where variations are taken to be of the form $\delta\xi = \dot{\eta} + [\xi, \eta]$, with $\eta(a) = \eta(b) = 0$.

Proof. Items i and ii are equivalent due to Theorem 7.1.

To demonstrate the equivalency of items ii and iv, one must show the virtual displacements $\delta g \in TG$ induce variations of the form $\delta\xi = \dot{\eta} + [\xi, \eta]$ when pulled back to the Lie algebra; but this follows from Lemma 9.3.

To show the equivalence of items iii and iv, we follow the computation

$$\begin{aligned}\delta \int_a^b l(\xi(t))\, dt &= \int_a^b \frac{\partial l}{\partial \xi} \delta\xi \, dt \\ &= \int_a^b \frac{\partial l}{\partial \xi} (\dot{\eta} + [\xi, \eta])\, dt \\ &= \int_a^b \frac{\partial l}{\partial \xi} (\dot{\eta} + \mathrm{ad}_\xi \eta)\, dt \\ &= \int_a^b \left(-\frac{d}{dt}\frac{\partial l}{\partial \xi} + \mathrm{ad}_\xi^* \frac{\partial l}{\partial \xi} \right) \eta \, dt,\end{aligned}$$

where the final term is obtained by integration by parts. The result follows since the components of η are algebraically independent. □

Remark. For Lie group Lagrangian systems under the influence of an applied force $F \in \mathfrak{X}^*(G)$, the Euler–Poincaré equation with forcing is given by

$$\frac{d}{dt}\frac{\partial l}{\partial \xi} - \mathrm{ad}_\xi^* \frac{\partial l}{\partial \xi} = f, \tag{9.4}$$

where $f \in \mathfrak{g}^*$ is the pullback of F to the dual of the Lie algebra under the left-translation map.

Kinematic Lie Group Integrators

The Euler–Lagrange equation, though classically viewed as a system of n second-order differential equations, is in actuality a system of first-order differential equations on the $2n$-dimensional tangent bundle TQ, in much the same way as Hamilton's equations constitute a system of first-order differential equations on the cotangent bundle T^*Q. The Euler–Lagrange equation is often seen as second-order due to the trivial relation between the time derivatives of the coordinates and the velocities, i. e., $d/dt(q^i) = \dot{q}^i$. Similarly, the Euler–Poincaré equation is to be considered a first-order system on the tangent bundle of the Lie group TG; however, as we have seen, the velocities are taken as the pullback of the velocity at $g \in G$ to the Lie algebra $\mathfrak{g}$. The kinematic relation that must be coupled with the Euler–Poincaré equation (9.4) is given by inverting the definition of body velocity (9.3) to obtain the matrix differential equation

$$\dot{g} = g\xi. \tag{9.5}$$

A naïve application of this equation could easily result in loss of the underlying geometric structures during numerical integration. The simple reason for this is the fact that all numerical integration schemes rely on finite time steps, and by approximating subsequent values of g, one is potentially stepping off the Lie group G. There are two ways around this. The first is to describe the Lie group G in terms of local coordinates, as one would on an ordinary manifold with no underlying group structure. *Exponential coordinates* are often used to achieve this, such as the Euler angles in the study of rigid-body mechanics. The problem with this approach is that local coordinates on Lie groups are never global, and one must constantly perform coordinate transformations and use multiple coordinate charts during long-term simulations. This is no worse than the difficulties faced with the regular Euler–Lagrange equation, but when the configuration manifold has a group structure, one can do better. The second method is to construct numerical methods directly using the result of Theorem 9.2. For example, the classical Euler method for integrating the first-order system (9.5) would yield

$$g_{k+1} = g_k + hg_k\xi_k,$$

where g_k and ξ_k are the approximated solutions at time t_k and $h = t_{k+1} - t_k$. Clearly, even if $g_k \in G$ and $g_k\xi_k \in T_{g_k}G$, as it should be, $g_{k+1} \notin G$. A vastly superior method that remedies this issue can easily be constructed based on (9.2); for example,

$$g_{k+1} = g_k \exp(\xi_k h).$$

We will see an application of this approach in §9.5.

9.4 Application: Rigid-Body Mechanics

In this section, we will see how these results on Lie groups can be applied to rigid-body mechanics. We will first discuss the kinematics of rigid-body motion and introduce the classical Euler angles.

Rigid-Body Kinematics

We begin our discussion on rigid-body mechanics with an examination of the relation between velocities on the Lie group SO(3) (Example 9.3) and rigid-body kinematics. In studying the dynamics and mechanics of rigid-body motion, one must introduce two

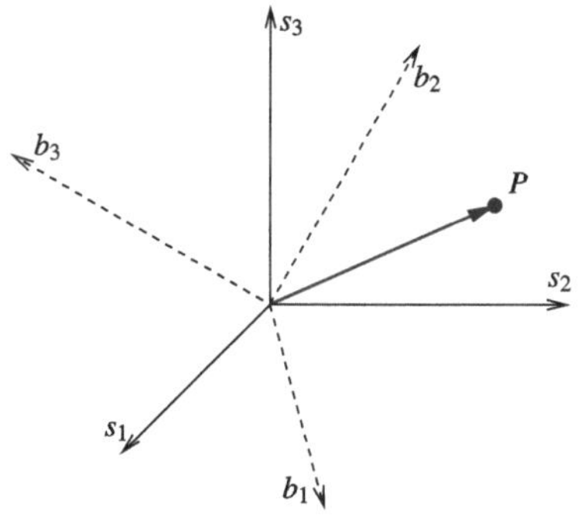

Figure 9.4: Spatial and body-fixed frames; position vector of a point *P* in a rigid body.

separate right-handed, orthogonal reference frames—an inertial, *spatial* frame and a *body-fixed* frame, as depicted in Figure 9.4. The body-fixed frame is considered to be fixed relative to the rigid body, so that its orientation equally describes the orientation of the rigid body. The spatial frame is determined by the orthonormal basis $\mathcal{S} = \{s_1, s_2, s_3\}$ and the body-fixed frame by the orthonormal basis $\mathcal{B} = \{b_1, b_2, b_3\}$. An arbitrary vector can be expressed using either *spatial coordinates* (x, y, z) or *body coordinates* (X, Y, Z) relative to the spatial and body bases $\mathcal{S}$ and $\mathcal{B}$, respectively. A given element $R \in \mathrm{SO}(3)$ determines the orientation of the rigid body as follows: the ith column of matrix R is comprised of the components of the basis vector b_i relative to the spatial coordinates $\mathcal{S}$; that is to say,

$$R = \begin{bmatrix} | & | & | \\ b_1 & b_2 & b_3 \\ | & | & | \end{bmatrix}.$$

It follows that $R \in \mathrm{SO}(3)$ since the basis $\mathcal{B}$ is a right-handed orthonormal basis. Matrix R can be further interpreted as the matrix that sends the vectors (s_1, s_2, s_3) into the vectors (b_1, b_2, b_3) when all six of these vectors are expressed relative to the basis $\mathcal{S}$, i. e., $b_i = R \cdot s_i$. A third and final way in which matrix R can be interpreted is as the matrix that transforms a vector expressed relative to the basis $\mathcal{B}$ into the equivalent vector expressed relative to $\mathcal{S}$. To see this, consider an arbitrary vector V and suppose that its components relative to the bases $\mathcal{B}$ and $\mathcal{S}$ are given by $V_{\mathcal{B}}$ and $V_{\mathcal{S}}$, respectively, i. e.,

$$V = V_{\mathcal{S}}^1 s_1 + V_{\mathcal{S}}^2 s_2 + V_{\mathcal{S}}^3 s_3 = V_{\mathcal{B}}^1 b_1 + V_{\mathcal{B}}^2 b_2 + V_{\mathcal{B}}^3 b_3.$$

Since $V_{\mathcal{B}}^i = b_i \cdot V_{\mathcal{S}}$, it follows that

$$V_{\mathcal{B}} = R^T \cdot V_{\mathcal{S}}$$

or, upon rearranging,

$$V_{\mathcal{S}} = R \cdot V_{\mathcal{B}}. \tag{9.6}$$

Now let P be a fixed point in a rigid body, X its (constant) position vector relative to the body basis $\mathcal{B}$, and x its position vector relative to the spatial basis $\mathcal{S}$. Let $R : \mathbb{R} \to SO(3)$ be the path of the body's motion through the configuration manifold SO(3). Then the constant position vector X of point P in the body is related to its position vector relative to the spatial frame by

$$x(t) = R(t) \cdot X.$$

Next, we explore the Lie algebra isomorphism between the Lie algebra $\mathfrak{so}(3)$ and the vector space $\mathbb{R}^3$. Consider the vector space isomorphism $\wedge : \mathbb{R}^3 \to \mathfrak{so}(3)$, defined by

$$\hat{\omega} = \begin{bmatrix} 0 & -\omega_3 & \omega_2 \\ \omega_3 & 0 & -\omega_1 \\ -\omega_2 & \omega_1 & 0 \end{bmatrix}, \tag{9.7}$$

for all $\omega = (\omega_1, \omega_2, \omega_3)^T \in \mathbb{R}^3$. The matrix product $\hat{\omega} \cdot v = \omega \times v$ for all $v \in \mathbb{R}^3$, where $\times : \mathbb{R}^3 \times \mathbb{R}^3 \to \mathbb{R}^3$ is the usual vector *cross product* in $\mathbb{R}^3$. One similarly defines a "down" operator $\vee : \mathfrak{so}(3) \to \mathbb{R}^3$ as the inverse $\vee = (\wedge)^{-1}$.

Exercise 9.6. Show that the vector space $\mathbb{R}^3$, coupled with the binary operation of the cross product $\times : \mathbb{R}^3 \times \mathbb{R}^3 \to \mathbb{R}^3$, is a Lie algebra.

Exercise 9.7. Show that the *Lie algebra* $\mathfrak{so}(3)$ is isomorphic to the *Lie algebra* $(\mathbb{R}^3, \times)$, i. e., show that

$$v \times w = [\hat{v}, \hat{w}]$$

for all $v, w \in \mathbb{R}^3$.

To connect this discussion with mechanics, consider a rigid body with a fixed point, described by the configuration manifold SO(3). Let $R : \mathbb{R} \to SO(3)$ represent the system's path through the Lie group SO(3). Following Definition 9.10, the *material velocity* of the rigid body is defined to be the derivative $\dot{R}(t) \in T_{R(t)}SO(3)$. Similarly, the *spatial* and *body* velocities are given by

$$\xi^s = \dot{R}(t)R(t)^T, \tag{9.8}$$
$$\xi^b = R(t)^T\dot{R}(t), \tag{9.9}$$

respectively. Notice that $\xi^s = \mathrm{Ad}_{R(t)}\xi^b$. Both the spatial and body velocities are curves through the Lie algebra $\mathfrak{so}(3)$.

Now, a rigid body has an *angular velocity vector*, which may be expressed relative to either the spatial or the body-fixed frames. These expressions are known as the *spatial angular velocity* $\omega : \mathbb{R} \to \mathbb{R}^3$ and the *body angular velocity* $\Omega : \mathbb{R} \to \mathbb{R}^3$, respectively. They are related to the spatial and body velocities of the system by the relations

$$\hat{\omega} = \dot{R}(t)R(t)^T = \xi^s,$$
$$\hat{\Omega} = R(t)^T\dot{R}(t) = \xi^b.$$

We will prove the validity of these definitions in a moment, after we introduce Euler angles.

Euler Angles

Euler angles are a classical tool used to describe the orientation of a rigid body in space. In a geometric setting, Euler angles represent a set of *local coordinates* for the configuration manifold SO(3). As such, one cannot describe every possible orientation of a rigid body using a single set of Euler angles. For a given set of Euler angles, configurations that lie outside the open set $U \subset \mathrm{SO}(3)$ on which the local coordinates are valid are known as *gimbal lock configurations*. They received this name because they correspond to configurations for which an aircraft's or spacecraft's gimbal assembly would literally "lock," resulting in the guidance and navigation computer's loss of information on the aircraft's or spacecraft's orientation.

Type I (Aircraft) Euler Angles

We will first focus our attention on a particular set of Euler angles commonly used in aviation, known as Type I (aircraft) Euler angles (ψ, θ, φ). This triple of angles describes the transformation from an *xyz* spatial axis that is fixed in inertial space to the *XYZ* body axis of the rigid body. To determine this transformation, one performs the following three (ordered) rotations. Beginning with the (x, y, z) inertial frame, one performs:

1. a rotation of ψ about the z-axis, resulting in the primed (x', y', z') coordinate system, followed by
2. a rotation of θ about the y'-axis, resulting in the double-primed (x'', y'', z'') coordinate system; and, finally, this is followed by
3. a rotation of φ about the x''-axis, resulting in the unprimed (X, Y, Z) body-fixed coordinate system.

See Figures 9.5 and 9.6 for an illustration of these three successive rotations.

The angular velocity of a rigid body is given by

$$\omega = \dot{\boldsymbol{\psi}} + \dot{\boldsymbol{\theta}} + \dot{\boldsymbol{\varphi}} = \Omega_x b_1 + \Omega_y b_2 + \Omega_z b_3.$$

The Euler angle rates are indicated by the angular velocity vectors in the direction of their line of nodes, i. e., $\dot{\boldsymbol{\psi}}$ about the Z-axis, $\dot{\boldsymbol{\theta}}$ about the y'-axis, and $\dot{\boldsymbol{\varphi}}$ about the x'' axis. The components of the body axis angular velocity Ω_x, Ω_y, and Ω_z, in terms

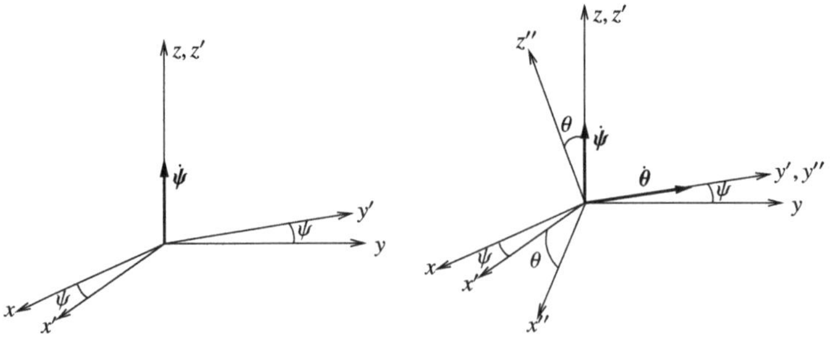

Figure 9.5: Type I (aircraft) Euler angles: first two intermediate rotations.

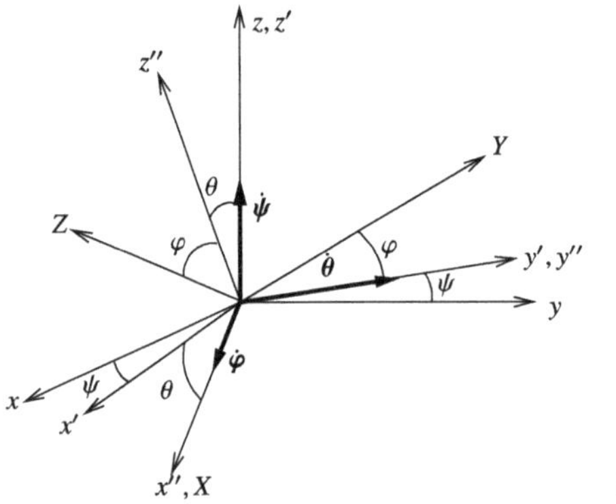

Figure 9.6: Type I (aircraft) Euler angles.

of aircraft and spacecraft orientations, are known as the *roll rate*, *pitch rate*, and *yaw rate*, respectively. (In aeronautical contexts, the spatial frame is conventionally taken with the z-axis pointing downward.)

To understand how this set of Euler angles may be used to describe the configuration manifold SO(3), consider the individual rotation matrices

$$\Phi = \begin{bmatrix} 1 & 0 & 0 \\ 0 & \cos\varphi & -\sin\varphi \\ 0 & \sin\varphi & \cos\varphi \end{bmatrix},$$

$$\Theta = \begin{bmatrix} \cos\theta & 0 & \sin\theta \\ 0 & 1 & 0 \\ -\sin\theta & 0 & \cos\theta \end{bmatrix}, \quad \text{and}$$

$$\Psi = \begin{bmatrix} \cos\psi & -\sin\psi & 0 \\ \sin\psi & \cos\psi & 0 \\ 0 & 0 & 1 \end{bmatrix},$$

which represent rotations of φ, θ, and ψ about the x-, y-, and z-axes, respectively. Working backward using (9.6), we have

$$v'' = \Phi V,$$

$$v' = \Theta v'' = \Theta\Phi V,$$
$$v = \Psi v' = \Psi\Theta\Phi V,$$

so that the rotation matrix R is given by $R = \Psi\Theta\Phi$. Here, the vectors V, v'', v', and v represent a single vector expressed relative to the XYZ, $x''y''z''$, $x'y'z'$, and xyz coordinates, respectively. Alternatively, we can obtain this formula by first applying a rotation of ψ about the z-axis, obtaining the primed coordinates. This transformation is represented by the matrix Ψ. A rotation of θ about the y'-axis is then represented by $\Psi\Theta\Psi^T$. (Note that Θ represents a rotation about the y-axis; to obtain a rotation about the y'-axis, we must transform the basis using the matrix Ψ.) Thus, the first two rotations composed together are represented by the matrix $(\Psi\Theta\Psi^T)\Psi = \Psi\Theta$. Finally, a rotation of φ about the x''-axis is represented by $\Psi\Theta\Phi\Theta^T\Psi^T$. Composing these matrices, we obtain $R = (\Psi\Theta\Phi\Theta^T\Psi^T)(\Psi\Theta\Psi^T)(\Psi)$, or

$$R = \Psi\Theta\Phi = \begin{bmatrix} c\psi c\theta & (-s\psi c\varphi + c\psi s\theta s\varphi) & (s\psi s\varphi + c\psi s\theta c\varphi) \\ s\psi c\theta & (c\psi c\varphi + s\psi s\theta s\varphi) & (-c\psi s\varphi + s\psi s\theta c\varphi) \\ -s\theta & c\theta s\varphi & c\theta c\varphi \end{bmatrix}, \tag{9.10}$$

where s and c represent sine and cosine, respectively. Matrix R constitutes a local parameterization of SO(3) in terms of local coordinates (ψ, θ, φ) on the domain

$$0 \le \psi < 2\pi, \quad -\frac{\pi}{2} < \theta < \frac{\pi}{2}, \quad 0 \le \varphi < 2\pi.$$

Singularities resulting in *gimbal lock* occur whenever any two Euler angles represent rotations about the same axis in space. For example, $\dot\psi$ and $\dot\varphi$ occur about the same vertical axis whenever $\theta = \pm\pi/2$. Such singularities are removed by restricting the Euler angles to lie on the specified domains.

Our next objective is to describe the spatial and body components of the angular velocity vector in terms of the Euler angle rates, thereby justifying the Definitions (9.8)–(9.9). Let $\mathbf{e}_\psi$, $\mathbf{e}_\theta$, and $\mathbf{e}_\varphi$ be unit vectors pointing toward the lines of nodes (i. e., the z-, y'-, and x''-axes, respectively). The z-axis is aligned with s_3. The transformation Ψ transforms the spatial axes to the primed axes; hence the direction of y' is the second column of matrix Ψ. Similarly, the transformation $\Psi\Theta$ transforms the spatial axes into the double-primed axes; hence the direction of the x'' axis is the first column of the product $\Psi\Theta$. The line-of-nodes vectors are therefore given, relative to the spatial basis, by the relations

$$\mathbf{e}_\psi = \begin{bmatrix} 0 \\ 0 \\ 1 \end{bmatrix}, \quad \mathbf{e}_\theta = \begin{bmatrix} -\sin\psi \\ \cos\psi \\ 0 \end{bmatrix}, \quad \mathbf{e}_\varphi = \begin{bmatrix} \cos\psi\cos\theta \\ \sin\psi\cos\theta \\ -\sin\theta \end{bmatrix}.$$

Since the angular velocity vector is $\boldsymbol{\omega} = \dot\psi\mathbf{e}_\psi + \dot\theta\mathbf{e}_\theta + \dot\varphi\mathbf{e}_\varphi = \omega_x s_1 + \omega_y s_2 + \omega_z s_3$, one easily finds the spatial components of the angular velocity to be

$$\omega_x = \dot\varphi\cos\psi\cos\theta - \dot\theta\sin\psi,$$

$$\omega_y = \dot{\varphi} \sin\psi \cos\theta + \dot{\theta} \cos\psi,$$
$$\omega_z = \dot{\psi} - \dot{\varphi} \sin\theta.$$

Similarly, the lines of nodes relative to the body frame are given by the vectors obtained by premultiplying $\mathbf{e}_\psi$, $\mathbf{e}_\theta$, and $\mathbf{e}_\varphi$ by R^T, i. e.,

$$(\mathbf{e}_\psi)_\mathcal{B} = \begin{bmatrix} -\sin\theta \\ \cos\theta \sin\varphi \\ \cos\theta\cos\varphi \end{bmatrix}, \quad (\mathbf{e}_\theta)_\mathcal{B} = \begin{bmatrix} 0 \\ \cos\varphi \\ -\sin\varphi \end{bmatrix}, \quad (\mathbf{e}_\varphi)_\mathcal{B} = \begin{bmatrix} 1 \\ 0 \\ 0 \end{bmatrix}.$$

Thus, the body axis components of the angular velocity are given by the relations

$$\Omega_x = \dot{\varphi} - \dot{\psi} \sin\theta, \tag{9.11}$$
$$\Omega_y = \dot{\psi} \cos\theta \sin\varphi + \dot{\theta} \cos\varphi, \tag{9.12}$$
$$\Omega_z = \dot{\psi} \cos\theta \cos\varphi - \dot{\theta} \sin\theta. \tag{9.13}$$

Finally, starting with (9.10), a simple, albeit tedious, computation reveals that

$$R^T \cdot \frac{\partial R}{\partial \psi} = \begin{bmatrix} 0 & -\cos\theta\cos\varphi & \cos\theta\sin\varphi \\ \cos\theta\cos\varphi & 0 & \sin\theta \\ -\cos\theta\sin\varphi & -\sin\theta & 0 \end{bmatrix},$$

$$R^T \cdot \frac{\partial R}{\partial \theta} = \begin{bmatrix} 0 & \sin\varphi & \cos\varphi \\ -\sin\varphi & 0 & 0 \\ -\cos\varphi & 0 & 0 \end{bmatrix},$$

$$R^T \cdot \frac{\partial R}{\partial \varphi} = \begin{bmatrix} 0 & 0 & 0 \\ 0 & 0 & -1 \\ 0 & 1 & 0 \end{bmatrix}.$$

Using the definition of the mapping $\wedge : \mathbb{R}^3 \to \mathfrak{so}(3)$ given by (9.7), it is thus clear that

$$\hat{\Omega} = R^T \dot{R} = R^T \left(\dot{\psi} \frac{\partial R}{\partial \psi} + \dot{\theta} \frac{\partial R}{\partial \theta} + \dot{\varphi} \frac{\partial R}{\partial \varphi} \right),$$

thereby justifying our definition of the body velocity (9.9). Similarly, one may compute

$$\frac{\partial R}{\partial \psi} \cdot R^T = \begin{bmatrix} 0 & -1 & 0 \\ 1 & 0 & 0 \\ 0 & 0 & 0 \end{bmatrix},$$

$$\frac{\partial R}{\partial \theta} \cdot R^T = \begin{bmatrix} 0 & 0 & \cos\psi \\ 0 & 0 & \sin\psi \\ -\cos\psi & -\sin\psi & 0 \end{bmatrix},$$

$$\frac{\partial R}{\partial \varphi} \cdot R^T = \begin{bmatrix} 0 & \sin\theta & \sin\psi\cos\theta \\ -\sin\theta & 0 & -\cos\psi\cos\theta \\ -\sin\psi\cos\theta & \cos\psi\cos\theta & 0 \end{bmatrix},$$

thereby demonstrating the relation $\hat{\omega} = \dot{R}R^T$ for the spatial angular velocity and hence justifying (9.8).

Example 9.7 (An Axisymmetric Top). As an example, let us consider the motion of an axisymmetric top with principal moments of inertia I_a about the axis of symmetry and I_t about the transversal axis. We will use aircraft Euler angles to specify the orientation of the top, with the spatial z-axis taken to be pointing downward and the body X-axis taken as the axis of symmetry. The angle θ may therefore be interpreted as the angle between the axis of symmetry of the top and the inertial x-y plane. Similarly, the angle ψ may be interpreted as the angle between the inertial x-axis and the projection of the symmetry axis of the body onto the x-y plane, as depicted in Figure 9.7.

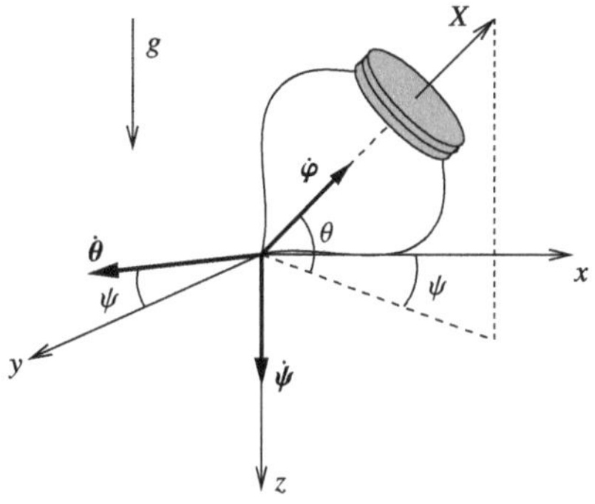

Figure 9.7: Axisymmetric top.

The kinetic energy of the top is simply $T = \frac{1}{2}I_a\Omega_x^2 + \frac{1}{2}I_t(\Omega_y^2 + \Omega_z^2)$. Suppose the top has mass m and its center of mass is a distance l from its tip, which is a fixed point attached to the origin. Then the Lagrangian may be written, with the aid of (9.11)–(9.13), in terms of Euler angles and Euler angle rates as

$$L = \frac{I_a}{2}\left(\dot{\varphi} - \dot{\psi}\sin\theta\right)^2 + \frac{I_t}{2}\left(\dot{\theta}^2 + \dot{\psi}^2\cos^2\theta\right) - mgl\sin\theta. \tag{9.14}$$

Though the kinetic energy expressed in terms of the body components of the angular velocity is decidedly simpler than this expression, the Euler–Lagrange equation is only valid when the Lagrangian is expressed in terms of the Euler angle rates. This is due to the fact that the body angular velocity components are not *true velocities*, i. e., they are not the perfect time derivative of any function of the configuration variables. We will explore this topic in greater depth in Chapter 10.

From (9.14), we may write the conjugate momenta as follows:

$$p_\psi = \frac{\partial L}{\partial \dot{\psi}} = -I_a\Omega_x\sin\theta + I_t\dot{\psi}\cos^2\theta,$$
$$p_\theta = \frac{\partial L}{\partial \dot{\theta}} = I_t\dot{\theta},$$

$$p_\varphi = \frac{\partial L}{\partial \varphi} = I_a \Omega_x,$$

where Ω_x is the *total spin*, defined by (9.11). From the Euler–Lagrange equation and the fact that the Lagrangian is independent of the coordinates ψ and φ, it follows that the conjugate momenta p_ψ and p_φ (or, equivalently, Ω_x) are integrals of motion for the axisymmetric top. Rearranging the p_ψ equation, we can solve for $\dot\psi$ in terms of the constants of motion and the angle θ, i. e.,

$$\dot\psi = \frac{p_\psi + I_a \Omega_x \sin\theta}{I_t \cos^2\theta}. \tag{9.15}$$

This is called the *precession rate*. Rather than integrating the final Euler–Lagrange equation

$$\frac{d}{dt} p_\theta - \frac{\partial L}{\partial \theta} = 0,$$

we will utilize the energy integral

$$H = \frac{I_a}{2}\Omega_x^2 + \frac{I_t}{2}\left(\dot\theta^2 + \dot\psi^2 \cos^2\theta\right) + mgl \sin\theta$$

as a final integral of motion. Since the first term is a constant, we will define a new function,

$$H' = H - \frac{I_a}{2}\Omega_x^2 = \frac{I_t}{2}\left(\dot\theta^2 + \dot\psi^2 \cos^2\theta\right) + mgl \sin\theta,$$

to be used as the third integral of motion in lieu of the straight-up energy H. Solving this equation for $\dot\theta$ yields the expression

$$\dot\theta^2 = \frac{2H'}{I_t} - \left(\frac{p_\psi + I_a \Omega_x \sin\theta}{I_t \cos\theta}\right)^2 - \frac{2mgl}{I_t} \sin\theta. \tag{9.16}$$

Hence, both the *precession rate* $\dot\psi$ and the *nutation rate* $\dot\theta$ may be written as a function of θ alone. The motion is therefore, in principle, solved by integrating the system (9.11), (9.15), and (9.16). The integrals of motion Ω_x, p_ψ, and H' are determined from the initial conditions (Euler angles and their rates). Additionally, the system depends on the parameters I_a, I_t, m, and l. See, e. g., [105] for a detailed treatment of the types of motion one can encounter in various regions of parameter space. □

Type II (Classic) Euler Angles

Type II (classic) Euler angles are formed by implementing three successive intermediate rotations as follows:

1. a rotation of φ about the z-axis, resulting in the primed (x', y', z') coordinate system, followed by

2. a rotation of θ about the x'-axis, resulting in the double-primed (x'', y'', z'') coordinate system; and, finally, this is followed by
3. a rotation of ψ about the z''-axis, resulting in the unprimed (X, Y, Z) body-fixed coordinate system.

The geometry of Type II Euler angles is illustrated in Figure 9.8. Type II Euler angles are local coordinates for SO(3) valid on the domain

$$0 \leq \varphi < 2\pi, \quad 0 < \theta < \pi, \quad 0 \leq \psi < 2\pi.$$

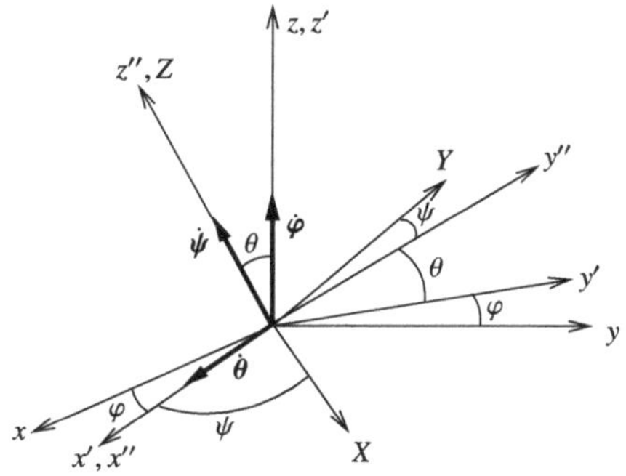

Figure 9.8: Type II (classic) Euler angles.

Exercise 9.8. Show that the body and spatial components of the angular velocity vector, expressed with classic Euler angles, are given by

$$\Omega_X = \dot{\varphi} \sin\theta \sin\psi + \dot{\theta} \cos\psi, \tag{9.17}$$
$$\Omega_Y = \dot{\varphi} \sin\theta \cos\psi - \dot{\theta} \sin\psi, \tag{9.18}$$
$$\Omega_Z = \dot{\psi} + \dot{\varphi} \cos\theta \tag{9.19}$$

and

$$\omega_x = \dot{\theta} \cos\varphi + \dot{\psi} \sin\varphi \sin\theta, \tag{9.20}$$
$$\omega_y = \dot{\theta} \sin\varphi - \dot{\psi} \cos\varphi \sin\theta, \tag{9.21}$$
$$\omega_z = \dot{\varphi} + \dot{\psi} \cos\theta, \tag{9.22}$$

respectively. Perform the computation in two separate ways:
(a) first, by determining the line of node vectors relative to both spatial and body coordinates and
(b) second, by directly computing the rotation matrix R and using Definitions (9.8) and (9.9).

Euler's Equation for Rigid-Body Mechanics

As an application of the Euler–Poincaré equation, we will derive the classic *Euler equations*, which govern the mechanical motion of a rigid body in terms of the body components of the body's angular velocity. Recall the Lie algebra vector-space isomorphism $\wedge : (\mathbb{R}^3, \times) \to \mathfrak{so}(3)$, defined by

$$\hat{\omega} = \begin{bmatrix} 0 & -\omega_3 & \omega_2 \\ \omega_3 & 0 & -\omega_1 \\ -\omega_2 & \omega_1 & 0 \end{bmatrix},$$

for $\omega = (\omega_1, \omega_2, \omega_3)^T \in \mathbb{R}^3$. Hence, for $A, B \in \mathbb{R}^3$, we have $\hat{A}, \hat{B} \in \mathfrak{so}(3)$ and

$$\mathrm{ad}_{\hat{A}} \hat{B} = \widehat{A \times B}.$$

Thus, with the identification $\mathfrak{so}(3) \cong (\mathbb{R}^3, \times)$, we can think of the *adjoint operator* as a binary operator on $\mathbb{R}^3$, so that

$$\mathrm{ad}_A B = A \times B.$$

Exercise 9.9. Use the vector identity $C \cdot (A \times B) = B \cdot (C \times A)$ to show that the *dual* adjoint operator is given by

$$\mathrm{ad}^*_A B = B \times A$$

for all $A, B \in \mathbb{R}^3$.

Now consider a rigid body with mass density ρ. Given a body-fixed XYZ-axis, the *moments of inertia* are defined as

$$I_{XX} = \int_V \rho(Y^2 + Z^2) dV,$$
$$I_{YY} = \int_V \rho(Z^2 + X^2) dV,$$
$$I_{ZZ} = \int_V \rho(X^2 + Y^2) dV,$$

where V is the volume of the rigid body, and the *products of inertia* are defined as

$$I_{XY} = I_{YX} = -\int_V \rho XY dV,$$
$$I_{XZ} = I_{ZX} = -\int_V \rho XZ dV,$$

$$I_{YZ} = I_{ZY} = -\int_V \rho YZ dV.$$

The moments and products of inertia for a rigid body constitute the components of its inertial matrix $\mathbb{I}$. Typically one chooses a *principal axis*, determined by the eigenvectors of this matrix. The moment of the inertia matrix becomes diagonal when expressed relative to the principal axis, and the moments of inertia (eigenvalues) are referred to as the *principal moments of inertia*. Now let $\omega, \Omega : \mathbb{R} \to \mathbb{R}^3$ represent the spatial and body-fixed angular velocities, respectively. If the origin of the body frame is either a fixed point or the center of mass, the kinetic energy may be expressed relative to the body components of the angular velocity as

$$l(\Omega) = \frac{1}{2}\Omega^T \cdot \mathbb{I} \cdot \Omega,$$

which may therefore equally be interpreted as the pullback of the Lagrangian to the Lie algebra. If we define the body axis *angular momentum vector* as $\Pi = \mathbb{I} \cdot \Omega$, then the Euler–Poincaré equation with forcing (9.4) yields *Euler's equation for rigid-body mechanics*, i. e.,

$$\dot{\Pi} - \Pi \times \Omega = m,$$

where m is the applied moment relative to the body-fixed frame. Relative to the principal axis, the components of Euler's equation are given by (3.18)–(3.20). Stability of the relative fixed points for the case of torque-free motion was previously analyzed in Section 3.7.

9.5 Application: Linearization of Hamiltonian Systems

In this final section of the chapter, we describe a surprising application of Lie group theory to the linearized flow of Hamiltonian systems. Given a smooth hyperregular Hamiltonian function $H : T^*Q \to \mathbb{R}$, recall that the Hamiltonian phase flow is determined by Hamilton's equations (8.4). Defining $x = (q^1, \ldots, q^n, p_1, \ldots, p_n) \in T^*Q$, we may represent these equations in the form

$$\dot{x} = J \cdot \frac{\partial H}{\partial x}.$$

The linearized equations that determine the dynamics of the state-transition matrix were given by (8.11), which we repeat here for convenience:

$$\dot{\Phi} = J \cdot \frac{\partial^2 H}{\partial x^2} \cdot \Phi. \tag{9.23}$$

Recall from Theorem 8.7 that the state-transition matrix, as defined along a given trajectory with a given set of initial conditions, is a symplectic matrix for all time,

i. e., $\Phi(t) \in \mathrm{Sp}(2n)$. One immediately realizes that the material velocity of the state-transition matrix, given by (9.23), is the right translation of a particular vector in the Lie algebra. The *spatial velocity* of the state-transition matrix for a Hamiltonian system is determined by the relation $\omega = \dot{\Phi} \cdot \Phi^{-1}$, which yields

$$\omega = J \cdot \frac{\partial^2 H}{\partial x^2} \in \mathfrak{sp}(2n). \tag{9.24}$$

Thus, instead of constructing numerical algorithms to directly integrate the matrix differential equation (9.23), which ultimately are likely to cause the numerically estimated state-transition matrix to wander away from the symplectic matrix Lie group, one can construct numerical algorithms using the result of Theorem 9.2. For example, one may integrate the variational equations using a Lie group form of the forward Euler or midpoint rule for flow along right-invariant vector fields, i. e.,

$$\Phi_{i+1} = \exp(h_i \omega_i) \cdot \Phi_i,$$
$$\Phi_{i+1} = \exp\left(h_i \frac{\omega_i + \omega_{i+1}}{2}\right) \cdot \Phi_i,$$

respectively. Since ω depends only on the position $x \in T^*Q$ along the phase flow and not upon the current value of the state-transition matrix, the Lie group midpoint rule constitutes an explicit method.

Example 9.8. For example, let us consider the Hénon–Heiles system (4.1). Recall that the Hénon–Heiles system is a Hamiltonian dynamical system with Hamiltonian function (8.5). The state-transition matrix is therefore a symplectic-matrix-valued function of time, $\Phi : \mathbb{R} \to \mathrm{Sp}(4)$, determined by the matrix initial value problem

$$\dot{\Phi} = \begin{bmatrix} 0 & 0 & 1 & 0 \\ 0 & 0 & 0 & 1 \\ -(1+2y) & -2x & 0 & 0 \\ -2x & -(1-2y) & 0 & 0 \end{bmatrix} \Phi \quad \text{and} \quad \Phi(0) = I_4,$$

so that $\Phi(t)$ is a solution curve along a right-invariant vector field on the Lie group $\mathrm{Sp}(4)$. The Lie algebra $\mathfrak{sp}(4)$ is 10-D and spanned by the basis vectors

$$e_{11} = \begin{bmatrix} 0 & 0 & 0 & 0 \\ 0 & 0 & 0 & 0 \\ -1 & 0 & 0 & 0 \\ 0 & 0 & 0 & 0 \end{bmatrix}, \qquad e_{22} = \begin{bmatrix} 0 & 0 & 0 & 0 \\ 0 & 0 & 0 & 0 \\ 0 & 0 & 0 & 0 \\ 0 & -1 & 0 & 0 \end{bmatrix},$$

$$e_{33} = \begin{bmatrix} 0 & 0 & 1 & 0 \\ 0 & 0 & 0 & 0 \\ 0 & 0 & 0 & 0 \\ 0 & 0 & 0 & 0 \end{bmatrix}, \qquad e_{44} = \begin{bmatrix} 0 & 0 & 0 & 0 \\ 0 & 0 & 0 & 1 \\ 0 & 0 & 0 & 0 \\ 0 & 0 & 0 & 0 \end{bmatrix},$$

$$e_{12} = \begin{bmatrix} 0 & 0 & 0 & 0 \\ 0 & 0 & 0 & 0 \\ 0 & -1 & 0 & 0 \\ -1 & 0 & 0 & 0 \end{bmatrix}, \quad e_{13} = \begin{bmatrix} 1 & 0 & 0 & 0 \\ 0 & 0 & 0 & 0 \\ 0 & 0 & -1 & 0 \\ 0 & 0 & 0 & 0 \end{bmatrix},$$

$$e_{14} = \begin{bmatrix} 0 & 0 & 0 & 0 \\ 1 & 0 & 0 & 0 \\ 0 & 0 & 0 & -1 \\ 0 & 0 & 0 & 0 \end{bmatrix}, \quad e_{23} = \begin{bmatrix} 0 & 1 & 0 & 0 \\ 0 & 0 & 0 & 0 \\ 0 & 0 & 0 & 0 \\ 0 & 0 & -1 & 0 \end{bmatrix},$$

$$e_{24} = \begin{bmatrix} 0 & 0 & 0 & 0 \\ 0 & 1 & 0 & 0 \\ 0 & 0 & 0 & 0 \\ 0 & 0 & 0 & -1 \end{bmatrix}, \quad e_{34} = \begin{bmatrix} 0 & 0 & 0 & 1 \\ 0 & 0 & 1 & 0 \\ 0 & 0 & 0 & 0 \\ 0 & 0 & 0 & 0 \end{bmatrix}.$$

With respect to this basis, the spatial velocity (9.24) of the state-transition matrix may be represented as

$$\omega = \sum_{i \le j} \frac{\partial^2 H}{\partial x^i \partial x^j} e_{ij} \in \mathfrak{sp}(4).$$

For the Hénon–Heiles system we obtain

$$\omega = (1 + 2y)e_{11} + (1 - 2y)e_{22} + e_{33} + e_{44} + 2xe_{12}.$$

The ten right-translation pushforward vector fields $E_{ij} = e_{ij} \cdot \Phi$ for $i \le j$ form a basis at each tangent space $T_\Phi \mathrm{Sp}(4)$. These vector fields constitute what is known as a *moving frame*, which will be the subject of our study in Chapter 10. Furthermore, note that the material velocity $\dot{\Phi}$ of the state-transition matrix is subject to a set of five nonholonomic constraints, in that its components along the noncommuting vector fields E_{13}, E_{14}, E_{23}, E_{24}, and E_{34} vanish identically. □

10 Moving Frames and Nonholonomic Mechanics

In Chapter 7, we introduced *nonholonomic* mechanical systems and discussed how the corresponding equations of motion arise due to the Lagrange–d'Alembert principle, i. e., by varying the action integral using variations whose virtual displacements are themselves kinematically admissible. For an n-degree-of-freedom system with a total of $m < n$ nonholonomic constraints, one obtains n second-order ordinary differential equations *and* m additional algebraic equations that must be solved to eliminate the m Lagrange multipliers used to enforce the constraints on the virtual displacement. Since each second-order ordinary differential equation may be replaced with two first-order ordinary differential equations, determining the motion is equivalent to solving a system of $(2n + m)$ first-order equations.

In this chapter, we discuss a technique that reduces the number of equations that one must solve when determining the ensuing dynamics of a given nonholonomic system. The technique utilizes a certain set of quantities known as *quasivelocities*. The idea is that a set of coordinates and quasivelocities $(q^1, \dots, q^n, v^1, \dots, v^n)$, when taken together, constitute a set of local coordinates for the tangent bundle TQ. The quasivelocities may then be selected so that $(q^1, \dots, q^n, v^1, \dots, v^{n-m})$ also constitute a complete set of local coordinates for the constraint distribution $\mathcal{D} \subset TQ$ and so that satisfaction of the constraint $\dot{q} \in \mathcal{D}$ is equivalent to the vanishing of the final m quasivelocities $v^{n-m+1} = 0, \dots, v^n = 0$. With appropriate care, one can derive an equation, known as *Hamel's equation*, that is multiplier-free and whose solution completely determines the resulting dynamics of the system. The components of Hamel's equation constitute a minimal set of $(2n - m)$ first-order, ordinary differential equations that completely govern the dynamics of nonholonomic mechanical systems. The derivation of Hamel's equation from a modern geometrical viewpoint will be the primary goal of this chapter.

10.1 Quasivelocities and Moving Frames

In this section, we define a set of quantities known as *quasivelocities*, which date back to the works of [116–118], [143], and [288]. For more recent, but classic, reviews on quasivelocity techniques, see [106] and [222]. The utility of these techniques was clearly demonstrated in these early works; however, the appropriate geometric setting was not understood until more recently. To uncover this geometric meaning, one makes use of a concept known as *moving frames*, which we will describe below.

Before proceeding, let us recall a few definitions and bits of terminology. A *nonholonomic system* is a triple $(Q, L, \mathcal{D})$, where Q is an n-dimensional configuration manifold, $L : TQ \to \mathbb{R}$ is a Lagrangian function, and $\mathcal{D} \subset TQ$ is an $(n - m)$-dimensional completely nonintegrable distribution. We say that a velocity $\dot{q}$ is *kinematically ad-*

https://doi.org/10.1515/9783110597806-010

missible at a point q if $\dot{q} \in \mathcal{D}_q$. The distribution $\mathcal{D}$ may be described *locally* as the null space of a set of m differential one-forms ω^a, for $a = (n - m + 1), \ldots, n$, so that $\dot{q} \in \mathcal{D}$ if and only if $\omega^a(\dot{q}) = 0$ for $a = (n - m + 1), \ldots, n$. These one-forms are referred to as *constraint one-forms*.

Now, the idea of quasivelocities is to use a linear transformation on the tangent space at each point to define a new set of velocity components

$$v^i = \phi^i_j \dot{q}^j, \tag{10.1}$$

where $\phi : U \to \mathrm{GL}(n; \mathbb{R})$ is a smooth, (invertible) matrix-valued function on some open set $U \subset Q$, with smooth inverse $\psi = \phi^{-1}$. The quantities $(v^1, \ldots, v^n)$ are known as *quasivelocities* since they, in general, may not be expressible as time derivatives of functions of q. The matrix-valued function ϕ is chosen so that the constraint one-forms ω^a may locally be described by the relation

$$\omega^a = \phi^a_j dq^j, \quad \text{for } a = (n - m + 1), \ldots, n,$$

i. e., so that $\dot{q} \in \mathcal{D}$ is equivalent to requiring the last m quasivelocities $v^{n-m+1} = 0, \ldots, v^n = 0$ to vanish. In this way, one needs only solve for the dynamics of the $(n - m)$ nonvanishing quasivelocities $v^1, \ldots, v^{n-m}$ and then use the kinematic relations

$$\dot{q}^i = \psi^i_j v^j \tag{10.2}$$

to solve for $q = q(t)$. When the nonholonomic constraints are given by $v^{n-m+1} = 0, \ldots, v^n = 0$, this equation reduces to $\dot{q}^i = \psi^i_\alpha v^\alpha$ for $\alpha = 1, \ldots, (n - m)$. Together, this will constitute a coupled system of $(2n - m)$ first-order differential equations known in vector form as Hamel's equation. These equations were recently generalized to infinite-dimensional nonholonomic systems in [269].

To describe quasivelocities in a geometric context, we will make use of the notion of a *moving frame*. The general definition of a moving frame was introduced by Cartan [44] and elaborated upon by Weyl [292]; see also, for example, [159, 160] and [285]. For the use of moving frames in nonholonomic mechanics, see, for example, [30], [80], and [206].

Definition 10.1. A *moving frame* on an open set U of a manifold Q is a collection $\mathscr{E} = \{E_1, \ldots, E_n\}$ of smooth vector fields that form a basis for the tangent space at each point in U. The *frame bundle*, $\mathcal{F}$, of manifold Q is the disjoint union of all frames on Q.

Remark. The frame bundle $\mathcal{F}$ may be represented *locally* as the direct product set $\mathcal{F} = Q \times \mathrm{GL}(n; \mathbb{R})$.

Definition 10.2. Let $U \subset Q$ be an open, connected subset of Q and let $\mathscr{E} = \{E_1, \ldots, E_n\}$ be a set of noncommuting vector fields that constitutes a moving frame on U. If $\dot{q} \in T_qU$ is the velocity of a curve at q, then the components of $\dot{q}$ with respect to the basis $\mathscr{E}$ are referred to as the *quasivelocities* of $\dot{q}$ at point q.

For a nonholonomic system with a constraint distribution $\mathcal{D}$, we say that a frame $\mathscr{E}$ on $U \subset Q$ is *consistent with the constraints* if its first $(n-m)$ vector fields $E_1, \ldots, E_{n-m}$ span the constraint distribution $\mathcal{D}|_U$.

Associated with a particular moving frame $\mathscr{E}$ is a *moving coframe*

$$\mathscr{W} = \{\omega^1, \ldots, \omega^n\},$$

that is, a set of differential forms that form a basis for each cotangent space $T_q^* U$ and that are dual to the vector fields that comprise the moving frame $\mathscr{E}$, i. e., we have

$$\omega^i(E_j) = \delta^i_j = \begin{cases} 1 & \text{if } i = j, \\ 0 & \text{if } i \neq j. \end{cases}$$

Exercise 10.1. Show that a moving frame $\mathscr{E}$ is *consistent with the distribution* $\mathcal{D}$ if and only if the distribution $\mathcal{D}$ is the null space of the last m differential forms $\omega^{n-m+1}, \ldots, \omega^n$ of the associated moving coframe $\mathscr{W}$.

Remark. Let $\mathscr{E}$ be a moving frame consistent with the nonintegrable constraint distribution $\mathcal{D}$. Then, in terms of the associated moving coframe $\mathscr{W}$, the *quasivelocities* of a velocity vector v are the quantities $v^1 = \omega^1(\dot{q}), \ldots, v^n = \omega^n(\dot{q})$.

We denote the components of the vector field E_i relative to the coordinate basis $\partial/\partial q^j$ by the symbols ψ^j_i, so that

$$E_i = \psi^j_i \frac{\partial}{\partial q^j}. \tag{10.3}$$

Exercise 10.2. Show that the components of the differential one-forms ω^i, dual to the moving frame $\mathscr{E}$, relative to the coordinate basis dq^j of $T_q Q$, are given by

$$\omega^i = \phi^i_j dq^j, \tag{10.4}$$

where $\phi = \psi^{-1}$. Equations (10.3) and (10.4) imply equations (10.1) and (10.2).

The ijth entry of matrix ψ is the ith component of vector field E_j relative to the coordinate basis $\partial/\partial q^i$. Similarly, the ijth entry of matrix ϕ is the jth component of the one-form ω^i relative to the coordinate basis dq^i, i. e., we have

$$[\psi] = \begin{bmatrix} | & | & & | \\ E_1 & E_2 & \cdots & E_n \\ | & | & & | \end{bmatrix} \quad \text{and} \quad [\phi] = \begin{bmatrix} - & \omega^1 & - \\ - & \omega^2 & - \\ & \vdots & \\ - & \omega^n & - \end{bmatrix}.$$

Furthermore, the condition that the one-forms ω^i are dual to the vector fields E_i, i. e., that $\omega^i(E_j) = \delta^i_j$, is equivalent to the condition that $\phi^i_k \psi^k_j = \delta^i_j$, or that the matrices ϕ and ψ are inverses of each other.

Typically, in a nonholonomic system, the constraint distribution is given in terms of the associated constraint forms $\omega^a = \phi^a_i dq^i$ for $a = (n-m+1), \dots, n$. These constraint forms dictate the final m rows of transformation matrix ϕ. The first $(n-m)$ rows $\omega^\alpha = \phi^\alpha_i dq^i$, for $\alpha = 1, \dots, (n-m)$, are then chosen in a convenient way, and the vector fields E_i constituting the moving frame are obtained by inverting matrix ϕ.

Once a moving frame (and dual moving coframe) is specified, one may define a mapping $\omega : TQ \to \mathbb{R}^n$ by the relation

$$\omega(X) = (\omega^1(X), \dots, \omega^n(X)), \tag{10.5}$$

so that $\omega(X(q))$ is the vector in $\mathbb{R}^n$ whose components are the components of the vector $X(q) \in T_qQ$ relative to the moving frame $\mathscr{E}$.

Let $\mathscr{E}$ be a moving frame in an open set U that contains a curve $c : [a,b] \to U$. Further, given a variation $\vartheta : [-\varepsilon, \varepsilon] \times [a,b] \to Q$, recall that the *virtual displacement*, defined in Definition 7.5, is denoted $\delta c : [a,b] \to TQ|_c$. As it turns out, the components of the virtual displacements δc relative to the moving frame $\mathscr{E}$ will be of great utility. We will denote these components by the symbols ζ^i, which may be precisely defined by the relation $\zeta = \omega(\delta q)$, so that

$$\delta c(t) = \zeta^i(t) E_i(c(t)). \tag{10.6}$$

As a result of this definition, the virtual displacement is kinematically admissible, i. e., $\delta c(t) \in \mathcal{D}_{c(t)}$, along curve c if and only if the components $\zeta^{n-m+1}(t) = 0, \dots, \zeta^n(t) = 0$ vanish identically for $t \in [a,b]$.

10.2 A Lie Algebra Bundle

When dealing with quasivelocities, it is practical to define a certain vector bundle, relative to the given moving frame, in which the quasivelocities live, as was introduced in [30] and further discussed in [206]. As we will see, this vector bundle can be endowed with a particular bracket structure that then constitutes a Lie algebra on each of its fibers, giving it the structure of a Lie algebra bundle. This sort of construction is motivated by the Euler–Poincaré equation on Lie groups, in which the velocity is described in terms of its pullback to the Lie algebra.

Definition 10.3. Given a moving frame $\mathscr{E}$ on the open set $U \subset Q$ and two elements $v, w \in \mathbb{R}^n$, we define the antisymmetric bracket operation $[\cdot,\cdot]_q : \mathbb{R}^n \times \mathbb{R}^n \to \mathbb{R}^n$ at the point $q \in U$ by the relation

$$[v,w]_q = v^i w^j \omega([E_i, E_j]), \tag{10.7}$$

where $[\cdot,\cdot]$ is the Jacobi–Lie bracket of vector fields on Q and $\omega : TQ \to \mathbb{R}^n$ is the mapping defined by relation (10.5).

Given this definition, each tangent space T_qU is therefore isomorphic to the Lie algebra $W_q = (\mathbb{R}^n, [\cdot,\cdot]_q)$ and the tangent bundle TU is diffeomorphic to the Lie algebra bundle W over U whose fibers are the Lie algebras W_q. Note that the mapping (10.5) maps vector fields on TU into the Lie algebra bundle, i. e., $\omega : TU \to W$, so that vectors in W_q correspond to the components of vectors in T_qU with respect to the moving frame.

The dual of the bracket (10.7) is defined as the operation $[\cdot,\cdot]_q^* : W_q \times W_q^* \to W_q^*$ given by

$$\langle [v,\alpha]_q^*, w\rangle = \langle \alpha, [v,w]_q\rangle . \tag{10.8}$$

Similarly to how we define a mapping $\omega : TQ \to W$ by relation (10.5), we can also define a mapping $E : T^*Q \to W^*$ by the relation

$$E[a] = (a[E_1],\dots,a[E_n]) \in W^*,$$

for $a \in T^*Q$. In local coordinates, if $a = a_i dq^i$ and the vector fields E_i are given by (10.3), we obtain

$$E[a] = \left(a_i\psi_1^i,\dots,a_i\psi_n^i\right).$$

The mapping E may be extended to act on functions of q in an obvious way; namely, one can define a related mapping $E : C^\infty(U) \to W^*$ by the relation $E[f] = E[df]$. Whether E is acting on functions or one-forms is understood from the context.

Exercise 10.3. Show that the definition $E[f] := E[df]$ is equivalent to

$$E[f] = (E_1[f],\dots,E_n[f]) \in W_q^*. \tag{10.9}$$

Recall that $df = \frac{\partial f}{\partial q^i} dq^i$.

Lemma 10.1. *Given a vector field $X \in TU$, a covector field $a \in T^*U$, and a moving frame $\mathscr{E}$, the following relation holds:*

$$\langle a, X\rangle = \langle E[a], \omega(X)\rangle , \tag{10.10}$$

where $E[a] \in W^*$ and $\omega(X) \in W$.

Proof. A simple computation yields

$$\langle a, X\rangle = a_i X^i = a_i\psi_k^i\phi_j^k X^j = a[E_k]\omega^k(X) = \langle E[a], \omega(X)\rangle ,$$

thereby demonstrating the result. □

Definition 10.4. The *Hamel coefficients* γ_{ij}^k are the structure coefficients of the moving frame $\mathscr{E}$, i. e.,

$$[E_i, E_j] = \gamma_{ij}^k E_k. \tag{10.11}$$

Relative to the Hamel coefficients, the bracket defined by (10.7) may be expressed locally as

$$[v,w]_q = (\gamma_{ij}^1 v^i w^j, \dots, \gamma_{ij}^n v^i w^j) \in W_q.$$

Proposition 10.1. *Relative to the components of the vector fields defining the moving frame, the Hamel coefficients are given by the relations*

$$\gamma_{ij}^k = -\left(\frac{\partial \phi_h^k}{\partial q^g} - \frac{\partial \phi_g^k}{\partial q^h}\right)\psi_i^g \psi_j^h. \tag{10.12}$$

Proof. A simple computation shows

$$\begin{aligned}
[E_i, E_j] &= E_i E_j - E_j E_i \\
&= \psi_i^g \frac{\partial}{\partial q^g}\left(\psi_j^h \frac{\partial}{\partial q^h}\right) - \psi_j^h \frac{\partial}{\partial q^h}\left(\psi_i^g \frac{\partial}{\partial q^g}\right) \\
&= \left(\psi_i^g \frac{\partial \psi_j^h}{\partial q^g}\phi_h^k - \psi_j^h \frac{\partial \psi_i^g}{\partial q^h}\phi_g^k\right) E_k = \gamma_{ij}^k E_k,
\end{aligned}$$

where the last equality holds because of the identity

$$\phi_j^k \frac{\partial \psi_i^j}{\partial q^h} = -\psi_i^j \frac{\partial \phi_j^k}{\partial q^h},$$

obtained by differentiating the identity $\phi_j^k \psi_i^j = \delta_i^k$ with respect to q^h. □

The coordinate form (10.12) of the Hamel coefficients is the classic form used in, for example, [106] and [222]. However, as we will see in several applications, it is usually simpler to compute these coefficients using the modern definition (10.11).

Following the preceding definitions, each $\dot{q} \in TU$ may be uniquely written as the sum

$$\dot{q} = \langle E(q), v^{\mathcal{D}} \rangle + \langle E(q), v^{\mathcal{U}} \rangle,$$

where $\langle E(q), v^{\mathcal{D}} \rangle \in \mathcal{D}_q$ and where $v = \omega(\dot{q}) \in W_q$. Whenever the moving frame is consistent with the constraint distribution, vectors $v^{\mathcal{D}}$ and $v^{\mathcal{U}}$ in W_q are literally just the vectors

$$\begin{aligned}
v^{\mathcal{D}} &= (v^1, \dots, v^{n-m}, 0, \dots, 0), \\
v^{\mathcal{U}} &= \left(0, \dots, 0, v^{n-m+1}, \dots, v^n\right).
\end{aligned}$$

Similarly, each $a \in T^*U$ can be uniquely decomposed as

$$a = \langle \alpha_{\mathcal{D}}, \omega(q) \rangle + \langle \alpha_{\mathcal{U}}, \omega(q) \rangle,$$

where $\langle \alpha_{\mathcal{D}}, \omega(q) \rangle$ is the component of α along the dual of $\mathcal{D}_q$, $\alpha = E(a)$ and where $\omega \in T^*U \times \cdots \times T^*U$ denotes the dual frame of $E \in TU \times \cdots \times TU$. Hence, the constraints read

$$v = v^{\mathcal{D}} \quad \text{or} \quad v^{\mathcal{U}} = 0.$$

Thus, for a virtual displacement δq that satisfies the constraints we have

$$\zeta = \zeta^{\mathcal{D}} \quad \text{or} \quad \zeta^{\mathcal{U}} = 0,$$

where $\zeta = \omega(\delta q) \in W_q$. In the case of nonholonomic mechanics, $v^{\mathcal{U}} = 0$ along the fiducial curve but not elsewhere in the variation (where $v = \omega(\dot{q})$). However, if one chooses the constrained variational principle, as one encounters in the calculus of variations, thereby obtaining the variational nonholonomic dynamics (vakonomic motions), one instead enforces $v^{\mathcal{U}} = 0$ throughout the variation using the condition $\delta v^{\mathcal{U}} = 0$. If this is selected, it is impossible, in general, to simultaneously enforce the condition $\zeta^{\mathcal{U}} = 0$.

10.3 Maggi's Equation

As an example of the utility of a moving-frame decomposition, we next devise a way to enforce the Lagrange–d'Alembert principle *without* the use of Lagrange multipliers. The resulting equation is known as *Maggi's equation.*

Theorem 10.1 (Maggi's Equation). *Let $(Q, L, \mathcal{D})$ be a nonholonomic system. Then a curve $c : [a, b] \to Q$ satisfies the Lagrange–d'Alembert principle* (7.8) *if and only if it is a solution of* Maggi's equation

$$E_{\mathcal{D}} \left[\frac{d}{dt} \frac{\partial L}{\partial \dot{q}} - \frac{\partial L}{\partial q} \right] = 0, \tag{10.13}$$

where $E_{\mathcal{D}} = (E_1, \ldots, E_{n-m})$. In terms of local coordinates and given a moving frame $\mathscr{E}$, consistent with the constraints, on an open set U with components $E_i = \psi_i^j \frac{\partial}{\partial q^j}$, this equation is equivalent to

$$\left(\frac{d}{dt} \frac{\partial L}{\partial \dot{q}^i} - \frac{\partial L}{\partial q^i} \right) \psi_\alpha^i = 0 \tag{10.14}$$

for $\alpha = 1, \ldots, (n - m)$.

Maggi's equation is comprised of only $(n - m)$ components, each of which represents a second-order ordinary differential equation. Those equations are coupled with the m constraint equations, forming a final set equivalent to $(2n - m)$ first-order differential equations.

Proof. Since the moving frame is consistent with the constraints, we have at each $q \in U$, $\mathcal{D}_q = \operatorname{span}\{E_1(q), \dots, E_{n-m}(q)\}$. By the Lagrange–d'Alembert principle, we require

$$\left\langle \frac{d}{dt}\frac{\partial L}{\partial \dot{q}} - \frac{\partial L}{\partial q}, \delta q \right\rangle = 0,$$

where the components of δq are not algebraically independent but subject to the condition that $\delta q \in \mathcal{D}$. From Lemma 10.1 we have

$$\left\langle E\left[\frac{d}{dt}\frac{\partial L}{\partial \dot{q}} - \frac{\partial L}{\partial q}\right], \zeta \right\rangle = 0,$$

where $\zeta = \omega(\delta q)$. However, the left-hand side may be replaced as follows:

$$\left\langle E_{\mathcal{D}}\left[\frac{d}{dt}\frac{\partial L}{\partial \dot{q}} - \frac{\partial L}{\partial q}\right], \zeta^{\mathcal{D}} \right\rangle + \left\langle E_{\mathcal{U}}\left[\frac{d}{dt}\frac{\partial L}{\partial \dot{q}} - \frac{\partial L}{\partial q}\right], \zeta^{\mathcal{U}} \right\rangle = 0.$$

Finally, recall that $\delta q \in \mathcal{D}$ if and only if $\zeta = \zeta^{\mathcal{D}}$ and $\zeta^{\mathcal{U}} = 0$, which demonstrates the result. □

Example 10.1 (Knife Edge on Inclined Plane). Consider again the system described in Example 7.7. The constraint form is given by

$$\omega^3 = -\sin\varphi dx + \cos\varphi dy.$$

The simplest way to span T^*Q with a moving coframe is to define the one-forms $\omega^1 = d\varphi$ and $\omega^2 = \cos\varphi dx + \sin\varphi dy$. Hence, matrix ϕ and its inverse are given by

$$[\phi] = \begin{bmatrix} 1 & 0 & 0 \\ 0 & \cos\varphi & \sin\varphi \\ 0 & -\sin\varphi & \cos\varphi \end{bmatrix} \quad \text{and} \quad [\psi] = \begin{bmatrix} 1 & 0 & 0 \\ 0 & \cos\varphi & -\sin\varphi \\ 0 & \sin\varphi & \cos\varphi \end{bmatrix},$$

respectively. We therefore choose the vector fields

$$E_1 = \frac{\partial}{\partial\varphi}, \quad E_2 = \cos\varphi\frac{\partial}{\partial x} + \sin\varphi\frac{\partial}{\partial y}, \quad E_3 = -\sin\varphi\frac{\partial}{\partial x} + \cos\varphi\frac{\partial}{\partial y}, \tag{10.15}$$

the column vectors of $[\psi]$, as a moving frame. Now, the variational derivative of the Lagrangian is the one-form

$$\frac{d}{dt}\frac{\partial L}{\partial \dot{q}} - \frac{\partial L}{\partial q} = I\ddot{\varphi}d\varphi + (m\ddot{x} - mg\sin\alpha)dx + m\ddot{y}dy$$

on the cotangent bundle. Recalling that, for $a \in T^*Q$, we have $E_{\mathcal{D}}[a] = (a(E_1), a(E_2), 0) \in W^*$, we find

$$E_{\mathcal{D}}\left[\frac{d}{dt}\frac{\partial L}{\partial \dot{q}} - \frac{\partial L}{\partial q}\right] = (I\ddot{\varphi}, m\ddot{x}\cos\varphi - mg\sin\alpha\cos\varphi + m\ddot{y}\sin\varphi, 0).$$

Maggi's equation therefore yields

$$I\ddot{\varphi} = 0,$$

$$m(\ddot{x}\cos\varphi + \ddot{y}\sin\varphi) - mg\sin\alpha\cos\varphi = 0.$$

Finally, the constraint equation $-\sin\varphi\dot{x} + \cos\varphi\dot{y} = 0$ can be differentiated to obtain

$$-\sin\varphi\ddot{x} + \cos\varphi\ddot{y} - \cos\varphi\dot{x}\dot{\varphi} - \sin\varphi\dot{y}\dot{\varphi} = 0.$$

Together, this equation and the two Maggi's equations yield a system that can be solved for $\ddot{\varphi}$, $\ddot{x}$, and $\ddot{y}$, which can then be integrated numerically to solve for the resulting trajectory as a function of time. □

Exercise 10.4. Write out Maggi's equation for the vertical rolling disc discussed in Example 7.6.

10.4 Hamel's Equation

In this section, we derive Hamel's classic equation from a modern geometric perspective. The components of this equation constitute a minimal set of $(2n - m)$ first-order ordinary differential equations that govern the ensuing dynamics of a nonholonomic mechanical system. These equations fully utilize the power of quasivelocities. One starts by expressing the Lagrangian in terms of quasivelocities, as follows.

Definition 10.5. Let $(Q, L, \mathcal{D})$ be a nonholonomic mechanical system, $\mathscr{E}$ a moving frame on the open set $U \subset Q$, and $W = (\mathbb{R}^n, [\cdot,\cdot]_q)$ its associated Lie algebra bundle. Then the *Lagrangian expressed in terms of quasivelocities* is the mapping $l : W \to \mathbb{R}$ defined by the relation $l(q, v) = L(q, v^i E_i)$.

Remark. A Lagrangian expressed in terms of quasivelocities must be taken as the *unconstrained* Lagrangian. Even if the moving frame $\mathscr{E}$ is consistent with the constraints (so that the constraints, in terms of quasivelocities, are given by the relations $v^{\mathcal{U}} = 0$), one must nevertheless leave these quasivelocities in the expression for l, as they could potentially affect the equations of motion via their derivatives, which may no longer necessarily depend on the vanishing quasivelocity components.

Since a Lagrangian expressed in terms of quasivelocities is a function on the Lie algebra bundle W and not the tangent bundle TQ, one must determine the appropriate formula to relate variations of the quasivelocities δv to the components of the time derivative of the virtual displacements relative to the moving frame. In other words, the quasivelocity analogy of the relation $\delta\dot{q} = d/dt(\delta q)$ is not simply $\delta v = \dot{\zeta}$, where $\zeta = \omega(\delta q)$, but $\delta v = \dot{\zeta} + \cdots$, a relation that contains additional terms. This relation is known as the *transpositional relation* and is a key ingredient to formulating Hamel's equations of motion in terms of a quasivelocity Lagrangian. To derive this relation, we start with the following lemma.

Lemma 10.2. *Let $X, Y \in \mathfrak{X}(U)$ and let $x = \omega(X)$ and $y = \omega(Y)$ be their representations in the Lie algebra bundle W with respect to the moving frame $\mathscr{E}$, respectively. Then*

$$[X, Y] = \left(X[y^i] - Y[x^i]\right) E_i + x^i y^j [E_i, E_j], \tag{10.16}$$

where the bracket $[\cdot, \cdot]$ represents the standard Jacobi–Lie bracket on TU.

Exercise 10.5. Prove Lemma 10.2.

As an immediate corollary to the preceding lemma, we obtain the following Hamel transpositional relation, which relates variations in quasivelocities to the time derivatives of components of the virtual displacement plus a bracket term.

Corollary 10.1 (Hamel Transpositional Relation). *Let $c : [a, b] \to U$ be a smooth curve, $\vartheta : [-\varepsilon, \varepsilon] \times [a, b] \to U$ an arbitrary C^2 variation of c, $\dot{q} = \vartheta_t(s, t)$ the velocity of the varied curves, and $\delta q = \vartheta_s(s, t)$ the virtual displacements. Then $v = \omega(\dot{q})$ and $\zeta = \omega(\delta q)$, the components of $\dot{q}$ and δq relative to the moving frame, represented as vectors in the Lie algebra bundle W, respectively, satisfy the* Hamel transpositional relation

$$\delta v = \dot{\zeta} + [v, \zeta]_q, \tag{10.17}$$

where $[\cdot, \cdot]_q$ is the bracket on W_q, defined by (10.7).

Proof. Since the variation ϑ is C^2, it follows from Clairaut's theorem that $[\dot{q}, \delta q] = 0$. Applying (10.16) to vector fields $\dot{q}$ and δq on $[-\varepsilon, \varepsilon] \times [a, b]$, one obtains

$$0 = \left(\dot{q}[\zeta^i] - \delta q[v^i]\right) E_i + v^i \zeta^j [E_i, E_j].$$

Recognizing that $\dot{q}[\zeta^i] = \dot{\zeta}^i$ and $\delta q[v^i] = \delta v^i$, by the chain rule, and applying the mapping $\omega : TU \to W$, we obtain our result. □

We are now in a position to derive Hamel's equation, which yields the equations of motion that the Lagrangian l, expressed in terms of quasivelocities, satisfies for a nonholonomic system.

Theorem 10.2 (Hamel). *Let $(Q, L, \mathcal{D})$ be a nonholonomic mechanical system, $\mathscr{E}$ a moving frame consistent with the constraints on an open set $U \subset Q$, W its associated Lie algebra bundle, and $l : W \to \mathbb{R}$ the Lagrangian expressed in terms of quasivelocities. Then a curve $c : [a, b] \to U$ satisfies the Lagrange–d'Alembert principle if and only if it is a solution of* Hamel's equation

$$\left(\frac{d}{dt}\frac{\partial l}{\partial v} - E[l] - \left[v^{\mathcal{D}}, \frac{\partial l}{\partial v}\right]_q^*\right)_{\mathcal{D}} = 0, \quad v^{\mathcal{U}} = 0, \quad \dot{q} = \langle E(q), v^{\mathcal{D}} \rangle. \tag{10.18}$$

In terms of local coordinates, this equation may be equivalently expressed as

$$\frac{d}{dt}\frac{\partial l}{\partial v^\alpha} - \frac{\partial l}{\partial q^i}\psi^i_\alpha - \frac{\partial l}{\partial v^i}\gamma^i_{\beta\alpha}v^\beta = 0, \quad v^a = 0, \quad \dot{q}^i = \psi^i_\alpha v^\alpha, \tag{10.19}$$

where $\alpha, \beta = 1, \ldots, (n - m)$, $a = (n - m + 1), \ldots, n$, and $i = 1, \ldots, n$.

Proof. First, $v^{\mathcal{U}} = 0$ is simply the imposition of the nonholonomic constraints and $\dot{q} = \langle E(q), v^{\mathcal{D}} \rangle$ is the kinematic relation (10.2). To prove Hamel's equation, let us begin with the Lagrange–d'Alembert principle,

$$\delta\mathcal{I}[c] = \delta \int_a^b l(q, v)\, dt = \int_a^b \left(\left\langle \frac{\partial l}{\partial q}, \delta q \right\rangle + \left\langle \frac{\partial l}{\partial v}, \delta v \right\rangle \right) dt,$$

where the virtual displacements $\delta q \in \mathcal{D}$ are taken to be kinematically admissible, as per the Lagrange–d'Alembert principle. Now, by Lemma 10.1 and by the Hamel transpositional relation (10.17), we obtain

$$\delta\mathcal{I}[c] = \int_a^b \left(\langle E[l], \zeta \rangle + \left\langle \frac{\partial l}{\partial v}, \dot{\zeta} + [v, \zeta]_q \right\rangle \right) dt.$$

Integrating by parts, while recalling that the variations vanish at the endpoints, we find

$$\delta\mathcal{I}[c] = \int_a^b \left\langle E[l] - \frac{d}{dt}\frac{\partial l}{\partial v} + \left[v, \frac{\partial l}{\partial v} \right]_q^*, \zeta \right\rangle dt.$$

Since requiring $\delta q \in \mathcal{D}$ is equivalent to the requirement that $\zeta^{\mathcal{U}} = 0$, we find that, to obtain an extreme value of this action subject to the condition that virtual displacements be kinematically admissible, the relationship

$$\left(\frac{d}{dt}\frac{\partial l}{\partial v} - E[l] - \left[v, \frac{\partial l}{\partial v} \right]_q^* \right)_{\mathcal{D}} = 0$$

must hold, similar to what we encountered in the proof of Maggi's equation. We thus obtain our result. □

Note that the components of Hamel's equation form a minimal set of $(2n - m)$ first-order ordinary differential equations. A related form of these equations was applied to optimal control problems in [203, 204]. Next we illustrate their benefit in the context of several applications.

Example 10.2. In this example, let us consider a dumbbell with a knife edge constraint, as depicted in Figure 10.1. The dumbbell consists of two weights, each of mass m, that are connected by a massless rod of length d. The first dumbbell is located at position (x, y) and has a knife edge constraint along the direction of the bar. The configuration manifold for the system is $Q = S^1 \times \mathbb{R}^2$, with local coordinates given by (φ, x, y), where θ is the angle between the bar and the x-axis.

The nonholonomic constraint is given by the equation

$$-\sin\varphi \dot{x} + \cos\varphi \dot{y} = 0.$$

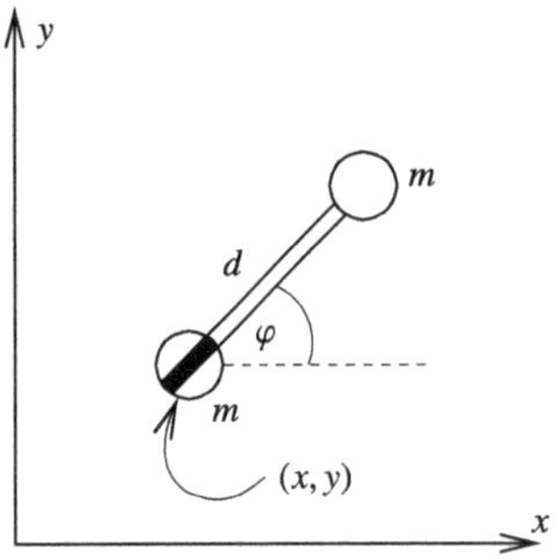

Figure 10.1: Diagram of dumbbell with knife edge constraint.

We may therefore choose vector fields (10.15) as our moving frame. The corresponding quasivelocities are given by

$$\begin{aligned} v_1 &= \dot{\varphi}, \\ v_2 &= \dot{x}\cos\varphi + \dot{y}\sin\varphi, \\ v_3 &= -\dot{x}\sin\varphi + \dot{y}\cos\varphi. \end{aligned}$$

The quasivelocity v_2 denotes the component of the dumbbell's velocity in the direction of the rod, whereas v_3 denotes the component of the velocity in a direction perpendicular to the rod. The nonholonomic constraint expressed in terms of quasivelocities simplifies to the relation $v_3 = 0$. The kinetic energy for this system is given in terms of quasivelocities by

$$\begin{aligned} T(q, v) &= \frac{1}{2}m(v_2^2 + v_3^2) + \frac{1}{2}m\left[v_2^2 + (v_3 + dv_1)^2\right] \\ &= m\left(\frac{1}{2}d^2v_1^2 + v_2^2 + v_3^3 + dv_1v_3\right). \end{aligned}$$

Now let us suppose that our dumbbell is located on an *inclined plane*, with angle α, as in Example 7.2, so that the system further possesses a potential energy of $U = -mg(2x + d\cos\varphi)\sin\alpha$, where g is the acceleration due to gravity. The Lagrangian, expressed in terms of quasivelocities, is therefore given by

$$l(q, v) = m\left(\frac{1}{2}d^2v_1^2 + v_2^2 + v_3^3 + dv_1v_3\right) + mg(2x + d\cos\varphi)\sin\alpha.$$

The next ingredient we require is the Hamel coefficients (10.11) for our moving frame (10.15). A simple computation reveals the nonzero commutation relations of the moving frame (10.15) to be

$$\begin{aligned} [E_1, E_2] &= E_3, \\ [E_1, E_3] &= -E_2. \end{aligned}$$

Of course, since the Jacobi–Lie bracket is antisymmetric, we also have $[E_2, E_1] = -E_3$ and $[E_3, E_1] = E_2$. Therefore, the nonzero Hamel coefficients are given by

$$\gamma_{12}^3 = \gamma_{31}^2 = 1, \quad \gamma_{21}^3 = \gamma_{13}^2 = -1.$$

Next, we begin computing the various terms of Hamel's equation (10.19). We find

$$
\begin{aligned}
\frac{d}{dt}\frac{\partial l}{\partial v_1} &= md^2\dot{v}_1,\\
\frac{d}{dt}\frac{\partial l}{\partial v_2} &= 2m\dot{v}_2,\\
E_1[l] &= -mgd\sin\varphi\sin\alpha,\\
E_2[l] &= 2mg\cos\varphi\sin\alpha,\\
\frac{\partial l}{\partial v_2}\gamma^2_{31}v_3 + \frac{\partial l}{\partial v_3}\gamma^3_{21}v_2 &= -mdv_1v_2,\\
\frac{\partial l}{\partial v_3}\gamma^3_{12}v_1 &= mdv_1^2,
\end{aligned}
$$

where we have set $v_3 = 0$ identically *after* taking the derivatives of l. (Notice that otherwise the last two equations would equal zero.) Combining these according to Hamel's equation, we obtain the following equations of motion:

$$
\begin{aligned}
d\dot{v}_1 + g\sin\varphi\sin\alpha + v_1v_2 &= 0,\\
\dot{v}_2 - g\cos\varphi\sin\alpha - \frac{d}{2}v_1^2 &= 0.
\end{aligned}
$$

These are then combined with the kinematic relations (10.2), from which one obtains

$$
\begin{aligned}
\dot{\varphi} &= v_1,\\
\dot{x} &= \cos\varphi v_2,\\
\dot{y} &= \sin\varphi v_2.
\end{aligned}
$$

Together, these equations form a minimum set of five first-order ordinary differential equations, which may be solved numerically to determine the ensuing motion of the system given a set of initial conditions. □

Exercise 10.6. Write out Hamel's equation for the knife edge on the inclined plane of Examples 7.7 and 10.1.

Exercise 10.7. Write out Hamel's equation for the vertical rolling disc of Example 7.6.

There are instances where quasivelocities, and thus Hamel's equation, are useful even in the absence of constraints. A famous example of this is Euler's equation for rigid-body mechanics. The body axis components of the angular velocity are not derivatives of functions of the rigid body's configuration; they therefore constitute a set of quasivelocities, as we will see in our next example.

Example 10.3 (Euler's Equation for Rigid-Body Mechanics). Let us consider again a rigid body with a fixed point, described by the configuration manifold SO(3). We will assume an *xyz* principal axis body-fixed coordinate system with the origin located at the body's center of mass. For local coordinates, let us choose Type I Euler angles (ψ, θ, φ). The quasivelocities are

$$\begin{aligned} v_1 &= \Omega_x = -\dot{\psi}\sin\theta + \dot{\varphi}, \\ v_2 &= \Omega_y = \dot{\psi}\cos\theta\sin\varphi + \dot{\theta}\cos\varphi, \\ v_3 &= \Omega_z = \dot{\psi}\cos\theta\cos\varphi - \dot{\theta}\sin\varphi, \end{aligned}$$

where $(\Omega_x, \Omega_y, \Omega_z)$ is the angular velocity relative to the body-fixed principal axis frame. The Lagrangian is given by the kinetic energy as

$$l = \frac{1}{2}\left(I_x\Omega_x^2 + I_y\Omega_y^2 + I_z\Omega_z^2\right),$$

where (I_x, I_y, I_z) are the principal moments of inertia. The transformation matrices are

$$[\phi] = \begin{bmatrix} -\sin\theta & 0 & 1 \\ \cos\theta\sin\varphi & \cos\varphi & 0 \\ \cos\theta\cos\varphi & -\sin\varphi & 0 \end{bmatrix}, \quad [\psi] = \begin{bmatrix} 0 & \sec\theta\sin\varphi & \sec\theta\cos\varphi \\ 0 & \cos\varphi & -\sin\varphi \\ 1 & \tan\theta\sin\varphi & \tan\theta\cos\varphi \end{bmatrix}.$$

It follows that the moving frame is given by

$$\begin{aligned} E_1 &= \frac{\partial}{\partial\varphi}, \\ E_2 &= \sec\theta\sin\varphi\frac{\partial}{\partial\psi} + \cos\varphi\frac{\partial}{\partial\theta} + \tan\theta\sin\varphi\frac{\partial}{\partial\varphi}, \\ E_3 &= \sec\theta\cos\varphi\frac{\partial}{\partial\psi} - \sin\varphi\frac{\partial}{\partial\theta} + \tan\theta\cos\varphi\frac{\partial}{\partial\varphi}. \end{aligned}$$

Exercise 10.8. Show that the nonzero commutators are given by $[E_1, E_2] = E_3$, $[E_1, E_3] = -E_2$, and $[E_2, E_3] = E_1$ (and their negatives). How is this related to the cross product on $\mathbb{R}^3$?

It follows that the nonzero Hamel coefficients are $\gamma_{12}^3 = 1$, $\gamma_{21}^3 = -1$, $\gamma_{13}^2 = -1$, $\gamma_{31}^2 = 1$, $\gamma_{23}^1 = 1$, and $\gamma_{32}^1 = -1$.

Exercise 10.9. Show that Hamel's equation for this system yields

$$\begin{aligned} I_x\dot{\Omega}_x + (I_z - I_y)\Omega_y\Omega_z &= 0, \\ I_y\dot{\Omega}_y + (I_x - I_z)\Omega_z\Omega_x &= 0, \\ I_z\dot{\Omega}_z + (I_y - I_x)\Omega_x\Omega_y &= 0. \end{aligned}$$

How are these related to Euler's equation?

This example shows that quasivelocities sometimes have applications even in unconstrained systems. □

Example 10.4 (Ball Rolling Inside a Cylinder). In this example, we will consider the case of a sphere rolling without slipping on the inside of a cylinder, as described in [271]. See [254] and [297] for a generalized version of this problem and also [49–52] for additional early references regarding the analysis of rolling balls. We will consider a homogeneous sphere of radius r to be rolling without slipping inside a hollow cylinder with radius R, as depicted in Figure 10.2. The sphere's center of mass is the point C, and the contact point between the sphere and the cylinder is P. The line segment CP makes an angle ρ with the positive x-axis. We will decompose the angular velocity using cylindrical coordinates, so that $\boldsymbol{\omega} = \omega_r \mathbf{e}_r + \omega_\rho \mathbf{e}_\rho + \omega_z \mathbf{e}_z$.

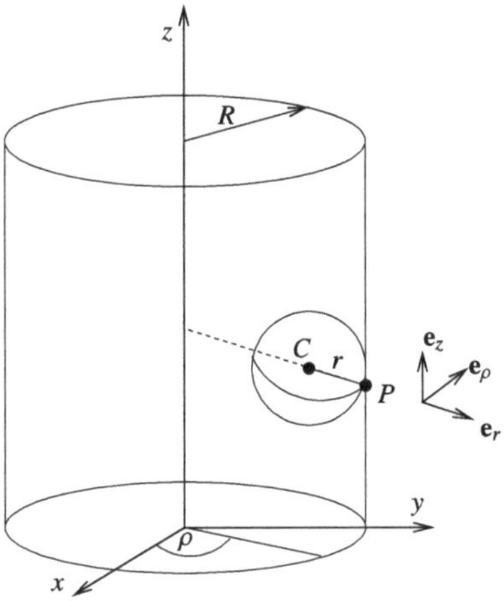

Figure 10.2: Rigid sphere rolling without slipping inside a hollow cylinder.

We will use the Type II (classical) Euler angles (φ, θ, ψ) as local coordinates on SO(3), so that the configuration manifold $Q = \mathrm{SO}(3) \times \mathbb{R} \times S^1$ may be described by the local coordinates $(\varphi, \theta, \psi, z, \rho)$. Recall that the spatial components of the angular velocity are given by (9.20)–(9.22). The spatial components (ω_x, ω_y) are related to the components of the angular velocity in cylindrical coordinates via the relations

$$\begin{aligned} \omega_x &= \omega_r \cos\rho - \omega_\rho \sin\rho, \\ \omega_y &= \omega_r \sin\rho + \omega_\rho \cos\rho, \end{aligned}$$

so that the quasivelocities $(\omega_r, \omega_\rho, \omega_z)$ are given by

$$\begin{aligned} \omega_r &= \dot\theta \cos(\varphi - \rho) + \dot\psi \sin\theta \sin(\varphi - \rho), \\ \omega_\rho &= \dot\theta \sin(\varphi - \rho) - \dot\psi \sin\theta \cos(\varphi - \rho), \\ \omega_z &= \dot\varphi + \dot\psi \cos\theta. \end{aligned}$$

The Lagrangian is given by

$$L = \frac{m}{2}\left(\dot z^2 + (R - r)^2 \dot\rho^2\right) + \frac{I}{2}\left(\dot\varphi^2 + \dot\theta^2 + \dot\psi^2 + 2\dot\varphi\dot\psi \cos\theta\right) - mgz.$$

The rotational kinetic-energy term is equivalent to $\omega_r^2 + \omega_\rho^2 + \omega_z^2$. Given the relative velocity of contact point P relative to the center of sphere C and the velocity of center C,

$$\mathbf{v}_{P/C} = r\omega_z \mathbf{e}_\rho - r\omega_\rho \mathbf{e}_z,$$
$$\mathbf{v}_C = (R - r)\dot{\rho}\mathbf{e}_\rho + \dot{z}\mathbf{e}_z,$$

respectively, we find the velocity of the contact point to be

$$\mathbf{v}_P = \mathbf{v}_{P/C} + \mathbf{v}_C = (r\omega_z + (R - r)\dot{\rho})\,\mathbf{e}_\rho + (\dot{z} - r\omega_\rho)\mathbf{e}_z.$$

The roll without slipping condition $\mathbf{v}_P = 0$ then yields the nonholonomic constraints

$$u_4 = -r\omega_\rho + \dot{z} = 0,$$
$$u_5 = \lambda\omega_z + \dot{\rho} = 0,$$

where we define the parameter λ as $\lambda = \frac{r}{R-r}$. Hence, the transformation matrix $[\phi]$: $(\dot{\varphi}, \dot{\theta}, \dot{\psi}, \dot{z}, \dot{\rho}) \to (\omega_r, \omega_\rho, \omega_z, u_4, u_5)$ is given by

$$[\phi] = \begin{bmatrix} 0 & \cos(\varphi - \rho) & \sin\theta\sin(\varphi - \rho) & 0 & 0 \\ 0 & \sin(\varphi - \rho) & -\sin\theta\cos(\varphi - \rho) & 0 & 0 \\ 1 & 0 & \cos\theta & 0 & 0 \\ 0 & -r\sin(\varphi - \rho) & r\sin\theta\cos(\varphi - \rho) & 1 & 0 \\ \lambda & 0 & \lambda\cos\theta & 0 & 1 \end{bmatrix},$$

and its inverse is given by

$$[\psi] = [\phi^{-1}] = \begin{bmatrix} -\cot\theta\sin(\varphi - \rho) & \cot\theta\cos(\varphi - \rho) & 1 & 0 & 0 \\ \cos(\varphi - \rho) & \sin(\varphi - \rho) & 0 & 0 & 0 \\ \csc\theta\sin(\varphi - \rho) & -\csc\theta\cos(\varphi - \rho) & 0 & 0 & 0 \\ 0 & r & 0 & 1 & 0 \\ 0 & 0 & -\lambda & 0 & 1 \end{bmatrix}.$$

The moving frame is therefore comprised of the vector fields

$$\begin{aligned} E_1 &= -\cot\theta\sin(\varphi - \rho)\frac{\partial}{\partial\varphi} + \cos(\varphi - \rho)\frac{\partial}{\partial\theta} + \csc\theta\sin(\varphi - \rho)\frac{\partial}{\partial\psi}, \\ E_2 &= \cot\theta\cos(\varphi - \rho)\frac{\partial}{\partial\varphi} + \sin(\varphi - \rho)\frac{\partial}{\partial\theta} - \csc\theta\cos(\varphi - \rho)\frac{\partial}{\partial\psi} + r\frac{\partial}{\partial z}, \\ E_3 &= \frac{\partial}{\partial\varphi} - \lambda\frac{\partial}{\partial\rho}, \\ E_4 &= \frac{\partial}{\partial z}, \\ E_5 &= \frac{\partial}{\partial\rho}. \end{aligned} \tag{10.20}$$

Exercise 10.10. Prove that $[E_1, E_2] = -E_3 - \lambda E_5$, $[E_5, E_1] = E_2 - rE_4$, $[E_5, E_2] = -E_1$, $[E_3, E_1] = -(1+\lambda)E_2 + (1+\lambda)rE_4$, and $[E_3, E_2] = (1+\lambda)E_1$. Conclude that the nonzero Hamel coefficients are given by

$$\begin{aligned}
\gamma_{51}^2 = \gamma_{25}^1 = \gamma_{21}^3 = 1 = -\gamma_{15}^2 = -\gamma_{52}^1 = -\gamma_{12}^3, \\
\gamma_{21}^5 = \lambda = -\gamma_{12}^5, \quad \gamma_{31}^4 = (1+\lambda)r = -\gamma_{13}^4, \\
\gamma_{51}^4 = -r = -\gamma_{15}^4, \quad \gamma_{13}^2 = \gamma_{32}^1 = (1+\lambda) = -\gamma_{31}^2 = -\gamma_{23}^1.
\end{aligned}$$

Further, note that the only *relevant* Hamel coefficients are $\gamma_{21}^3, \gamma_{12}^3, \gamma_{21}^5, \gamma_{12}^5, \gamma_{31}^4, \gamma_{13}^4, \gamma_{13}^2, \gamma_{31}^2, \gamma_{32}^1$, and γ_{23}^1 since only the coefficients $\gamma_{\beta\alpha}^i$ appear in Hamel's equation.

The Lagrangian written in terms of quasivelocities is given by

$$l = \frac{m}{2}\left((r\omega_\rho + u_4)^2 + (R-r)^2(-\lambda\omega_z + u_5)^2\right) + \frac{I}{2}\left(\omega_r^2 + \omega_\rho^2 + \omega_z^2\right) - mgz.$$

The derivatives with respect to the quasivelocities $u^1 = \omega_r$, $u^2 = \omega_\rho$, $u^3 = \omega_z$, u^4, and u^5, with u^4 and u^5 set to zero *after* the differentiation, are given by

$$\begin{aligned}
\frac{\partial l}{\partial u^1} = I\omega_r, \quad \frac{\partial l}{\partial u^2} = (I + mr^2)\omega_\rho, \quad \frac{\partial l}{\partial u^3} = (I + mr^2)\omega_z, \\
\frac{\partial l}{\partial u^4} = mr\omega_\rho, \quad \frac{\partial l}{\partial u^5} = -mr(R-r)\omega_z.
\end{aligned}$$

Each of these five derivatives will come into play for computing the bracket term

$$\left(\left[v^{\mathcal{D}}, \frac{\partial l}{\partial v}\right]_{\mathcal{T}}^*\right)_{\mathcal{D}} = \frac{\partial l}{\partial v^i}\gamma_{\beta\alpha}^i v^\beta \omega^\alpha$$

in Hamel's equation. Note again the *extreme* importance of not setting u^4 and u^5 to zero until *after* the operations of differentiation have been performed.

We next proceed to systematically calculating the various terms of Hamel's equation, obtaining

$$\begin{aligned}
\frac{d}{dt}\frac{\partial l}{\partial u^1} &= I\dot{\omega}_r, \\
\frac{d}{dt}\frac{\partial l}{\partial u^2} &= (I + mr^2)\dot{\omega}_\rho, \\
\frac{d}{dt}\frac{\partial l}{\partial u^3} &= (I + mr^2)\dot{\omega}_z, \\
E_1[l] &= 0, \\
E_2[l] &= -mgr, \\
E_3[l] &= 0.
\end{aligned}$$

Note that the gravitational force leads to acceleration of the rotational degree of freedom corresponding to ω_ρ. We will come back to this important point later.

Exercise 10.11. Show that the bracket terms of Hamel's equation yield

$$\frac{\partial l}{\partial u^i}\gamma^i_{\beta 1}u^\beta = -I\lambda\omega_\rho\omega_z,$$
$$\frac{\partial l}{\partial u^i}\gamma^i_{\beta 2}u^\beta = \lambda I\omega_r\omega_z,$$
$$\frac{\partial l}{\partial u^i}\gamma^i_{\beta 3}u^\beta = 0,$$

where the summation over $\beta = 1,\ldots,3$ and $i = 1,\ldots,5$ is implied.

Combining the foregoing nine equations, as prescribed by Hamel's equation, we obtain the following set of differential equations of motion:

$$I\dot{\omega}_r + I\lambda\omega_\rho\omega_z = 0, \tag{10.21}$$
$$(I + mr^2)\dot{\omega}_\rho + mgr - \lambda I\omega_r\omega_z = 0, \tag{10.22}$$
$$(I + mr^2)\dot{\omega}_z = 0. \tag{10.23}$$

These are then combined with the kinematic relations (10.2) in the form $\dot{q}^i = \psi^i_\alpha v^\alpha$ for $\alpha = 1,\ldots,3$, i. e.,

$$\dot{\varphi} = -\cot\theta\sin(\varphi - \rho)\omega_r + \cot\theta\cos(\varphi - \rho)\omega_\rho + \omega_z,$$
$$\dot{\theta} = \cos(\varphi - \rho)\omega_r + \sin(\varphi - \rho)\omega_\rho,$$
$$\dot{\psi} = \csc\theta\sin(\varphi - \rho)\omega_r - \csc\theta\cos(\varphi - \rho)\omega_\rho,$$
$$\dot{z} = r\omega_\rho, \tag{10.24}$$
$$\dot{\rho} = -\lambda\omega_z, \tag{10.25}$$

yielding, in total, a minimal set of eight coupled first-order ordinary differential equations that can be solved to determine the ensuing motion.

The story, however, does not quite stop here. With a little further analysis, we uncover a very counterintuitive result. First, note from (10.23) that ω_z is constant; more importantly, $\dot{\rho}$ is also constant due to (10.25). Next, observe that (10.21)–(10.22) may be cast into the form

$$\frac{d}{dt}\begin{bmatrix}\omega_r\\ \omega_\rho\end{bmatrix} = \begin{bmatrix}0 & -A\\ B & 0\end{bmatrix}\begin{bmatrix}\omega_r\\ \omega_\rho\end{bmatrix} + \begin{bmatrix}0\\ -C\end{bmatrix},$$

where

$$A = \lambda\omega_z,\quad B = \frac{\lambda\omega_z I}{I + mr^2},\quad C = \frac{mgr}{I + mr^2}.$$

A *particular solution* to this nonhomogeneous system is immediately identified as

$$x_p = \begin{bmatrix}C/B\\ 0\end{bmatrix}.$$

Hence, due to the structure of these equations, a torque on ω_ρ actually affects the solution for ω_r but not the solution for ω_ρ! The *homogeneous solution* has purely imaginary eigenvalues. The final, *total* solution yields

$$\omega_\rho = c_1 \cos\left(\sqrt{\frac{I}{I + mr^2}}\dot{\rho}t\right) + c_2 \sin\left(\sqrt{\frac{I}{I + mr^2}}\dot{\rho}t\right).$$

However, $\dot{z}$ is connected to ω_ρ by (10.24). It follows that the solution for $z(t)$ is also *sinusoidal* in time. This implies that the ball never falls; its vertical height simply oscillates in time! An example trajectory is plotted in Figure 10.3. This example demonstrates the counterintuitive way in which even simple mechanical systems can behave when acted upon by nonholonomic constraints. □

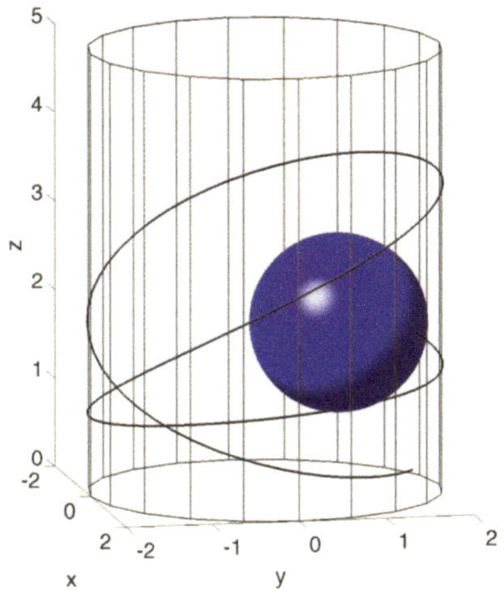

Figure 10.3: Example trajectory of a sphere rolling without slipping inside of a hollow cylinder; acceleration due to gravity is in the negative z-direction.

Exercise 10.12. Set up a Lagrangian and determine Hamel's equation for a sphere rolling without slipping on an inclined plane.

10.5 Relation between the Hamel and Euler–Poincaré Equations

We will now explore the relationship between the Euler–Poincaré equation for mechanics on Lie groups and Hamel's equation. In particular, we will show that the Euler–Poincaré equation is a special case of Hamel's equation for systems with symmetry and that, for such cases, the Lie algebra bundle on TU defined by a moving frame is equivalent to the Lie algebra of the Lie group for a particular choice of moving frame. A similar discussion that utilizes exponential coordinates on Lie groups is presented in [202].

Let G be a Lie group and $\mathfrak{g}$ its associated Lie algebra. Furthermore, suppose that the vectors $\{e_1, \ldots, e_n\} \subset \mathfrak{g}$ form a basis for the Lie algebra and define the left-invariant vector fields $\mathscr{E} = \{E_i\}_{i=1}^n$ by the relations

$$E_i(g) = d(L_g)_e(e_i) = ge_i,$$

where $E_i(g) \in T_gG$. The left-invariant vector fields $\mathscr{E}$ therefore constitute a moving frame on G. As usual, we may define quasivelocities $v = \omega(\dot{g}) \in W_g$ for tangent vectors $\dot{g} \in T_gG$, so that the velocity $\dot{g}$ can be represented as

$$\dot{g} = v^i E_i = v^i g e_i.$$

The *body velocity* (9.3) in the Lie algebra $\mathfrak{g}$ may therefore be described in terms of the quasivelocities as

$$\xi = d(L_{g^{-1}})_g \dot{g} = g^{-1}\dot{g} = v^i e_i.$$

Thus, relative to the basis $(e_1, \ldots, e_n)$ of $\mathfrak{g}$, vectors in the Lie algebra may be represented as the vectors in $\mathbb{R}^n$ with the components $(v^1, \ldots, v^n) \in \mathbb{R}^n$, i. e., $\check{\xi} = v \in \mathbb{R}^n$, where $\vee : \mathfrak{g} \to \mathbb{R}^n$ yields the components of a vector in $\mathfrak{g}$ relative to the preceding basis. The mapping $\omega \circ d(L_g)_e \circ \wedge : \mathbb{R}^n \to \mathbb{R}^n$ is therefore the identity transformation, as depicted in Figure 10.4.

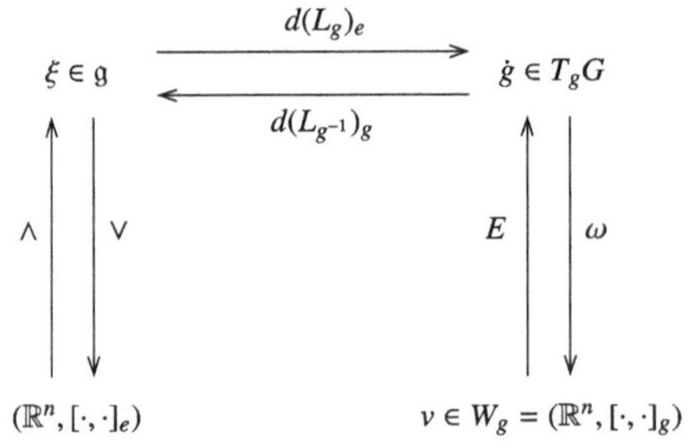

Figure 10.4: Relation between the Lie algebra and quasivelocities; $\omega \circ d(L_g)_e \circ \wedge = Id$.

Next, we must show that the bracket structure $[\cdot,\cdot]_e$ on the Lie algebra $\mathfrak{g}$ is the same as the bracket $[\cdot,\cdot]_g$ obtained on W_g by the moving frame $\mathscr{E}$, as defined by (10.7). However, this is precisely the statement of Corollary 9.1. Suppose the structure coefficients on the Lie algebra are given by $[e_i, e_j]_e = \gamma_{ij}^k e_k$. It therefore follows from Corollary 9.1 that the structure coefficients for the moving frame $\mathscr{E}$ are given by

$$[E_i, E_j](g) = g(e_i e_j - e_j e_i) = \gamma_{ij}^k g e_k = \gamma_{ij}^k E_k.$$

Therefore,

$$\operatorname{ad}_\xi^* \frac{\partial l}{\partial \xi} = \left[v, \frac{\partial l}{\partial v}\right]_g^*,$$

where $\xi = \hat{v} = v^i e_i$ and the Euler–Poincaré equation is thus a special case of Hamel's equation for left-invariant Lagrangians on Lie groups.

10.6 Application: Constrained Optimal Control

In this application, we discuss constrained kinematic optimal control, paying particular attention to *driftless* systems, in which a particularly nice simplification results. We will closely follow the discussion of [204], using the more rigorous formalism introduced by Bloch et al. [30] and Maruskin et al. [206]. A different approach, in which constrained optimal control problems are related to Pontryagin's maximum principle, is discussed in [25]. We begin with the following definition.

Definition 10.6. A *constrained driftless kinematic control system* is a triple $(Q, \mathcal{D}, \mathcal{X})$, where Q is an n-dimensional manifold, $\mathcal{D}$ is an $(n-m)$-dimensional, completely nonintegrable distribution, and $\mathcal{X} = \{X_\alpha\}_{\alpha=1}^{n-m}$ is a set of $(n-m)$ kinematically admissible vector fields that span the distribution $\mathcal{D}$ at each point on Q.

Given a constrained driftless kinematic control system, the *controls* are a set of $(n-m)$ functions $u^\alpha : [t_1, t_2] \to \mathbb{R}$ for $\alpha = 1, \ldots, n-m$ that, when coupled with an initial condition $q(0) = q_0$, determine a *resultant trajectory* by the differential equation

$$\dot{q}(t) = X_\alpha(q(t))u^\alpha(t). \tag{10.26}$$

The system is *driftless* due to the absence of a *drift vector field* that would determine the flow when all of the controls are set to zero.

As is typical in nonholonomic systems, the constraint distribution $\mathcal{D}$ may be described as the null space of a set of m nonintegrable constraint one-forms, ω^a, for $a = (n-m+1), \ldots, n$. One may then select a set of $(n-m)$ additional vector fields $\{X_a\}_{a=n-m+1}^{n}$ that are dual to the constraint one-forms, so that $\omega^a(X_b) = \delta_b^a$ for $a, b = (n-m+1), \ldots, n$. This selection is not unique, and one usually chooses the simplest vector fields that possess this property. A moving frame $\mathscr{E} = \{X_1, \ldots, X_n\}$ that is consistent with the constraints is therefore given on the tangent space. The nonvanishing quasivelocities $v^{\mathcal{D}}$ exactly equal the controls u since the first $(n-m)$ vector fields of the moving frame were chosen so as to coincide with the control vector fields. For affine systems with a drift vector field, a set of affine quasivelocities may be chosen so that it is still the case; see [106] for a detailed treatment of affine quasivelocities in various contexts.

An *optimal control problem* is given on a constrained driftless kinematic control system by specifying two fixed endpoints $q_1, q_2 \in Q$ and a cost functional

$$\mathcal{I}[c] = \int_{t_1}^{t_2} l(c(t), u(t))dt, \tag{10.27}$$

where $l : Q \times \mathbb{R}^{n-m} \to \mathbb{R}$ is a smooth function. The problem is to determine a set of controls $u(t)$ that steer the system from q_1 to q_2 while minimizing the cost functional (10.27) relative to the class of admissible trajectories. The optimal controls may be computed with the aid of the following theorem.

Theorem 10.3. *Extremal curves in the constrained driftless kinematic control system* $(q, \mathcal{D}, \mathcal{X})$ *of cost function* (10.27) *satisfy the* modified Hamel equation for constrained optimal control, *i. e.*,

$$\frac{d}{dt}\frac{\partial l}{\partial v^{\mathcal{D}}} - E_{\mathcal{D}}[l] - \left[v, \frac{\partial l}{\partial v^{\mathcal{D}}}\right]^*_{\mathcal{D}} - [v, \mu]^*_{\mathcal{D}} = 0, \tag{10.28}$$

$$\dot{\mu} - E_{\mathcal{U}}[l] - \left[v, \frac{\partial l}{\partial v^{\mathcal{D}}}\right]^*_{\mathcal{U}} - [v, \mu]^*_{\mathcal{U}} = 0, \tag{10.29}$$

as well as the kinematic equation (10.26), *where* $\mu \in W_q^*$ *(such that* $\mu_{\mathcal{D}} = 0$*) constitutes a set of m Lagrange multipliers whose initial conditions must be chosen so as to enforce the boundary condition* $q(t_2) = q_2$ *on curve c.*

When carrying out variations of action (10.27), one may not select the Lagrange–d'Alembert principle, given by equation (7.8); rather, one must select the nonholonomic form of Hamilton's principle, given by equation (7.9). As such, the virtual displacements δc of the fiducial curve c cannot be kinematically admissible. Without further ado, we proceed to the proof.

Proof. We begin by taking (unconstrained) variations of the action functional (10.27). Identifying $u = v^{\mathcal{D}}$, we obtain the relation

$$\delta \mathcal{I} = \int_{t_1}^{t_2} \left(\left\langle \frac{\partial l}{\partial q}, \delta q \right\rangle + \left\langle \frac{\partial l}{\partial v^{\mathcal{D}}}, \delta v^{\mathcal{D}} \right\rangle \right) dt.$$

To enforce the condition that varied curves must be kinematically admissible, which we may alternatively state with the relation $\delta v^{\mathcal{U}} = 0$, we append a set of m Lagrange multipliers $\{\mu_a\}_{a=n-m+1}^n$—written collectively as an element $\mu = \mu_a \omega^a \in W_q^*$ such that $\mu_{\mathcal{D}} = 0$—to the integrand, thereby obtaining the expression

$$\delta \mathcal{I} = \int_{t_1}^{t_2} \left(\left\langle \frac{\partial l}{\partial q}, \delta q \right\rangle + \left\langle \frac{\partial l}{\partial v^{\mathcal{D}}}, \delta v^{\mathcal{D}} \right\rangle + \left\langle \mu, \delta v^{\mathcal{U}} \right\rangle \right) dt.$$

Recalling the algebraic relation (10.9) and the Hamel transpositional relation (10.17), we may rewrite $\delta \mathcal{I}$ as

$$\delta \mathcal{I} = \int_{t_1}^{t_2} \left(\langle E[l], \zeta \rangle + \left\langle \frac{\partial l}{\partial v^{\mathcal{D}}}, \dot{\zeta}^{\mathcal{D}} + [v, \zeta]_q^{\mathcal{D}} \right\rangle + \left\langle \mu, \dot{\zeta}^{\mathcal{U}} + [v, \zeta]_q^{\mathcal{U}} \right\rangle \right) dt.$$

Integrating by parts and using the definition of the dual bracket given by equation (10.8), we obtain

$$\delta \mathcal{I} = \int_{t_1}^{t_2} \left(\left\langle E_{\mathcal{D}}[l] - \frac{d}{dt}\frac{\partial l}{\partial v^{\mathcal{D}}} + \left[v, \frac{\partial l}{\partial v^{\mathcal{D}}}\right]^*_{\mathcal{D}} + [v, \mu]^*_{\mathcal{D}}, \zeta^{\mathcal{D}} \right\rangle \right.$$

$$+ \left\langle E_{\mathcal{U}}[l] - \dot{\mu} + \left[v, \frac{\partial l}{\partial v^{\mathcal{D}}}\right]^*_{\mathcal{U}} + [v, \mu]^*_{\mathcal{U}}, \zeta^{\mathcal{U}} \right\rangle \Bigg) dt.$$

Since the components $\zeta^{\mathcal{D}}$ and $\zeta^{\mathcal{U}}$ are independent, we obtain our result. □

Remark. If we define the augmented Lagrangian $\Lambda : W \to \mathbb{R}$ on the Lie algebra bundle W by the relation $\Lambda(q, v) = l(q, u) + \langle \mu, v^{\mathcal{U}} \rangle$, where we identify $u = v^{\mathcal{D}}$, the modified Hamel equation for constrained optimal control (10.28)–(10.29) is equivalent to the equation

$$\frac{d}{dt}\frac{\partial \Lambda}{\partial v} - E[\Lambda] - \left[v, \frac{\partial \Lambda}{\partial v}\right]^*_q = 0,$$

i. e., the augmented Lagrangian satisfies an unconstrained version of Hamel's equation (10.18) on W^*.

Remark. The modified Hamel equation for constrained optimal control differs from Hamel's equation for nonholonomic mechanics in two fundamental ways. First, it is obtained using a true variational principle in which the varied curves, and not the virtual displacements, are kinematically admissible. Second, the function $l : W \to \mathbb{R}$ that appears in the integrand of action (10.27) is not a mechanical Lagrangian but rather the integrand of a performance measure that might even be defined differently for the same physical system depending on one's performance needs (e. g., minimum fuel problems, minimum displacement, minimum control expenditure).

Example 10.5 (Vertical Rolling Disc). As an example, we will consider control of the vertical rolling disc discussed in Example 7.6. Recall that the configuration manifold is $Q = S^1 \times S^1 \times \mathbb{R}^2$, which may be described using local coordinates (φ, θ, x, y) and nonholonomic constraints (7.4) and (7.5). Suppose that we have direct control over the heading angle φ and the rolling angle θ. The kinematically admissible control vector fields are then given by

$$X_1 = \frac{\partial}{\partial \varphi} \quad \text{and} \quad X_2 = \frac{\partial}{\partial \theta} + R\cos\varphi \frac{\partial}{\partial x} + R\sin\varphi \frac{\partial}{\partial y}$$

and the corresponding control system (10.26) by

$$\dot{q} = u^1 X_1 + u^2 X_2.$$

We must next append to the control vector fields two additional vector fields that are dual to the constraint one-forms

$$\omega^3 = -R\cos\varphi d\theta + dx,$$
$$\omega^4 = -R\sin\varphi d\theta + dy.$$

As mentioned previously, this selection is not unique but is often chosen for simplicity. Such being the case, we select the additional vector fields

$$X_3 = \frac{\partial}{\partial x} \quad \text{and} \quad X_4 = \frac{\partial}{\partial y}.$$

The set $\mathcal{E} = \{X_1, X_2, X_3, X_4\}$ constitutes a moving frame on TQ whose dual coframe is given by $\omega^1 = d\varphi$, $\omega^2 = d\theta$, ω^3, and ω^4. The quasivelocities of the system are given by $v^i = \omega^i(\dot{q})$; as expected, we obtain $v^1 = \dot{\varphi}$ and $v^2 = \dot{\theta}$, the velocities over which we have direct control, and $v^3 = -R\cos\varphi\dot{\theta} + \dot{x}$ and $v^4 = -R\sin\varphi\dot{\theta} + \dot{y}$, which are constrained to vanish identically.

Now consider the cost function

$$\mathcal{I} = \int_{t_1}^{t_2} \frac{1}{2}\left(\dot{x}^2 + \dot{y}^2 + \frac{R^2\dot{\varphi}^2}{4} + \frac{R^2\dot{\theta}^2}{2}\right) dt,$$

which represents the integral of the specific kinetic energy of the disc. (*Specific* means per unit of mass.) This cost function is motivated by the following physical consideration: the controls cost energy to implement. This energy is converted into the kinetic energy of the rolling disc. By integrating the kinetic energy of the disc over time, we obtain a measure that is proportional to the total energy required to perform the controls; minimizing this measure minimizes the control cost.

Rewriting the cost function in terms of quasivelocities, setting the constrained quasivelocities v^3 and v^4 to zero, and dividing out a common factor of R^2 from the result, we obtain a cost function of the form (10.27), i. e.,

$$\mathcal{I} = \frac{1}{2}\int_{t_1}^{t_2} \left(\frac{v_1^2}{4} + \frac{3v_2^2}{2}\right) dt.$$

Unlike in the case of nonholonomic mechanics, we are free to enforce the constraints $v^3 = v^4 = 0$ before computing Hamel's equation. This follows due to the selection of the nonholonomic form of Hamilton's principle, in which we seek extremal curves among the class of kinematically admissible curves, as opposed to the selection of the Lagrange–d'Alembert principle of mechanics.

Next, when considering the vector fields that constitute the moving frame, one finds a single, nonzero commutator, i. e.,

$$[X_1, X_2] = -R\sin\varphi X_3 + R\cos\varphi X_4.$$

The nonzero Hamel coefficients are therefore given by

$$\gamma_{12}^3 = -\gamma_{21}^3 = -R\sin\varphi \quad \text{and} \quad \gamma_{12}^4 = -\gamma_{21}^4 = R\cos\varphi.$$

To determine the governing equations that produce the optimal trajectory, one next computes the modified Hamel equation. Equation (10.28) yields the differential equations

$$\frac{1}{2}\dot{v}_1 = (\mu_3\sin\varphi - \mu_4\cos\varphi)Rv_2, \tag{10.30}$$

$$\frac{3}{2}\dot{v}_2 = (-\mu_3 \sin\varphi + \mu_4 \cos\varphi)\, R v_1, \tag{10.31}$$

whereas (10.29) yields $\dot{\mu}_3 = 0$ and $\dot{\mu}_4 = 0$. To solve for the optimal trajectory, one appends the kinematic relations (10.26), thereby obtaining

$$\dot{\theta} = v_1, \qquad \dot{x} = R\cos\varphi v_2,$$
$$\dot{\varphi} = v_2, \qquad \dot{y} = R\sin\varphi v_2.$$

Exercise 10.13. Show that the modified Hamel equations (10.28) and (10.29) yield the differential equations (10.30) and (10.31) and $\dot{\mu}_3 = \dot{\mu}_4 = 0$.

Of course, determining the differential equations that are satisfied by optimal trajectories of our cost functional is only part of the story. As is customarily the case in control problems, one has desired initial and terminal conditions between which one wants to steer the system. The optimal trajectory is therefore determined not by a full set of initial conditions but by a set of boundary conditions on the system state and the multipliers. Techniques for solving such boundary value problems are beyond the scope of the present text, which is primarily focused on using geometry and variational principles to derive the differential equations satisfied by optimal curves; for details on numerical solutions to boundary value problems for ordinary differential equations, we refer the interested reader to, for example, [14]. □

Exercise 10.14. In this exercise, we consider the kinematic control of the Heisenberg system, as introduced by Brockett [35] and further discussed in [24]. The system is comprised of a particle in $\mathbb{R}^3$ whose motion is constrained by the one-form

$$\omega^3 = -y dx + x dy + dz,$$

so that $\omega^3(\dot{q}) = 0$. Kinematically admissible control vector fields are given by

$$X_1 = \frac{\partial}{\partial x} + y\frac{\partial}{\partial z} \quad \text{and} \quad X_2 = \frac{\partial}{\partial y} - x\frac{\partial}{\partial z}.$$

The motion of the controlled particle is determined by the differential equation

$$\dot{q} = u^1(t)X_1(q(t)) + u^2(t)X_2(q(t)).$$

The optimal control problem is to steer the particle from $(0,0,0)$ to $(0,0,a)$, for a given $a > 0$, during the fixed interval $t \in [0,T]$ while minimizing the cost functional

$$\mathcal{I} = \frac{1}{2}\int_0^T (u_1^2 + u_2^2)\, dt.$$

(a) Use the modified Hamel equation to show that the optimal trajectory satisfies the system

$$\dot{u}_1 = 2\mu v_2 \quad \text{and} \quad \dot{u}_2 = -2\mu v_1,$$

where $\dot{\mu} = 0$ (i. e., μ is a constant).

(b) Integrate to obtain

$$\begin{bmatrix} \dot{x}(t) \\ \dot{y}(t) \end{bmatrix} = \begin{bmatrix} \cos(2\mu t) & \sin(2\mu t) \\ -\sin(2\mu t) & \cos(2\mu t) \end{bmatrix} \begin{bmatrix} \dot{x}(0) \\ \dot{y}_0 \end{bmatrix}.$$

(c) Integrate again to obtain

$$\begin{bmatrix} x(t) \\ y(t) \end{bmatrix} = \frac{1}{2\mu} \begin{bmatrix} \sin(2\mu t) & 1 - \cos(2\mu t) \\ \cos(2\mu t) - 1 & -\sin(2\mu t) \end{bmatrix} \begin{bmatrix} \dot{x}(0) \\ \dot{y}(0) \end{bmatrix}.$$

Use the boundary conditions to conclude that $\mu = n\pi/T$ for $n \in \mathbb{Z}$.

(d) Use the constraint equation to determine $z(t)$ and use the boundary conditions on z to show that

$$\dot{x}(0)^2 + \dot{y}(0)^2 = -\frac{2\pi n a}{T^2}.$$

Determine the value of n that minimizes the cost functional.

11 Fiber Bundles and Nonholonomic Mechanics

Fiber bundles are certain mathematical structures that naturally arise in the fields of nonholonomic mechanics and control. Their conceptual origin actually preceded the rigorous, mathematical development of fiber bundles that is known today. Early references of the classic treatment of nonholonomic mechanics date back to such works as [186], [258], and [275, 276]. The authors of this literature actually formed a different school of thought, known as the *Suslov approach*, from their peer contemporaries (such as [116–118], [143], and [288]) who advocated for the *Hamel approach* (i. e., use of quasivelocities) discussed in Chapter 10. For a time in the early development of nonholonomic mechanics, the Suslov approach enjoyed widespread acceptance, as did the belief that the quasivelocity approach held by Hamel was erroneous. Hamel contradicted this belief in [118], where he showed that both viewpoints were equally valid descriptions of nonholonomic mechanics [222].

This discrepancy arose due to different applications of the *general transpositional relation*. In Suslov's approach, velocity was no longer understood as a tangent vector to the variation but rather as a kinematically admissible vector in the constraint distribution. This distinction thus altered the geometric meaning of the variations of the velocity. It is therefore necessary to properly define an *extended velocity field* before carrying out computation of the transpositional relation, as shown in [206]. The geometry and integrability of the Suslov problem was recently discussed in [31].

The approach advocated by Suslov and his contemporaries was the classical equivalent of the fiber bundle approach to nonholonomic mechanics developed in [28]. The approach entails dividing the generalized coordinates into two preferred sets called *independent* and *dependent* variables. The idea is that one may regard certain velocity components as dependent upon the other (independent) components by the nonholonomic constraint relations. These dependent velocities may therefore be eliminated from the equations of motion by judicious use of the constraints. Since the constraints are nonintegrable, the net change of the dependent variables depends on the precise path the independent variables traverse to get from one point to another: closed paths traversed in the space of independent variables do not constitute closed paths in the space of dependent variables. In relation to fiber bundles, the space of independent variables is known as the *base space*, whereas the dependent variables constitute the *fibers*. In this chapter, we discuss the modern formulation of the early viewpoint adopted by Suslov and his contemporaries and finally discuss a combination of the two approaches in which one is able to utilize quasivelocities on the base space of a fiber bundle, as discussed in [206].

https://doi.org/10.1515/9783110597806-011

11.1 Fiber Bundles

In this paragraph we introduce the mathematical concept of *fiber bundles*, discuss connections on fiber bundles and their curvature, and relate these concepts to the discussion of affine connections dealt with previously. Classic references include [3], [265], and [273]; Bloch [24] provides a concise overview and discusses applications of fiber bundle structures to nonholonomic mechanics and control.

A fiber bundle is a common structure encountered in differential geometry. In fact, we saw several instances of this structure employed in our previous discussions: *tangent bundles*, *cotangent bundles*, *frame bundles*, *Lie algebra bundles*, and *principal bundles* are all examples of a fiber bundle.

Definition 11.1. A *fiber bundle* is a quadruple (Q, R, F, π), where Q, R, and F are topological spaces and $\pi : Q \to R$ is a continuous, locally trivial surjection called a *projection map* or *bundle projection*. Space R is called the *base space*, Q is called the *total space*, and F is called the *fiber*. We require $\pi^{-1}(r)$ to be homeomorphic to fiber F for any $r \in R$. A *smooth fiber bundle* is one in which spaces Q, R, and F are differentiable manifolds.

By *locally trivial* we mean that the total space is locally homeomorphic to the product space $R \times F$. If the fibers are homeomorphic to their own structure group, then Q is a principal bundle. If the fibers are homeomorphic to a vector space, Q is a *vector bundle*.

Example 11.1. Let M be an n-dimensional, differentiable manifold and consider the tangent bundle $Q = TM$, which constitutes a smooth fiber bundle. The bundle projection $\pi : TM \to M$ is the canonical projection given by $\pi(q, v) = q$, and the fibers $\pi^{-1}(q) = T_qM$ are the tangent spaces, each of which is diffeomorphic to the real vector space $\mathbb{R}^n$. Therefore, TM is in fact a vector bundle.

Now let $\mathcal{F}$ be the frame bundle of M. Again, M is the base space. Each fiber $F_q = \pi^{-1}(q)$ consists of an ordered basis for the tangent space T_qM at a point q. The fibers are therefore homeomorphic to their structure group $\mathrm{GL}(n; \mathbb{R})$. Therefore, all frame bundles are principal bundles. □

Definition 11.2. A *section* of a fiber bundle is a continuous selection of a point in the fiber for every point in the base.

To define the operation of differentiation, one must make a particular choice of section that will be regarded as *constant*. This particular section is known as the *horizontal section* of the bundle. There is no unique way to define a horizontal section. As we will see, defining a horizontal section is equivalent to defining a certain structure on a fiber bundle known as an *Ehresmann connection*. Before we introduce this structure, we will require one more definition.

Definition 11.3. Let (Q, R, F, π) be a smooth fiber bundle. The *vertical space* at any point q, which will be denoted by V_q, is the kernel of the mapping $d_q\pi : T_qQ \to T_{\pi(q)}R$.

Given a fiber bundle, the vertical space is always uniquely defined, and it always lies tangent to the fibers. We now define a particular connection on the fiber bundle, first introduced by Ehresmann [81].

Definition 11.4. Given a fiber bundle (Q, R, F, π), an *Ehresmann connection* is a smooth mapping $A : TQ \to V$ with the property that $A(v) = v$ for every $v \in V_q$ and $q \in Q$.

Hence, one may consider an Ehresmann connection to be a *vector-valued one-form* on Q that is *vertical-valued* and, moreover, a projection onto the vertical space V. As we will see next, defining an Ehresmann connection on a fiber bundle is equivalent to defining a *horizontal space*.

Definition 11.5. Let (Q, R, F, π) be a fiber bundle and $A : TQ \to V$ an Ehresmann connection on Q. Then the *horizontal space* at any point q, which we denote as H_q, is the kernel of the Ehresmann connection A_q.

Exercise 11.1. Given a fiber bundle and Ehresmann connection, prove that the tangent space at any $q \in Q$ can be decomposed as the direct sum decomposition $T_qQ = V_q \oplus H_q$.

Thus, any tangent vector to Q can be decomposed as a sum of its *horizontal* and *vertical parts*. The Ehresmann connection may be used to project a tangent vector onto the vertical space.

When applying a fiber bundle approach to nonholonomic mechanics, it is customary to choose the connection so that the constraint distribution $\mathcal{D}$ is the horizontal space of the connection. We will see examples of this later.

Local coordinates on space Q are denoted $q = (r, s)$, where $r \in R$ and $s \in \pi^{-1}(r) \cong F$. In what follows, we will take the dimension of the base space to be p, so that the dimension of the fiber is $(n - p)$. An Ehresmann connection may be represented locally as a vector-valued one-form. In particular, we write

$$A(v) = \omega^a(v)\frac{\partial}{\partial s^a}, \quad \text{where } \omega^a = ds^a + A^a_\alpha(r, s)dr^\alpha \tag{11.1}$$

for $a = (n - p + 1), \ldots, n$ and $\alpha = 1, \ldots, (n - p)$. The one-forms ω^a are chosen such that connection A is a projection onto the vertical space (which, as you recall, lies tangent to the fibers). In particular, consider a tangent vector $\dot{q} \in TQ$ given by

$$\dot{q} = \dot{r}^\beta \frac{\partial}{\partial r^\beta} + \dot{s}^b \frac{\partial}{\partial s^b} \tag{11.2}$$

for $\beta = 1, \ldots, (n - p)$ and $b = (n - p + 1), \ldots, n$. The one-forms ω^a acting on $\dot{q}$ therefore yield $\omega^a(\dot{q}) = \dot{s}^a + A^a_\alpha \dot{r}^\alpha$ for $\alpha = 1, \ldots, (n - p)$ and $a = (n - p + 1), \ldots, n$ (we will assume this same summation convention for the remainder of the discussion). Therefore, the Ehresmann connection sends the velocity vector $\dot{q}$ to the vertical-valued vector

$$A(\dot{q}) = \left(\dot{s}^a + A^a_\alpha \dot{r}^\alpha\right) \frac{\partial}{\partial s^a}. \tag{11.3}$$

This is literally the *vertical projection* (the *vertical part*) of vector $\dot{q}$ onto the vertical space V_q. Due to the direct sum decomposition of TQ, the *horizontal projection* (the *horizontal part*) of vector $\dot{q}$ onto the horizontal space H_q is given uniquely as

$$\mathrm{hor}(\dot{q}) = \dot{q} - A(\dot{q}) = \dot{r}^\alpha \frac{\partial}{\partial r^\alpha} - A^a_\alpha \dot{r}^\alpha \frac{\partial}{\partial s^a},$$

so that $\dot{q} = A(\dot{q}) + \mathrm{hor}(\dot{q})$. For the special case of a tangent vector to the base space, say $\dot{r} = \dot{r}^\alpha \frac{\partial}{\partial r^\alpha}$, the horizontal part is also known as the *horizontal lift*. The *horizontal part* (relative to a given connection) of a vector $\dot{q}$ is determined *uniquely* by its components that lie tangent to the base space, i. e., for a general $\dot{q}$ given by (11.2) we have $\mathrm{hor}(\dot{q}) = \mathrm{hor}(\dot{r})$. This will be an important observation later.

Definition 11.6. Given a fiber bundle (Q, R, F, π) and an Ehresmann connection $A : TQ \to V$, the *curvature*, B, of connection A is a vertical vector-valued two-form $B : TQ \times TQ \to V$, defined by the operation

$$B(X, Y) = A([\mathrm{hor}(X), \mathrm{hor}(Y)]),$$

for all $X, Y \in \mathfrak{X}(Q)$, where $[\cdot, \cdot]$ is the Jacobi–Lie bracket on Q.

Proposition 11.1. *The Ehresmann curvature may be represented by*

$$B(X, Y) = B^a_{\alpha\beta} X^\alpha Y^\beta \frac{\partial}{\partial s^a}, \tag{11.4}$$

where

$$B^a_{\alpha\beta} = \frac{\partial A^a_\alpha}{\partial r^\beta} - \frac{\partial A^a_\beta}{\partial r^\alpha} + A^b_\alpha \frac{\partial A^a_\beta}{\partial s^b} - A^b_\beta \frac{\partial A^a_\alpha}{\partial s^b}, \tag{11.5}$$

for $\alpha, \beta = 1, \dots, (n-p)$ *and* $a, b = (n-p+1), \dots, n$.

Exercise 11.2. Prove the statement of Proposition 11.1.

Relation to Affine Connection

At first pass, admittedly, these definitions seem hopelessly incongruous with the definitions of affine connections and their curvature as discussed in §6.7. However, there is a deep connection between these two concepts, as outlined in [24] and [40]. In fact, the tangent bundle of a manifold constitutes an archetypical and intuitive example of a fiber bundle. In this paragraph, we will examine the tangent bundle and affine connection as special cases of fiber bundles with an Ehresmann connection. We will analyze the tangent bundle $Q = TR$ of a manifold R as a fiber bundle with projection operator $\pi : TR \to R$. However, since the horizontal spaces live in $TQ = T^2R$, we must first make a few important remarks regarding *second tangent bundles*.

First, recall that the tangent bundle of R is the set

$$TR = \{(r,s) : r \in R, s \in T_rR\},$$

where the tangent spaces to R are viewed as the fibers of TR and the fiber variables s as elements of the tangent space at r. Elements of the *second tangent bundle*, $T^2R = T(TR)$, are therefore identified with the quadruples $(r,s,\dot{r},\dot{s})$, where $(r,s) \in TR$ and $(\dot{r},\dot{s})$ represent an equivalence class of curves through TR passing through the point (r,s) with velocity $(\dot{r},\dot{s})$, i. e.,

$$(\dot{r},\dot{s}) = \dot{r}^\alpha \frac{\partial}{\partial r^\alpha} + \dot{s}^a \frac{\partial}{\partial s^a} \in T_{(r,s)}(TR).$$

There is no requirement that $s = \dot{r}$.

The differential of the projection operator $\pi : TR \to R$ is a mapping $d\pi : T^2R \to TR$ whose kernel consists of elements of the second tangent bundle of the form $(r,s,0,\dot{s})$. One therefore identifies the vertical space with the acceleration components of the tangent vectors to TR. The horizontal space, which can be defined by an affine connection, as we will shortly demonstrate, then defines a covariant notion of acceleration-free paths.

Now consider an affine connection on R with Christoffel symbols Γ^a_{bc}. Recall that the covariant derivative of vector field X with respect to Y is given by

$$\nabla_Y X = \left(Y^b \frac{\partial X^a}{\partial r^b} + \Gamma^a_{bc} X^b Y^c \right) \frac{\partial}{\partial r^a}.$$

Given this Levi-Civita connection, we now define an Ehresmann connection at the point $(r,s) \in Q = TR$ by the relation

$$A(r,s) = \omega^a \frac{\partial}{\partial s^a}, \quad \omega^a = ds^a + \Gamma^a_{b\alpha} s^b dr^\alpha. \tag{11.6}$$

Such connections (with $A^\alpha_a(r,s) = \Gamma^a_{b\alpha}(r)s^b$) are known as *linear connections*, as they are linear in the fiber variable s. Next, we define the lift at the point $(r, X(r))$ of the vector field $Y \in TR$ to the second tangent bundle $\hat{Y}_X \in T_{(r,X)}TR$ by the relation

$$\hat{Y}_X = Y^\alpha \frac{\partial}{\partial r^\alpha} + Y^\alpha \frac{\partial X^a}{\partial r^\alpha} \frac{\partial}{\partial s^a}.$$

Evaluating the Ehresmann connection at the point $(r,s) = (r,X) \in TR$ and letting it act on $\hat{Y}_X$, we obtain

$$A(r,X)(\hat{Y}_X) = \left(Y^\alpha \frac{\partial X^a}{\partial r^\alpha} + \Gamma^a_{b\alpha} X^b Y^\alpha \right) \frac{\partial}{\partial s^a},$$

which is equivalent to $\nabla_Y X$.

Finally, we may define the *acceleration bundle*, AR, of R as the vector subbundle of T^2R, i. e.,

$$AR = \{(r, s, \dot{r}, \dot{s}) \in T^2R : s = \dot{r}\}.$$

One can easily show that the elements of the acceleration bundle may be equivalently thought of as equivalence classes of curves through the point $r \in R$ with identical velocities and accelerations when expressed in an arbitrary set of local coordinates, i. e., an equivalence subclass of the tangent bundle. Thus, points in AR are equally well described by triples of the form $(r, \dot{r}, \ddot{r})$. Applying the Ehresmann connection (11.6) to the vector field $X = \dot{r}$ and its lift

$$\hat{X}_X = \dot{r}^\alpha \frac{\partial}{\partial r^\alpha} + \ddot{r}^a \frac{\partial}{\partial s^a},$$

we obtain

$$A(r, X)(\hat{X}_X) = (\ddot{r}^a + \Gamma^a_{bc}\dot{r}^b\dot{r}^c)\frac{\partial}{\partial s^a}.$$

But we immediately identify the components of this equation with the geodesic equations. Therefore, the horizontal paths of the Ehresmann connection (11.6) in the tangent bundle are precisely the geodesics relative to the affine connection ∇. For more details on geodesics in nonholonomic geometries, see [279].

11.2 The Transpositional Relation and Suslov's Principle

The general transpositional relation provides greater flexibility as to how one defines variations of velocity. This freedom arises since, a priori, one only has a definition of velocity along a fiducial curve, whereas the concept of velocity elsewhere in the variation remains to be defined. This definition of velocity, as we will see, need not coincide with the tangent vectors to the varied curve, as is understood in the Hamel approach. After first defining the concept of an extended velocity field, we will then derive the generalized transpositional relation that will be used to derive new forms of the equations of motion from this conceptual framework.

Variations of Velocity

Recall that in Definition 7.6 we used the concept of the *canonical lift* of a curve and its variation to properly define the notion of the variation of the curve's velocity. Since the velocity of a curve is, a priori, defined only on the curve itself and not elsewhere in the curve's variation, there is no unique way to assign a meaning to the symbol δV. To resolve this dilemma, we will make use of the concept of an *extended velocity field*, as discussed in [206].

Definition 11.7. Given a curve $c : [a,b] \to Q$ on a manifold Q and an open set $U \subset Q$ containing c, an *extended velocity* is a fiber-preserving mapping $\mathbb{V} : TU \to TU$ that satisfies $\mathbb{V}(c(t), \dot{c}(t)) = \dot{c}(t)$ for all $t \in [a,b]$ and that defines a rule by which one can assign an associated *extended velocity vector field* on the image of an arbitrary variation ϑ by restricting this mapping to the image of ϑ, i. e., $V(s,t) = \mathbb{V}(\vartheta(s,t), \vartheta_t(s,t)) \in TU$.

Given a variation $\vartheta : [-\varepsilon, \varepsilon] \times [a,b] \to U$ of curve c, define $\hat{\vartheta}_V : [-\varepsilon, \varepsilon] \times [a,b] \to TU$ by the relation $\hat{\vartheta}_V(s,t) = (\vartheta(s,t), V(s,t))$. We may now define the variation of the extended velocity field as

$$(\delta c(t), \delta V(t)) = \left.\frac{\partial \hat{\vartheta}_V(s,t)}{\partial s}\right|_{s=0}.$$

One typically considers two primary choices of extended velocity. As we have seen, when deriving Hamel's equation, one implicitly selects the mapping $\mathbb{V}(q, \dot{q}) = \dot{q}$. Hence, we will refer to this choice as the *Hamel extended velocity*. When using the Hamel extended velocity, the velocity in the variation is literally taken as the velocity of the varied curves. Since, to satisfy the Lagrange–d'Alembert principle, one requires the virtual displacements $\delta c(t)$ to be kinematically admissible, the nonintegrability of the constraints coupled with the smoothness of the variations implies that it is not possible in general for the extended velocity field V to be kinematically admissible, except along the fiducial curve c. Note that it was the Hamel extended velocity that was used when defining the *canonical lift* in Definition 7.6.

The second possible choice for an extended velocity is the one used by Suslov [275, 276] and Voronets [288], in which one defines $V(q, \dot{q}) = \operatorname{hor}(\dot{q})$, where the horizontal distribution is taken to coincide with the nonholonomic constraint distribution $\mathcal{D}$. We will refer to this choice as the *Suslov extended velocity*. When using the Suslov extended velocity, the velocity in the variation is taken to be the horizontal part of the tangent velocity to the varied curve. Hence, δV is understood as a variation in the kinematically admissible extended velocity field V and not a variation of the tangent vectors of the varied curve. When one utilizes this extended velocity, one is literally applying the constraints before taking variations. Hence, additional terms arise in the variational principle, as we will discuss in our next paragraph.

Transpositional Relation

As we have seen, given an extended velocity mapping $\mathbb{V} : TU \to TU$, one may define an extended velocity field $V : [-\varepsilon, \varepsilon] \times [a,b] \to TU$ on the image of an arbitrary variation to a curve c. Taken together with the virtual displacements $\delta q : [-\varepsilon, \varepsilon] \times [a,b] \to TU$, one obtains two vector fields on the (s,t)-space for each arbitrary variation of curve c. Given a moving frame $\mathscr{E}$, each of these vector fields may be expressed relative to $\mathscr{E}$ as vector fields in the Lie algebra bundle, i. e., one obtains $v = \omega(V)$ and

$\zeta = \omega(\delta q)$. The components of these vector fields relative to the moving frame satisfy the *general transpositional relation*

$$\delta v = \dot{\zeta} + [v, \zeta]_q - \omega([V, \delta q]). \tag{11.7}$$

The transpositional relation follows immediately by applying (10.16) to the vector fields $V = v^i E_i$ and $\zeta = \zeta^i E_i$. It should be emphasized that the transpositional relation is defined on the (s,t)-space for each particular, arbitrary variation of curve c. The transpositional relation may also be expressed in terms of coordinates in its classic form as

$$\delta v^i = \dot{\zeta}^i + \gamma^i_{jk} v^j \zeta^k - \left(\frac{\partial \delta q^j}{\partial t} - \frac{\partial V^j}{\partial s} \right) \phi^i_j, \tag{11.8}$$

where γ^i_{jk} are the Hamel coefficients (10.12). This classic form appears in, for example, [106] and [221, 222].

Note that the *Hamel transpositional relation* (10.17) is just a special case of the general transpositional relation (11.7), obtained by selecting the Hamel extended velocity $\mathbb{V}(q, \dot{q}) = \dot{q}$. This follows from the C^2 differentiability of the variation and Clairaut's theorem, $\theta_{st} - \theta_{ts} = 0$, which in turn implies the vanishing of $[V, \delta q]$.

Suslov's Principle

In this paragraph, we continue our discussion of *extended velocity fields*. As you will recall from the last chapter, it was of paramount importance to write out the full, *unconstrained* Lagrangian when employing Hamel's equation. The purpose of Suslov's approach to nonholonomic mechanics is to enable one to work directly with the *constrained Lagrangian*, which may be formed by making a particular choice of extended velocity.

Definition 11.8. Let $\mathbb{V} : TU \to TU$ be an extended velocity mapping of a curve $c : [a, b] \to U$. Then the *modified Lagrangian* on U associated with $\mathbb{V}$ is the function $L_{\mathbb{V}} : TU \to \mathbb{R}$ defined by the relation

$$L_{\mathbb{V}}(q, \dot{q}) = L(q, \mathbb{V}(q, \dot{q})). \tag{11.9}$$

As should be obvious, one cannot immediately use the modified Lagrangian (11.9) directly in the standard variational principles of nonholonomic mechanics. The difference between taking variations of the unconstrained and constrained Lagrangians is quantified by *Suslov's principle*, which we state in the modern form following [206].

Theorem 11.1 (Suslov's principle). *Let $(Q, L, \mathcal{D})$ be a nonholonomic mechanical system and $\mathbb{V} : TU \to TU$ an extended velocity. For an arbitrary C^2 variation ϑ of a curve c, the* *variations of the Lagrangian L and the modified Lagrangian $L_{\mathbb{V}}$ are related to each other along the fiducial curve by Suslov's principle, which states that*

$$\delta L = \delta L_{\mathbb{V}} + \left\langle \frac{\partial L}{\partial \dot{q}}, [V, \delta q] \right\rangle, \tag{11.10}$$

where

$$\delta L = \frac{\partial L}{\partial q}\delta q + \frac{\partial L}{\partial \dot{q}}\delta \dot{q} \quad and \quad \delta L_{\mathbb{V}} = \frac{\partial L}{\partial q}\delta q + \frac{\partial L}{\partial \dot{q}}\delta V.$$

Proof. This follows from the formula for the Lie bracket of vector fields V and δq evaluated along the fiducial curve. Explicitly,

$$[V, \delta q] = \left(V^j \frac{\partial \delta q^i}{\partial q^j} - \delta q^j \frac{\partial V^i}{\partial q^j} \right) \frac{\partial}{\partial q^i} = \frac{d\delta q}{dt} - \delta V = \delta \dot{q} - \delta V.$$

The second equality holds due to the chain rule and the fact that $V(0,t) = \dot{q}(t)$; the third equality holds due to the fact that the variation ϑ is C^2 and Clairaut's theorem. □

11.3 Voronets' Equation

In this section, we will derive a set of equations of motion that are satisfied by the *constrained Lagrangian*, which may be defined as the modified Lagrangian associated with a particular choice of extended velocity. Let $(Q, L, \mathcal{D})$ be a nonholonomic system and suppose we can provide a fiber bundle structure on the configuration manifold Q such that it may locally be represented as a direct product of an $(n-p)$-dimensional base space R and a p-dimensional fiber space. In the classic language of Suslov, the base variables are referred to as *independent* variables, whereas fiber variables are *dependent* variables [275, 276]. Now let us choose an Ehresmann connection whose horizontal space coincides with the constraint distribution, i. e., $\mathcal{D}_q = H_q$. This can be accomplished using the Ehresmann connection (11.1), where the one-forms ω^a coincide with the constraint one-forms whose kernel coincides with $\mathcal{D}$.

Definition 11.9. The *Suslov extended velocity* is the fiber-preserving mapping given by

$$\mathbb{V}(q, \dot{q}) = \operatorname{hor}(\dot{q}) = \dot{r}^\alpha \frac{\partial}{\partial r^\alpha} - A^a_\alpha \dot{r}^\alpha \frac{\partial}{\partial s^a}, \tag{11.11}$$

where $\alpha = 1, \dots, (n-p)$ and $a = (n-p+1), \dots, n$. The modified Lagrangian associated with this choice of extended velocity is known as the *constrained Lagrangian* $L_c : \mathcal{D} \to \mathbb{R}$ and is defined by $L_c(q, \dot{q}) = L(q, \operatorname{hor}(\dot{q}))$.

By construction, this extended velocity field is kinematically admissible. Hence, the constrained Lagrangian L_c is literally the Lagrangian rewritten so as to enforce the constraints.

Following [206], we will next derive the transpositional relation for the Suslov-type extended velocity field, given by (11.11), and the particular moving frame

$$E_\alpha = \operatorname{hor}\left(\frac{\partial}{\partial r^\alpha}\right) = \frac{\partial}{\partial r^\alpha} - A_\alpha^a \frac{\partial}{\partial s^a}, \tag{11.12}$$

$$E_a = \frac{\partial}{\partial s^a}, \tag{11.13}$$

which was introduced in [30].

The first $(n-p)$ vector fields of the moving frame are the horizontal parts of the basis tangent vectors to the base space, whereas the remaining p vector fields lie tangent to the fibers.

Exercise 11.3. Show that the quasivelocities of the Suslov extended velocity field (11.11) relative to the moving frame (11.12)–(11.13) are given by $v^{\mathcal{D}} = \dot{r}$ and $v^{\mathcal{U}} = 0$. Similarly, conclude that the components of kinematically admissible virtual displacements $\delta q \in \mathcal{D}$ relative to this moving frame are simply $\zeta^{\mathcal{D}} = \delta r$ and $\zeta^{\mathcal{U}} = 0$.

Remark. The constrained Lagrangian $L_c : \mathcal{D} \to \mathbb{R}$, defined by $L_c(q, \dot{q}) = L(q, \operatorname{hor}(\dot{q}))$, is a function on the constraint distribution $\mathcal{D}$. The velocities $\dot{r}$ lie tangent to the base space, and not the constraint distribution. Exercise 11.3 shows, however, that the velocities $\dot{r}$ constitute a set of quasivelocities for the kinematically admissible velocity vectors in $\mathcal{D}$. Hence, one may define the constrained Lagrangian in terms of these quasivelocities as $l_c(r, s, \dot{r}) = L(r, s, \dot{r}, -A(\dot{r}))$.

Exercise 11.4. Prove that the commutation relations for the moving frame (11.12)–(11.13) are given by

$$[E_\alpha, E_\beta] = B_{\alpha\beta}^a \frac{\partial}{\partial s^a}, \tag{11.14}$$

$$[E_\alpha, E_b] = \frac{\partial A_\alpha^a}{\partial s^b} \frac{\partial}{\partial s^a},$$

where the coefficients $B_{\alpha\beta}^a$ are the coefficients of the curvature of the Ehresmann connection, given by (11.5).

It follows that the nonzero Hamel coefficients of the frame (11.12)–(11.13) are given by

$$\gamma_{\alpha\beta}^a = B_{\alpha\beta}^a \quad \text{and} \quad \gamma_{\alpha b}^a = -\gamma_{b\alpha}^a = \frac{\partial A_\alpha^a}{\partial s^b}.$$

We next demonstrate a lemma that relates the Jacobi–Lie bracket of the kinematically admissible Suslov extended velocity and virtual displacements with the curvature of the Ehresmann connection. This lemma lies at the heart of connecting the classic Suslov principle with the modern fiber bundle form of Voronets' equations of motion.

Lemma 11.1. *Let V be the Suslov-type extended velocity* (11.11) *and B the curvature of the Ehresmann connection. Then, for $\delta q \in \mathcal{D}$, one has the relation*

$$[V, \delta q] = B(\dot{r}, \delta r). \tag{11.15}$$

In local bundle coordinates (r^α, s^a), this relation may be expressed as

$$(\delta\dot{s}^a - \delta V^a)\frac{\partial}{\partial s^a} = B^a_{\alpha\beta}\dot{r}^\alpha \delta r^\beta \frac{\partial}{\partial s^a}.$$

Proof. As noted earlier, the components of V and δq relative to the moving frame (11.12)–(11.13) are simply $v = v^{\mathcal{D}} = \dot{r}$ and $\zeta = \zeta^{\mathcal{D}} = \delta r$, i. e., $V = \dot{r}^\alpha E_\alpha$ and $\delta q = \delta r^\alpha E_\alpha$. Now, applying relation (10.16) and noting that $d/dt(\delta r) = \delta\dot{r}$, we obtain

$$[V, \delta q] = \dot{r}^\alpha \delta r^\beta [E_\alpha, E_\beta] = B(\dot{r}, \delta r),$$

where the second relation follows from (11.14). The result follows. □

Remark. Even though the proof of Lemma 11.1 utilizes the particular moving frame (11.12)–(11.13), the resulting relationship (11.15) does not depend on the choice of frame, as neither the quantities $[V, \delta q]$ nor $B(\dot{r}, \delta r)$ depend on the choice of frame.

We now state the main result of this paragraph, which dates back to [288] and is further described from a geometric, fiber bundle perspective in [28].

Theorem 11.2 (Voronets' Equation). *Let $(Q, L, \mathcal{D})$ be a nonholonomic mechanical system and $l_c(r, s, \dot{r}) = L(r, s, \dot{r}, -A(\dot{r}))$ the constrained Lagrangian. Then a curve satisfies the Lagrange–d'Alembert principle for nonholonomic mechanics if and only if it satisfies* Voronets' equation

$$\frac{d}{dt}\frac{\partial l_c}{\partial \dot{r}} - E_{\mathcal{H}}[l_c] = \left\langle \frac{\partial L}{\partial \dot{s}}, \mathbf{i}_{\dot{r}} B \right\rangle, \tag{11.16}$$

where $E_{\mathcal{H}}[l_c] := (E_1[l_c], \ldots, E_{n-p}[l_c])$. In local coordinates, this equation is given by

$$\frac{d}{dt}\frac{\partial l_c}{\partial \dot{r}^\alpha} - \frac{\partial l_c}{\partial r^\alpha} + A^a_\alpha \frac{\partial l_c}{\partial s^a} = \frac{\partial L}{\partial \dot{s}^a} B^a_{\beta\alpha}\dot{r}^\beta.$$

Proof. Using Suslov's principle (11.10), we write the Lagrange–d'Alembert principle in terms of the constrained Lagrangian as

$$\int_a^b \left(\delta l_c + \left\langle \frac{\partial L}{\partial \dot{q}}, [V, \delta q] \right\rangle\right) dt = 0, \tag{11.17}$$

where the virtual displacements must satisfy $\delta q \in \mathcal{D}$, i. e., $\delta s^a = -A^a_\alpha \delta r^\alpha$. Taking variations of the constrained Lagrangian we obtain

$$\delta l_c = \left\langle \frac{\partial l_c}{\partial q}, \delta q \right\rangle + \left\langle \frac{\partial l_c}{\partial \dot{r}}, \delta\dot{r} \right\rangle.$$

Now from (10.10) and noting that $\zeta^{\mathcal{D}} = \delta r$ and $\zeta^{\mathcal{U}} = 0$, we have

$$\delta l_c = \langle E_{\mathcal{H}}[l_c], \delta r\rangle + \left\langle \frac{\partial l_c}{\partial \dot{r}}, \delta \dot{r} \right\rangle .$$

Integrating the second term by parts, we find

$$\int_a^b \delta l_c \, dt = \int_a^b \left\langle E_{\mathcal{H}}[l_c] - \frac{d}{dt}\frac{\partial l_c}{\partial \dot{r}}, \delta r \right\rangle .$$

Finally, from Lemma 11.1, we find

$$\left\langle \frac{\partial L}{\partial \dot{q}}, [V, \delta q] \right\rangle = \left\langle \frac{\partial L}{\partial \dot{s}}, B(\dot{r}, \delta r) \right\rangle = \left\langle \frac{\partial L}{\partial \dot{s}}, \mathbf{i}_{\dot{r}} B \right\rangle \delta r, \tag{11.18}$$

where $\langle \partial_{\dot{s}} L, \mathbf{i}_{\dot{r}} B\rangle$ is viewed as a linear functional from the tangent to the base space to $\mathbb{R}$. Combining the preceding equations, we obtain the statement of the theorem. □

Example 11.2 (Vertical Rolling Disc). Consider again the vertical rolling disc of Example 7.6. Recall that the configuration manifold is given by $Q = S^1 \times S^1 \times \mathbb{R}^2$, with local coordinates (φ, θ, x, y), Lagrangian

$$L = \frac{m}{2}(\dot{x}^2 + \dot{y}^2) + \frac{I}{2}\dot{\theta}^2 + \frac{J}{2}\dot{\varphi}^2,$$

and nonholonomic constraints, given by (7.4)–(7.5). Hence, we may decompose Q as a fiber bundle with base space $R = S^1 \times S^1$ and fibers $F = \mathbb{R}^2$, so that $(r^1, r^2, s^3, s^4) = (\varphi, \theta, x, y)$. The Ehresmann connection (11.1) is therefore defined given

$$\omega^3 = -R\cos\varphi d\theta + dx,$$
$$\omega^4 = -R\sin\varphi d\theta + dy,$$

with coefficients A_α^a given by $A_2^3 = -R\cos\varphi$ and $A_2^4 = -R\sin\varphi$. The moving frame (11.12)–(11.13) is thus given by the vector fields

$$E_1 = \frac{\partial}{\partial\varphi}, \quad E_2 = \frac{\partial}{\partial\theta} + R\cos\varphi\frac{\partial}{\partial x} + R\sin\varphi\frac{\partial}{\partial y}, \quad E_3 = \frac{\partial}{\partial x}, \quad E_4 = \frac{\partial}{\partial y}.$$

A simple computation shows that $[E_1, E_2] = -R\sin\varphi E_3 + R\cos\varphi E_4$, from which it immediately follows that

$$B_{21}^3 = R\sin\varphi = -B_{12}^3 \quad \text{and} \quad B_{21}^4 = -R\cos\varphi = -B_{12}^4.$$

Exercise 11.5. Use (11.5) and connection A to compute the components of the Ehresmann curvature directly and show that they are consistent with those obtained previously.

Therefore, it follows that

$$\frac{\partial L}{\partial \dot{s}^a} B^a_{\beta 1} \dot{r}^\beta = mR \sin\varphi \dot{x}\dot{\theta} - mR \cos\varphi \dot{y}\dot{\theta},$$
$$\frac{\partial L}{\partial \dot{s}^a} B^a_{\beta 2} \dot{r}^\beta = -mR \sin\varphi \dot{x}\dot{\varphi} + mR \cos\varphi \dot{y}\dot{\varphi}.$$

Now we write the constrained Lagrangian as

$$l_c(q, \dot{r}) = \frac{1}{2}(I + mR^2)\dot{\theta}^2 + \frac{J}{2}\dot{\varphi}^2.$$

Therefore,

$$\frac{d}{dt}\frac{\partial l_c}{\partial \dot{r}^1} = J\ddot{\varphi},$$
$$\frac{d}{dt}\frac{\partial l_c}{\partial \dot{r}^2} = (I + mR^2)\ddot{\theta}.$$

Since the constrained Lagrangian is independent of the position variables, Voronets' equation yields the equations

$$J\ddot{\varphi} = mR \sin\varphi \dot{x}\dot{\theta} - mR \cos\varphi \dot{y}\dot{\theta},$$
$$(I + mR^2)\ddot{\theta} = -mR \sin\varphi \dot{x}\dot{\varphi} + mR \cos\varphi \dot{y}\dot{\varphi},$$

which are coupled with the constraint equations and integrated to determine the motions of the system. □

11.4 Combined Hamel–Suslov Approach

To conclude our discussion on the use of fiber bundles in nonholonomic mechanics, we next present a combined approach, as described in [206]. The idea is to begin with a fiber bundle reduction and then choose a moving frame on the base space. With such an approach, it is possible, for example, to eliminate some of the constraints with the fiber bundle reduction and others with a judicious choice of quasivelocities on the base space.

We begin with the fiber bundle reduction. First, we define the Suslov-type extended velocity field given by equation (11.11) and the constrained Lagrangian $L_c(q, \dot{q}) = L(q, \mathbb{V}(q, \dot{q})) = L(q, \mathrm{hor}(\dot{q}))$, as was done in the derivation of the Voronets equation. Next, we define a moving frame $\{e_\alpha\}_{\alpha=1}^{n-p}$ on the base space, which can be described in terms of local coordinates as

$$e_\alpha = \psi^\beta_\alpha \frac{\partial}{\partial r^\beta},$$

where $\psi_\alpha^\beta = \psi_\alpha^\beta(r)$. Next, define an associated moving frame on the fiber bundle Q by the relations

$$E_\alpha = \mathrm{hor}(e_\alpha) = \psi_\alpha^\beta \frac{\partial}{\partial r^\beta} - A_\beta^a \psi_\alpha^\beta \frac{\partial}{\partial s^a}, \tag{11.19}$$

$$E_a = \frac{\partial}{\partial s^a}. \tag{11.20}$$

Note that, as previously, the vector fields $E_\alpha = \mathrm{hor}(e_\alpha)$ span the horizontal constraint distribution $\mathcal{D}$, whereas vector fields $\{E_a\}_{a=n-p+1}^n$ span the space tangent to the fibers (compare with vector fields (11.12)–(11.13)). The dual one-forms to the above frame on Q are given by

$$\omega^\alpha = \phi_\beta^\alpha dr^\beta,$$
$$\omega^a = A_\alpha^a dr^\alpha + ds^a,$$

where the $(n-p) \times (n-p)$ matrix $[\phi_\beta^\alpha]$ is the inverse of $[\psi_\beta^\alpha]$, so that $\phi_\beta^\alpha \psi_\gamma^\beta = \delta_\gamma^\alpha$. It follows that the quasivelocities of the extended velocity relative to the moving frame (11.19)–(11.20) are given by $v^\alpha = \phi_\beta^\alpha \dot{r}^\beta$ and $v^a = 0$. Finally, we define the *constrained Lagrangian in terms of quasivelocities* by the relation

$$l_c(q, v^{\mathcal{D}}) = L_c(q, v^\alpha e_\alpha).$$

As usual, one can define a Lie algebra bundle over Q with fibers $W_q = (\mathbb{R}^n, [\cdot,\cdot]_q)$, where $[\cdot,\cdot]_q$ is the bracket associated with the quasivelocities relative to the frame (11.19)–(11.20). We will, however, obtain a simplification of the transpositional relation by considering a similar Lie algebra bundle over the base space R with fibers $w_r = (\mathbb{R}^{n-p}, [\cdot,\cdot]_r)$, where $[v, w]_r = v^\alpha w^\beta \omega^{\mathcal{D}}([e_\alpha, e_\beta])$ for $v, w \in w_r$. Note that $\omega^{\mathcal{D}}([e_\alpha, e_\beta])$ are just the components of the structure coefficients of the frame $\{e_1, \ldots, e_{n-p}\}$ on TR relative to the frame $\{e_1, \ldots, e_{n-p}\}$. Also, in general, $\omega^{\mathcal{U}}([e_\alpha, e_\beta]) \neq 0$; however, these components belong to the larger space W_q.

Theorem 11.3. *Let (Q, R, F, π) be a fiber bundle, A an Ehresmann connection with horizontal distribution $H = \mathrm{ker}(A)$, and $(e_\alpha)_{\alpha=1}^{n-p} \subset TR$ a moving frame on R, with horizontal parts $E_\alpha = \mathrm{hor}(e_\alpha) \in H$. Then*

$$\mathrm{hor}([E_\alpha, E_\beta]) = \mathrm{hor}([e_\alpha, e_\beta]). \tag{11.21}$$

Alternatively, this relation may be expressed as

$$[E_\alpha, E_\beta] = [e_\alpha, e_\beta] - A([e_\alpha, e_\beta]) + B(e_\alpha, e_\beta), \tag{11.22}$$

where $B(e_\alpha, e_\beta) = A([E_\alpha, E_\beta])$.

Proof. We first note that $E_\alpha = e_\alpha - A(e_\alpha)$, for $\alpha = 1, \ldots, (n-p)$. Next, we compute

$$\begin{aligned}[E_\alpha, E_\beta] &= (e_\alpha - A(e_\alpha))(e_\beta - A(e_\beta)) - (e_\beta - A(e_\beta))(e_\alpha - A(e_\alpha)) \\ &= (e_\alpha e_\beta - e_\beta e_\alpha) + (A(e_\beta)e_\alpha - A(e_\alpha)e_\beta) \\ &\quad + (e_\beta A(e_\alpha) - e_\alpha A(e_\beta) + A(e_\alpha)A(e_\beta) - A(e_\beta)A(e_\alpha)).\end{aligned}$$

Now, $A(e_\beta)e_\alpha$ vanishes identically since connection A is vertical-valued and vector fields (e_α) are vector fields on the base space. Recognizing that $[e_\alpha, e_\beta] = e_\alpha e_\beta - e_\beta e_\alpha$, we find

$$[E_\alpha, E_\beta] = [e_\alpha, e_\beta] + e_\beta A(e_\alpha) - e_\alpha A(e_\beta) + A(e_\alpha)A(e_\beta) - A(e_\beta)A(e_\alpha).$$

The result follows by taking the horizontal part of this equation, as the last four terms on the right-hand side are vertical-valued. □

Remark. Note that $[e_\alpha, e_\beta] \in TR$, whereas $B(e_\alpha, e_\beta), A([e_\alpha, e_\beta]) \in V$. Hence, (11.22) shows how one can decompose the bracket $[E_\alpha, E_\beta]$ relative to the local direct sum decomposition $TQ = TR \oplus V$ of the fiber bundle.

Proposition 11.2. *The transpositional relation for the Suslov extended velocity* (11.11) *and the moving frame* (11.19)–(11.20) *is given by*

$$\delta v - \dot{\zeta} = [v, \zeta]_r, \tag{11.23}$$

where $v^{\mathcal{U}} = \zeta^{\mathcal{U}} = 0$.

Proof. We first relate the bracket on W_q to the bracket on w_r by computing the corresponding structure coefficients. It follows immediately from (11.22) that

$$\begin{aligned}\omega^{\mathcal{D}}([E_\alpha, E_\beta]) &= \omega^{\mathcal{D}}([e_\alpha, e_\beta]), \\ \omega^{\mathcal{U}}([E_\alpha, E_\beta]) &= \omega^{\mathcal{U}}(B(e_\alpha, e_\beta)).\end{aligned}$$

The second relation follows since $\mathrm{hor}([e_\alpha, e_\beta])$ is horizontal and therefore contained in the null space of $\omega^{\mathcal{U}}$. Therefore, for vectors $v, w \in W_q$ with $v^{\mathcal{U}} = w^{\mathcal{U}} = 0$ it follows that

$$[v, w]_q = [v, w]_r + v^\alpha w^\beta \omega(B(e_\alpha, e_\beta)) = [v, w]_r + \omega(B(V, \delta q)).$$

The result therefore follows due to the general transpositional relation (11.7) and relation (11.15), which holds whenever one chooses the Suslov extended velocity (11.11). □

Finally, we define a related curvature on quasivelocities by the relation

$$\tilde{B}(x, y) = x^\alpha y^\beta B(e_\alpha, e_\beta)$$

for $x, y \in w_r$, where the forms B and $\tilde{B}$ depend on (r, s). In coordinates, one easily obtains

$$\tilde{B}^a_{\alpha\beta}\frac{\partial}{\partial s^a} = \psi^\gamma_\alpha \psi^\delta_\beta B^a_{\gamma\delta}\frac{\partial}{\partial s^a}.$$

The advantage of this definition is that one obtains the simple expression

$$B(e_\alpha, e_\beta) = A([E_\alpha, E_\beta]) = \tilde{B}^a_{\alpha\beta}\frac{\partial}{\partial s^a}, \tag{11.24}$$

where vector fields E_α are given by (11.19). This formula provides a simpler approach to computing the components of the curvature relative to the given moving frame on the base space, as opposed to first computing (11.5) and then applying a coordinate transformation.

We now state our main theorem, as given in [206].

Theorem 11.4. *Let $l_c(q, v) = L(q, v^\alpha E_\alpha)$ be the constrained Lagrangian expressed in terms of the $(n-p)$ nonzero quasivelocities of the Suslov extended velocity* (11.11) *relative to the moving frame* (11.19)–(11.20). *Then the equations of motion for nonholonomic mechanics are equivalent to*

$$\frac{d}{dt}\frac{\partial l_c}{\partial v} - E_{\mathcal{H}}[l_c] - \left[v, \frac{\partial l_c}{\partial v}\right]^*_r - \left\langle \frac{\partial L}{\partial \dot{s}}, \mathbf{i}_v \tilde{B}\right\rangle = 0, \tag{11.25}$$

where $E_{\mathcal{H}} := (E_1, \dots, E_{n-p})$ and $E_\alpha = \mathrm{hor}(e_\alpha)$. In local coordinates, this equation is given by

$$\frac{d}{dt}\frac{\partial l_c}{\partial v^\alpha} - \frac{\partial l_c}{\partial r^\beta}\psi^\beta_\alpha + \frac{\partial l_c}{\partial s^a}A^a_\beta\psi^\beta_\alpha - \frac{\partial l_c}{\partial v^\delta}\gamma^\delta_{\beta\alpha}v^\beta - \frac{\partial L}{\partial \dot{s}^a}\tilde{B}^a_{\beta\alpha}v^\beta = 0.$$

Proof. We will use the extended velocity field (11.11) and moving frame (11.19)–(11.20). Suslov's principle (11.10) and the Lagrange–d'Alembert principle yield

$$\int \left(\delta l_c + \left\langle \frac{\partial L}{\partial \dot{q}}, [V, \delta q]\right\rangle\right) dt = 0,$$

where $\delta q \in \mathcal{D}$. By Lemma 11.1 we have $[V, \delta q] = B(\dot{r}, \delta r)$, so that Suslov's principle is equivalent to

$$\int \left(\delta l_c + \left\langle\left\langle \frac{\partial L}{\partial \dot{s}}, \mathbf{i}_{\dot{r}} B\right\rangle, \delta r\right\rangle\right) dt = \int \left(\delta l_c + \left\langle\left\langle \frac{\partial L}{\partial \dot{s}}, \mathbf{i}_v \tilde{B}\right\rangle, \zeta\right\rangle\right) dt = 0.$$

Using (10.10) and the transpositional relation (11.23), we obtain

$$\delta l_c = \langle E_{\mathcal{H}}[l_c], \zeta\rangle + \left\langle \frac{\partial l_c}{\partial v}, \dot{\zeta} + [v, \zeta]_r\right\rangle.$$

Integrating by parts we obtain

$$\int \delta l_c dt = \int \left\langle E_{\mathcal{H}}[l_c] - \frac{d}{dt}\frac{\partial l_c}{\partial v} + \left[v, \frac{\partial l_c}{\partial v}\right]^*_r, \zeta\right\rangle dt = 0.$$

Combining this with Suslov's principle yields our result. ☐

The Lie algebra bundle structure on the tangent bundle, as one encounters in Hamel's equation, is thus equally represented as a reduced Lie algebra bundle structure on the base space plus a curvature structure on the fibers. The true benefit of the preceding equation is one of typography; it is easier for the eye to separate the structure coefficients on the base and fiber, easier for the wit to compute two smaller commutators than synergistically more complicated ones on a larger dimensional space, and easier for one's aesthetic assessment to appreciate hidden mathematical structures in complex equations.

Now suppose that the constraint distribution $\mathcal{D}$, regarded as the null space of the p one-forms $\omega^a = A^a_\alpha dr^\alpha + ds^a$, does not represent all of the constraints (for instance, when not all constraints are given in this form). Then there exists a constraint subbundle $\mathscr{D} \subset \mathcal{D}$ that describes these additional constraints. We enforce these constraints by selecting a frame on the base space so that any vector $v \in w_r$ can be decomposed as $v = v^{\mathscr{D}} + v^{\mathscr{U}}$, with $v \in \mathscr{D}$ if and only if $v = v^{\mathscr{D}}$ or, equivalently, $v^{\mathscr{U}} = 0$.

Theorem 11.5. *Consider a nonholonomic system in which p constraints are described by identifying the $(n-p)$-dimensional constraint distribution $\mathcal{D}$ with the horizontal space associated with an Ehresmann connection A, and an additional m constraints are described by the $(n-p-m)$-dimensional constraint subbundle $\mathscr{D} \subset \mathcal{D}$. Let $\{e_1, \ldots, e_{n-p}\}$ be a moving frame on the base space R such that a vector $v \in \mathscr{D}$ if and only if $v = v^{\mathscr{D}}$. Then the mechanical equations of motion that satisfy the Lagrange–d'Alembert principle are equivalent to the equations*

$$\left(\frac{d}{dt}\frac{\partial l_c}{\partial v} - E_{\mathcal{H}}[l_c] - \left[v, \frac{\partial l_c}{\partial v}\right]_r^* - \left\langle \frac{\partial L}{\partial \dot{s}}, \mathbf{i}_{v^{\mathscr{D}}} \tilde{B} \right\rangle\right)_{\mathscr{D}} = 0,$$

where

$$v^{\mathscr{U}} = 0 \quad \textit{and} \quad \dot{q} = \langle E(q), v^{\mathscr{D}} \rangle.$$

Proof. These equations follow directly by restricting (11.25) to the distribution $\mathscr{D}$. □

11.5 Application: Rolling-Without-Slipping Constraints

Arising commonly in systems with rolling-without-slipping constraints, fiber bundle geometries are abundant in mechanical systems. In this paragraph, we analyze several such examples.

Example 11.3 (Ball Rolling Inside a Cylinder). As an example to Theorem 11.4, let us consider again the case of the rigid ball rolling without slipping inside a hollow cylinder, as given in Example 10.4. Recall that the configuration manifold is given by $Q = \mathrm{SO}(3) \times \mathbb{R} \times S^1$, with local coordinates given by $(\varphi, \theta, \psi, z, \rho)$. The precise construction of a moving frame and its dual coframe is precisely the structure of an

Ehresmann connection, with a moving frame on the base space. In this vein, we take $r^1 = \varphi$, $r^2 = \theta$, and $r^3 = \psi$ as local coordinates (Euler angles) for the base and $s^4 = z$ and $s^5 = \rho$ for local (cylindrical) coordinates for the fibers. The transformation matrices can be partitioned so that their base and fiber components are apparent if we write

$$[\phi] = \left[\begin{array}{ccc|cc} 0 & \cos(\varphi-\rho) & \sin\theta\sin(\varphi-\rho) & 0 & 0 \\ 0 & \sin(\varphi-\rho) & -\sin\theta\cos(\varphi-\rho) & 0 & 0 \\ 1 & 0 & \cos\theta & 0 & 0 \\ \hline 0 & -r\sin(\varphi-\rho) & r\sin\theta\cos(\varphi-\rho) & 1 & 0 \\ \lambda & 0 & \lambda\cos\theta & 0 & 1 \end{array}\right]$$

and its inverse

$$[\psi] = [\phi^{-1}] = \left[\begin{array}{ccc|cc} -\cot\theta\sin(\varphi-\rho) & \cot\theta\cos(\varphi-\rho) & 1 & 0 & 0 \\ \cos(\varphi-\rho) & \sin(\varphi-\rho) & 0 & 0 & 0 \\ \csc\theta\sin(\varphi-\rho) & -\csc\theta\cos(\varphi-\rho) & 0 & 0 & 0 \\ \hline 0 & r & 0 & 1 & 0 \\ 0 & 0 & -\lambda & 0 & 1 \end{array}\right].$$

Therefore, the Ehresmann connection (11.1) is defined by the one-forms

$$\omega^4 = -r\sin(\varphi-\rho)d\theta + r\sin\theta\cos(\varphi-\rho)d\psi + dz,$$
$$\omega^5 = \lambda d\varphi + \lambda\cos\theta d\psi + d\rho,$$

with nonzero components A^a_α given by

$$A^4_2 = -r\sin(\varphi-\rho),\ A^4_3 = r\sin\theta\cos(\varphi-\rho),\ A^5_1 = \lambda,\ \text{and}\ A^5_3 = \lambda\cos\theta.$$

Exercise 11.6. Show that the nonzero components of the curvature of this Ehresmann connection are given by

$$B^4_{12} = r(1+\lambda)\cos(\varphi-\rho) = -B^4_{21}, \quad B^5_{23} = \lambda\sin\theta = -B^5_{32},$$
$$B^4_{32} = r(1+\lambda)\cos\theta\cos(\varphi-\rho) = -B^4_{23},$$
$$B^4_{13} = r(1+\lambda)\sin\theta\sin(\varphi-\rho) = -B^4_{31}.$$

It now follows from Exercise 10.10 and equation (11.24) that the nonzero components $\tilde{B}^a_{\alpha\beta} = B^a_{\gamma\delta}\psi^\gamma_\alpha\psi^\delta_\beta$ are given by

$$\tilde{B}^4_{31} = r(1+\lambda) = -\tilde{B}^4_{13} \quad \text{and} \quad \tilde{B}^5_{21} = \lambda = -\tilde{B}^5_{12}. \tag{11.26}$$

The moving frame on the base space consists of the vector fields

$$e_1 = -\cot\theta\sin(\varphi-\rho)\frac{\partial}{\partial\varphi} + \cos(\varphi-\rho)\frac{\partial}{\partial\theta} + \csc\theta\sin(\varphi-\rho)\frac{\partial}{\partial\psi},$$

$$e_2 = \cot\theta\cos(\varphi - \rho)\frac{\partial}{\partial\varphi} + \sin(\varphi - \rho)\frac{\partial}{\partial\theta} - \csc\theta\cos(\varphi - \rho)\frac{\partial}{\partial\psi},$$
$$e_3 = \frac{\partial}{\partial\varphi}.$$

Notice that the horizontal parts of these vector fields are precisely the vector fields E_1, E_2, and E_3 of the equation set (10.20). It follows from the results of Exercise 10.10 and Theorem 11.3 that

$$[e_1, e_2] = -e_3, \quad [e_1, e_3] = (1 + \lambda)e_2, \quad \text{and} \quad [e_3, e_2] = (1 + \lambda)e_1. \tag{11.27}$$

Next, we write the constrained Lagrangian in terms of quasivelocities as

$$l_c = \frac{1}{2}(I\omega_r^2 + (I + mr^2)\omega_\rho^2 + (I + mr^2)\omega_z^2) - mgz.$$

Writing out the components for the terms of (11.25) we obtain

$$\frac{d}{dt}\frac{\partial l_c}{\partial\omega_r} = I\dot{\omega}_r,$$
$$\frac{d}{dt}\frac{\partial l_c}{\partial\omega_\rho} = (I + mr^2)\dot{\omega}_\rho,$$
$$\frac{d}{dt}\frac{\partial l_c}{\partial\omega_z} = (I + mr^2)\dot{\omega}_z.$$

Next, note that $E_{\mathcal{H}}[l_c] = (0, -mgr, 0) \in w_r^*$. From (11.27) we compute the bracket terms as

$$\frac{\partial l_c}{\partial v^\delta}\gamma_{\beta 1}^\delta v^\beta = -\lambda(I + mr^2)\omega_\rho\omega_z,$$
$$\frac{\partial l_c}{\partial v^\delta}\gamma_{\beta 1}^\delta v^\beta = \lambda I\omega_r\omega_z - mr^2\omega_r\omega_z,$$
$$\frac{\partial l_c}{\partial v^\delta}\gamma_{\beta 1}^\delta v^\beta = (1 + \lambda)mr^2\omega_r\omega_\rho.$$

Next, note that

$$\frac{\partial L}{\partial\dot{z}} = m\dot{z} = mr\omega_\rho \quad \text{and} \quad \frac{\partial L}{\partial\dot{\rho}} = (R - r)^2 m\dot{\rho} = -(R - r)^2 m\lambda\omega_z.$$

Finally, using (11.26), we obtain for our final terms

$$\frac{\partial L}{\partial\dot{s}^a}\tilde{B}_{\beta 1}^a v^\beta = \lambda mr^2\omega_\rho\omega_z,$$
$$\frac{\partial L}{\partial\dot{s}^a}\tilde{B}_{\beta 2}^a v^\beta = mr^2\omega_r\omega_z,$$
$$\frac{\partial L}{\partial\dot{s}^a}\tilde{B}_{\beta 3}^a v^\beta = -mr^2(1 + \lambda)\omega_r\omega_\rho.$$

Combining the preceding terms as prescribed by (11.25), we obtain precisely the same equations of motion as given by (10.21)–(10.23). □

Example 11.4 (Falling Rolling Disc). We now consider a disc that rolls without slipping on a horizontal plane, free to fall as it rolls. This system was discussed classically in [287] and more recently in [132], [228], [63], and [298] and is a generalization of the vertical rolling disc that was discussed in Examples 7.6 and 11.2. The configuration manifold is $Q = SO(3) \times \mathbb{R}^2 \subset SE(3)$; the factor SO(3) describes the configuration of the disc in space, whereas the factor $\mathbb{R}^2$ describes the position of the contact point C on the horizontal plane. We choose Type II (classic) Euler angles as local coordinates on SO(3), yielding the set $(\varphi, \theta, \psi, x, y)$ of local coordinates on Q. The body-fixed frame is selected so that the Z-axis is perpendicular to the plane of the disc and the X-axis points toward a fixed point P on the rim of the disc; the Euler angles, their rates, and the lines of nodes are depicted in Figure 11.1. Vectors $\mathbf{e}_\varphi$ and $\mathbf{e}_\theta$ are vertical and horizontal, respectively, and vector $\mathbf{e}_\psi$ is normal to the plane of the disc.

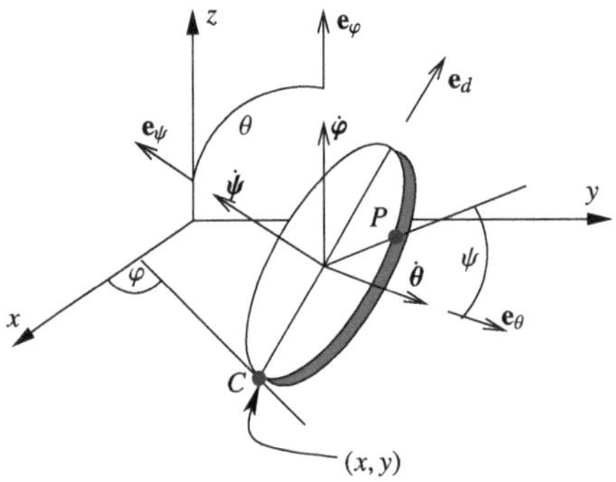

Figure 11.1: Schematic of a falling rolling disc.

The constraint that the disc must roll without slipping is enforced by setting

$$\dot{x} + r\dot{\psi}\cos\varphi = 0,$$
$$\dot{y} + r\dot{\psi}\sin\varphi = 0.$$

We immediately recognize Q as a fiber bundle with base space $R = SO(3)$, fibers $\mathbb{R}^2$, and Ehresmann connection

$$A = \omega^4 \frac{\partial}{\partial x} + \omega^5 \frac{\partial}{\partial y},$$

where the differential constraint forms ω^4 and ω^5 are defined by

$$\omega^4 = dx + r\cos\varphi d\psi,$$
$$\omega^5 = dy + r\sin\varphi d\psi.$$

The body axis components of the angular velocity are given in terms of the Euler angles by relations (9.17)–(9.19). The rotational kinetic energy of the disc in terms of these body axis components of the angular velocity is

$$T_{\text{rot}} = \frac{I_a}{2}\Omega_z^2 + \frac{I_t}{2}(\Omega_x^2 + \Omega_y^2),$$

where I_a and I_t are the axial and transversal moments of inertia of the disc, respectively. However, a simplification to the relation between Euler angle rates and quasivelocities can be made by defining $\mathbf{e}_d = \mathbf{e}_\psi \times \mathbf{e}_\theta$ and replacing the quasivelocities Ω_x and Ω_y with

$$\Omega_\theta = \dot{\theta}, \tag{11.28}$$

$$\Omega_d = \dot{\varphi}\sin\theta, \tag{11.29}$$

the components of the angular velocity relative to $\mathbf{e}_\theta$ and $\mathbf{e}_d$, respectively. Thus, the total angular velocity may be written in terms of quasivelocities as

$$\boldsymbol{\omega} = \Omega_\theta \mathbf{e}_\theta + \Omega_d \mathbf{e}_d + \Omega_z \mathbf{e}_\psi.$$

As the position of the center of mass relative to contact point C is simply $r\mathbf{e}_d$, the relative velocity of the center of mass with respect to the contact point is

$$\mathbf{v}_{\text{rel}} = \boldsymbol{\omega} \times (r\mathbf{e}_d) = -r\Omega_z \mathbf{e}_\theta + r\Omega_\theta \mathbf{e}_\psi.$$

Furthermore, the *unconstrained* velocity of the rim of the disc at contact point C is

$$\begin{aligned}\mathbf{v}_C &= (\dot{x} + r\cos\varphi\dot{\psi})\mathbf{i} + (\dot{y} + r\sin\varphi\dot{\psi})\mathbf{j}\\ &= (\dot{x} - r\cos\varphi\cot\theta\Omega_d + r\cos\varphi\Omega_z)\mathbf{i} + (\dot{y} - r\sin\varphi\cot\theta\Omega_d + r\sin\varphi\Omega_z)\mathbf{j}.\end{aligned}$$

(The constraint that the disc must roll without slipping is equivalent to $\mathbf{v}_C = \mathbf{0}$.) Computing the components of the lines of node relative to the inertial frame, we determine the total, *unconstrained* velocity of the center of mass as follows:

$$\begin{aligned}\mathbf{v} = \mathbf{v}_C + \mathbf{v}_{\text{rel}} &= (\dot{x} - r\cos\varphi\cot\theta\Omega_d + r\sin\varphi\sin\theta\Omega_\theta)\mathbf{i}\\ &\quad + (\dot{y} - r\sin\varphi\cot\theta\Omega_d - r\cos\varphi\sin\theta\Omega_\theta)\mathbf{j} + r\cos\theta\Omega_\theta\mathbf{k}.\end{aligned}$$

The total Lagrangian in terms of quasivelocities is therefore

$$\begin{aligned}l = {}&\frac{m}{2}(\dot{x}^2 + \dot{y}^2) + \frac{1}{2}(I_t + mr^2)\Omega_\theta^2 + \frac{1}{2}(I_t + mr^2\cot^2\theta)\Omega_d^2 + \frac{I_a}{2}\Omega_z^2\\ &- mr\cos\varphi\cot\theta\dot{x}\Omega_d + mr\sin\varphi\sin\theta\dot{x}\Omega_\theta\\ &- mr\sin\varphi\cot\theta\dot{y}\Omega_d - mr\cos\varphi\sin\theta\dot{y}\Omega_\theta - mgr\sin\theta.\end{aligned}$$

The *constrained* Lagrangian in terms of quasivelocities is thus given by

$$l_c = \frac{1}{2}(I_t + mr^2)\Omega_\theta^2 + \frac{I_t}{2}\Omega_d^2 + \frac{1}{2}(I_a + mr^2)\Omega_z^2 - mgr\sin\theta.$$

The comoving frame on the base space is given directly by converting the quasivelocities (11.28), (11.29), and (9.19) into the differential forms

$$\omega^1 = d\theta,$$
$$\omega^2 = \sin\theta d\varphi,$$
$$\omega^3 = \cos\theta d\varphi + d\psi,$$

so that $v = \omega(\dot{r})$, with $\omega^\alpha = \phi^\alpha_\beta \dot{r}^\beta$. The corresponding moving frame on the base space is found by computing $[\psi] = [\phi]^{-1}$ and considering its columns, i. e.,

$$e_1 = \frac{\partial}{\partial\theta},$$
$$e_2 = \csc\theta\frac{\partial}{\partial\varphi} - \cot\theta\frac{\partial}{\partial\psi},$$
$$e_3 = \frac{\partial}{\partial\psi}.$$

One easily verifies that $\dot{r} = v^\alpha e_\alpha$, where $v = (\Omega_\theta, \Omega_d, \Omega_z) \in w_r$.

Exercise 11.7. Show that the nonzero commutators of the moving frame vector fields on the base space are

$$[e_1, e_2] = -\cot\theta e_2 + e_3.$$

Then show that the nonzero components of $\tilde{B}$ are

$$\tilde{B}(e_2, e_3) = r\sin\varphi\csc\theta\frac{\partial}{\partial x} - r\cos\varphi\csc\theta\frac{\partial}{\partial y}.$$

Conclude that $\gamma^2_{21} = -\gamma^2_{12} = \cot\theta$, $\gamma^3_{12} = -\gamma^3_{21} = 1$, $\tilde{B}^4_{23} = -\tilde{B}^4_{32} = r\sin\varphi\csc\theta$, and $\tilde{B}^5_{32} = -\tilde{B}^5_{23} = r\cos\varphi\csc\theta$.

We now possess the necessary ingredients to obtain

$$\frac{d}{dt}\left[\frac{\partial l_c}{\partial v}\right] = ((I_t + mr^2)\dot{\Omega}_\theta, I_t\dot{\Omega}_d, (I_a + mr^2)\dot{\Omega}_z),$$
$$E_{\mathcal{H}}[l_c] = (-mgr\cos\theta, 0, 0),$$
$$\left[v, \frac{\partial l_c}{\partial v}\right]^*_r = (I_t\cot\theta\Omega_d^2 - (I_a + mr^2)\Omega_d\Omega_z$$
$$- I_t\cot\theta\Omega_\theta\Omega_d + (I_a + mr^2)\Omega_\theta\Omega_z, 0),$$
$$\left\langle\frac{\partial l}{\partial\dot{s}}, \mathbf{i}_v\tilde{B}\right\rangle = (0, -mr^2\Omega_\theta\Omega_z, mr^2\Omega_\theta\Omega_d).$$

Exercise 11.8. Verify the preceding four equations.

Finally, following (11.25), the equations of motion are given by

$$
\begin{aligned}
(I_t + mr^2)\dot{\Omega}_\theta + mgr\cos\theta - I_t\cot\theta\Omega_d^2 + (I_a + mr^2)\Omega_d\Omega_z &= 0,\\
I_t\dot{\Omega}_d + I_t\cot\theta\Omega_\theta\Omega_d - I_a\Omega_\theta\Omega_z &= 0,\\
(I_a + mr^2)\dot{\Omega}_z - mr^2\Omega_\theta\Omega_d &= 0,
\end{aligned}
$$

which may be integrated along with the kinematic relations

$$
\dot{\varphi} = \csc\theta\Omega_d, \quad \dot{\theta} = \Omega_\theta, \quad \text{and} \quad \dot{\psi} = -\cot\theta\Omega_d + \Omega_z
$$

to yield the motions of the system. □

Bibliography

[1] Abad, A., R. Barrio, and Á. Dena [2011] Computing periodic orbits with arbitrary precision, *Phys. Rev. E* **84**, 016701.

[2] Abraham, R. and J. E. Marsden [1978] *Foundations of Mechanics*, Reading, MA: Benjamin-Cummings; reprinted in (2008) Second Edition, Providence RI: AMS-Chelsea.

[3] Abraham, R., J. E. Marsden, and T. S. Ratiu [1988] *Manifolds, Tensor Analysis, and Applications*, Second Edition, New York: Springer.

[4] Agrachev, A. and Y. Sachkov [2004] *Control Theory from the Geometric Viewpoint*, New York: Springer.

[5] Anderson, J. D. [2004] *Modern Compressible Flow: With Historical Perspective*, Third Edition, New York: McGraw-Hill.

[6] Andronov, A. A. and A. Witt [1930] Sur la theórie mathematiques des autooscillations, *C. R. Acad. Sci. Paris* **190**, 256–258.

[7] Arnold, V. I. [1973] *Ordinary Differential Equations*, Cambridge MA: MIT Press.

[8] Arnold, V. I. [1989] *Mathematical Methods of Classical Mechanics*, New York: Springer.

[9] Arnold, V. I. and A. Avez [1967] *Problèmes ergodiques de la mécanique classique*, Paris: Gauthier-Villars; translated in (1968) *Ergodic Problems of Classical Mechanics*, New York: Benjamin.

[10] Arnold, V. I. and A. Givental [2001] Symplectic topology; in Arnold, V. I. and S. P. Novikov (eds.), *Dynamical Systems IV: Symplectic Geometry and Its Applications*, Second Edition, New York: Springer.

[11] Arnold, V. I., V. V. Kozlov, and A. I. Neishtadt [2006] *Dynamical Systems III: Mathematical Aspects of Classical and Celestial Mechanics*, Third Edition, New York: Springer.

[12] Artin, E. [1924] Ein mechanisches System mit quasiergodischen Bahnen, *Abh. Math. Semin. Univ. Hamb.* **3**, 170–175.

[13] Artin, M. [1995] *Algebra*, Englewood Cliffs NJ: Prentice Hall.

[14] Ascher, U. M., R. M. M. Mattheij, and R. D. Russell [1995] *Numerical Solution of Boundary Value Problems for Ordinary Differential Equations*, Philadelphia: SIAM.

[15] Audin, M., A. C. da Silva, and E. Lerman [2003] *Symplectic Geometry of Integrable Hamiltonian Systems*, Boston: Birkhäuser.

[16] Baillieul, J. [1975] *Some Optimization Problems in Geometric Control Theory*, Ph.D. Thesis, Harvard University.

[17] Baker, H. F. [1905] Note on the integration of linear differential equations, *Proc. Lond. Math. Soc.* **s2-2**(1), 293–296.

[18] Barrar, R. B. [1963] An analytic proof that the Hohmann-type transfer is the true minimum two-impulse transfer, *Acta Astronaut.* **9**, 1–11.

[19] Bendixson, I. [1901] Sur les courbes définies par des équations différentielles, *Acta Math.* **24**, 1–88.

[20] Benettin, G., L. Galgani, A. Giorgilli, and J. M. Strelcyn [1980] Lyapunov characteristic exponents for smooth dynamical systems and for Hamiltonian systems; a method for computing all of them, Part 1: Theory, *Meccanica* **15**(1), 9–20.

[21] Benettin, G., L. Galgani, A. Giorgilli, and J. M. Strelcyn [1980] Lyapunov characteristic exponents for smooth dynamical systems and for Hamiltonian systems; a method for computing all of them, Part 2: Numerical application, *Meccanica* **15**(1), 21–30.

[22] Betts, J. T. [2010] *Practical methods for optimal control and estimation using nonlinear programming*, Second Edition, Philadelphia: SIAM.

https://doi.org/10.1515/9783110597806-012

[23] Binney, J. and S. Tremaine [2008] *Galactic Dynamics*, Second Edition, Princeton NJ: Princeton University Press.

[24] Bloch, A. M. [2003] *Nonholonomic Mechanics and Control*, New York: Springer.

[25] Bloch, A. M. and P. E. Crouch [1994] Reduction of Euler Lagrange problems for constrained variational problems and relation with optimal control problems, *Proc. 33rd IEEE Conf. Decis. Control*, 2584–2590.

[26] Bloch, A. and A. Rojo [2016] The optical mechanical analogy and nonlinear nonholonomic systems, *Phys. Rev. E* **93**, 023005.

[27] Bloch, A. M., P. E. Crouch, and T. S. Ratiu [1994] Sub-Riemannian optimal control problems; in Bloch, A. M. (ed.), *Hamiltonian and Gradient Flows, Algorithms and Control*, Providence RI: AMS, 35–48.

[28] Bloch, A. M., P. S. Krishnaprasad, J. E. Marsden, and R. M. Murray [1996] Nonholonomic mechanical systems with symmetry, *Arch. Ration. Mech. Anal.* **136**, 21–99.

[29] Bloch, A. M., P. S. Krishnaprasad, J. E. Marsden, and T. S. Ratiu [1996] The Euler–Poincaré equations and double bracket dissipation, *Commun. Math. Phys.* **175**, 1–42.

[30] Bloch, A. M., J. E. Marsden, and D. V. Zenkov [2009] Quasivelocities and symmetries in nonholonomic systems, *Dyn. Syst.* **24**, 187–222.

[31] Bloch, A., O. Fernandez and D. Zenkov [2014] The geometry and integrability of the Suslov problem, *J. Math. Phys.* **55**, 4901754.

[32] Bloch, A., L. Colombo, R. Gupta and T. Ohsawa [2018] Optimal control problems with symmetry breaking cost functions, *SIAM J. Appl. Algebra Geom.*, **1**(1) 626–646.

[33] Boltzmann, L. [1872] Weitere Studien über das Wärmegleichgewicht unter Gasmolekülen, *Wien. Ber.* **66**, 275–370; in Hasenöhrl, F. (ed.), *Wissenschaftliche Abhandlungen* **I**, Leipzig: Barth, paper 23; reprinted in (1968), New York: Chelsea.

[34] Boyce, W. E. and R. C. DiPrima [2008] *Elementary Differential Equations*, Ninth Edition, New York: Wiley.

[35] Brockett, R. W. [1981] Control theory and singular Riemannian geometry; in Hilton, P. J. and G. S. Young (eds.), *New Directions in Applied Mathematics*, New York: Springer, 11–27.

[36] Brouwer, D. and G. M. Clemence [1961] *Methods of Celestial Mechanics*, New York: Academic Press.

[37] Bullo, F. and A. D. Lewis [2005] *Geometric Control of Mechanical Systems*, New York: Springer.

[38] Bunimovich, L. A. [1979] On the ergodic properties of nowhere dispersing billiards, *Commun. Math. Phys.* **65**(3), 295–312.

[39] Bunimovich, L. A. and Ya. G. Sinai [1980] Markov partitions for dispersed billiards, *Commun. Math. Phys.* **78**(2), 247–280.

[40] Burke, W. L. [1985] *Applied Differential Geometry*, Cambridge UK: Cambridge University Press.

[41] Cardin, F. and M. Favretti [1996] On nonholonomic and vakonomic dynamics of mechanical systems with nonintegrable constraints, *J. Geom. Phys.* **18**, 295–325.

[42] Carr, J. [1981] *Applications of Center Manifold Theory*, New York: Springer.

[43] Cartan, É. [1922] *Leçons sur les Invariants Intégraux*, Paris: Hermann.

[44] Cartan, É. [1937] *La théorie des groupes finis et continus et la géométrie différentielle traitées par la méthode du repére mobile*, Paris: Gauthier-Villars.

[45] Cartwright, J., V. Eguiluz, E. Hernandez-Garcia, and O. Piro [1999] Dynamics of elastic excitable media, *Int. J. Bifurc. Chaos Appl. Sci. Eng.* **9**, 2197–2202.

[46] Cayley, A. [1858] A memoir on the theory of matrices, *Philos. Trans. R. Soc. Lond.* **148**, 17–37.

[47] Cendra, H., J. E. Marsden, and T. S. Ratiu [2001] Geometric mechanics, Lagrangian reduction, and nonholonomic systems; in Engquist, B. and W. Schmid (eds.), *Mathematics Unlimited: 2001 and Beyond*, New York: Springer, 221–273.

[48] Chance, B., E. K. Pye, A. K. Ghosh, and B. Hess (eds.) [1973] *Biological and Biochemical Oscillators*, New York: Academic Press.

[49] Chaplygin, S. A. [1897] On the motion of a heavy body of revolution on a horizontal plane (in Russian), *Physics Section of the Imperial Society of Friends of Physics, Anthropology and Ethnographics* **9**, 10–16; translated in (2002) *Regul. Chaotic Dyn.* **7**(2), 119–130.

[50] Chaplygin, S. A. [1897] On some feasible generalization of the theorem of area, with an application to the problem of rolling spheres (in Russian), *Mat. Sb.* **20**, 1–32.

[51] Chaplygin, S. A. [1903] On a rolling sphere on a horizontal plane (in Russian), *Mat. Sb.* **24**, 139–168; translated in (2002) *Regul. Chaotic Dyn.* **7**(2), 131–148.

[52] Chaplygin, S. A. [1911] On the theory of the motion of nonholonomic systems: the reducing-multiplier theorem (in Russian), *Mat. Sb.* **28**, 303–314; translated in Getling, A. V. (trans.) (2008), *Regul. Chaotic Dyn.* **13**(4), 369–376.

[53] Chatayev, N. G. [1955] *The Stability of Motion (in Russian)*, Second Edition, Moscow: Gosudarstv. Izdat. Tehn.-Teor. Lit.; translated in Babister, A. W. and J. Burlak (eds.) (1961) *The Stability of Motion*, Oxford: Pergamon.

[54] Chow, S. N. and J. K. Hale [1982] *Methods of Bifurcation Theory*, New York: Springer.

[55] Clebsch, A. [1866] Über die simultane Integration linearer partieller Differentialgleichungen, *J. Reine Angew. Math.* **65**, 257–268.

[56] Coddington, E. A. and N. Levinson [1984] *Theory of Ordinary Differential Equations*, Krieger.

[57] Conway, B. (ed.) [2010] *Spacecraft Trajectory Optimization*, Cambridge UK: Cambridge University Press.

[58] Cortés, J., M. De Leon, and D. De Diego [2003] Geometric description of vakonomic and nonholonomic dynamics, comparison of solutions, *SIAM J. Control Optim.* **41**, 1389–1412.

[59] Crassadis, J. L. and J. L. Junkins [2004] *Optimal Estimation of Dynamical Systems*, Boca Rotan FL: CRC Press.

[60] Crouch, P. E. and R. Grossman [1993] Numerical integration of ordinary differential equations on manifolds, *J. Nonlinear Sci.* **3**, 1–33.

[61] Crouch, P. E. and F. Silva Leite [1995] The dynamic interpolation problem: on Riemannian manifolds, Lie groups, and symmetric spaces, *J. Dyn. Control Syst.* **1**(2), 177–202.

[62] Curtis, C. W. [1974] *Linear Algebra*, New York: Springer.

[63] Cushman, R., J. Hermans, and D. Kemppainen [1996] The rolling disc; in *Nonlinear Dynamical Systems and Chaos (Groningen, 1995)*, Basel: Birkhäuser, 21–60.

[64] de Fermat, P. [1662] *Letter to Coreau de la Chambre*.

[65] de Leon, M., J. C. Marrero, and D. M. De Diego [2000] Vakonomic mechanics versus nonholonomic mechanics: a unified geometrical approach, *J. Geom. Phys.* **35**, 126–144.

[66] Deahna, F. [1840] Über die Bedingungen der Integrabilitat, *J. Reine Angew. Math.* **20**, 340–350.

[67] Deprit, A. and J. Henrard [1967] Natural families of periodic orbits, *Astron. J.* **72**(2), 158–172.

[68] Deryabin, M. V. and L. D. Pustyl'nikov [2007] Nonequilibrium gas and generalized billiards, *J. Stat. Phys.* **126**(1), 117–132.

[69] Dieci, L., R. D. Russell, and E. S. Van Vleck [1997] On the computation of Lyapunov exponents for continuous dynamical systems, *SIAM J. Numer. Anal.* **34**, 402–423.

[70] Do Carmo, M. D. [1992] *Riemannian Geometry*, Boston: Birkhäuser.

[71] Dowell, D. H. [1975] *Aeroelasticity of Plates and Shells*, Leyden: Noordhoff.

[72] Dowell, D. H. [1980] Nonlinear aeroelasticity; in Holmes, P. (ed.), *New Approaches to Nonlinear Problems in Dynamics*, Philadelphia: SIAM, 147–172.

[73] Dowell, E. H. and M. Ilgamova [1988] *Studies in Nonlinear Aeroelasticity*, New York: Springer.

[74] Drazin, P. G. and W. H. Reid [1981] *Hydrodynamic Stability*, Cambridge UK: Cambridge University Press.

[75] Dubrovin, B. A., A. T. Fomenko, and S. P. Novikov [1992] *Modern Geometry – Methods and Applications, Part I: The Geometry of Surfaces, Transformation Groups, and Fields*, New York: Springer.

[76] Dunham, D. W. and R. W. Farquhar [2002] Libration-point missions 1978–2000, in *Libration Point Orbits and Applications*, Parador d'Aiguablava, Girona, Spain, June 2002.

[77] Edgar, G. [2008] *Measure, Topology, and Fractal Geometry*, Second Edition, New York: Springer.

[78] Efroimsky, M. [2001] Relaxation of wobbling asteroids and comets—theoretical problems, perspectives of experimental observation, *Planet. Space Sci.* **49**(9), 937–955.

[79] Efroimsky, M. [2002] Euler, Jacobi, and missions to comets and asteroids, *Adv. Space Res.* **29**(5), 725–734.

[80] Ehlers, K., J. Koiller, R. Montgomery, and P. Rios [2005] Nonholonomic systems via moving frames: Cartan equivalence and Chaplygin hamiltonization; in Marsden, J. and Ratiu, T. S. (eds.), *The Breadth of Symplectic and Poisson Geometry*, Boston: Birkhäuser, 75–120.

[81] Ehresmann, C. [1950] Les connexions infinitésimales dans un espace fibré différentiable, in *Colloque de Topologie*, Bruxelles, 29–55.

[82] Ekeland, I. and H. Hofer [1990] Symplectic topology and Hamiltonian dynamics, *Math. Z.* **200**, 355–378.

[83] Ekeland, I. and H. Hofer [1990] Symplectic topology and Hamiltonian dynamics, *Math. Z.* **203**, 553–567.

[84] Euclid [280 BCE] *Catoptrics*, Alexandria: np.

[85] Euler, L. [1744] Methodus inveniendi lineas curvas, in *Lausannae, Genevae, apud Marcum-Michaelem Bosquet and Socias.*

[86] Euler, L. [1752] Decouverte d'un nouveau principe de Mecanique, *Mem. Acad. Sci. Berlin* **6**, 185–217.

[87] Falconer, K. J. [1985] *The Geometry of Fractal Sets*, Cambridge UK: Cambridge University Press.

[88] Farquhar, R. W. [1968] The Control and Use of Libration-Point Satellites, Ph.D. Thesis, Stanford University.

[89] Favretti, M. [1998] Equivalence of dynamics for nonholonomic systems with transverse constraints, *J. Dyn. Differ. Equ.* **10**, 511–535.

[90] Feigenbaum, M. J. [1979] The universal metric properties of nonlinear transformations, *J. Stat. Phys.* **21**, 669–706.

[91] Fernandez, O. E. and A. M. Bloch [2008] Equivalence of the dynamics of nonholonomic and variational nonholonomic systems for certain initial data, *J. Phys. A* **41**.

[92] Ferraz-Mello, S. [2007] *Canonical Perturbation Theories: Degenerate Systems and Resonance*, New York: Springer.

[93] FitzHugh, R. [1961] Impulses and physiological states in theoretical models of nerve membranes, *Biophys. J.* **1**, 445–466.

[94] Floquet, G. [1883] Sur les équations différentialles linéaires à coefficients périodiques, *Ann. Sci. Éc. Norm. Supér. Sér. 2*, **12**, 47–88; available from http://www.Numdam.Org/item?id=ASENS_1883_2_12__47_0.

[95] Folland, G. B. [1999] *Real Analysis: Modern Techniques and Their Applications*, New York: Wiley.

[96] Frobenius, G. [1877] Über das Pfaffsche Probleme, *J. Reine Angew. Math.* **82**, 230–315.

[97] Gauss, C. F. [1829] Über ein neues allgemeines Grundgesatz der Mechanik, *J. Reine Angew. Math.* **4**, 232–235.

[98] Gelfand, I. M. and S. V. Fomin [1963] *Calculus of Variations*, Englewood Cliffs NJ: Prentice Hall; reprinted in (2000), New York: Dover.

[99] Gibbs, J. W. [1902] *Elementary Principles in Statistical Mechanics: Developed with Especial Reference to the Rational Foundation of Thermodynamics*, New York: Scribner; reprinted in (1960), New York: Dover.

[100] Gilat, A. [2008] *MATLAB: An Introduction with Applications*, Third Edition, New York: Wiley.

[101] Goldbeter, A. [1980] Models for oscillations and excitability in biochemical systems; in Segel, L. A. (ed.), *Mathematical Models in Molecular and Cellular Biology*, Cambridge UK: Cambridge University Press.

[102] Goldstein, H., C. P. Poole, and J. L. Safko [2001] *Classical Mechanics*, Third Edition, San Francisco: Addison Wesley.

[103] Gracia, X., J. Marin-Solano, and M. C. Munoz-Lecanda [2003] Some geometric aspects of variational calculus in constrained systems, *Rep. Math. Phys.* **51**, 127–147.

[104] Greenwood, D. T. [1977] *Classical Dynamics*, Englewood Cliffs NJ: Prentice Hall; reprinted in (1997), New York: Dover.

[105] Greenwood, D. T. [1987] *Principles of Dynamics*, Second Edition, Englewood Cliffs NJ: Prentice Hall.

[106] Greenwood, D. T. [2003] *Advanced Dynamics*, Cambridge UK: Cambridge University Press.

[107] Griffel, D. H. [1981] *Applied Functional Analysis*, New York: Dover.

[108] Grobman, D. [1959] Homeomorphisms of systems of differential equations, *Dokl. Akad. Nauk SSSR* **128**, 880–881.

[109] Gromov, M. [1985] Pseudoholomorphic curves in symplectic manifolds, *Invent. Math.* **82**, 307–347.

[110] Guckenheimer, J. and P. Holmes [1983] *Nonlinear Oscillations, Dynamical Systems, and Bifurcations of Vector Fields*, New York: Springer.

[111] Guckenheimer, J., M. Myers, and B. Sturmfels [1997] Computing Hopf bifurcations I, *SIAM J. Numer. Anal.* **34**(1), 1–22.

[112] Hadamard, J. [1893] Résolution d'une question relative aux déterminants, *Bull. Sci. Math.* **17**, 240–246.

[113] Hadamard, J. [1898] Les surfaces à curbures opposés et leurs lignes géodesiques, *J. Math. Pures Appl.* **4**, 27–73.

[114] Hairer, E., C. Lubich, and G. Wanner [2002] *Geometric Numerical Integration*, New York: Springer.

[115] Hale, J. and H. Koçak [1991] *Dynamics and Bifurcations*, New York: Springer.

[116] Hamel, G. [1904] Die lagrange-eulersche Gleichungen der Mechanik, *Z. Math. Phys.* **50**, 1–57.

[117] Hamel, G. [1904] Über die virtuellen Verschiebungen in der Mechanik, *Math. Ann.* **59**, 416–434.

[118] Hamel, G. [1949] *Theoretische Mechanik. Eine Einheitliche Einfuehrung in die gesamte Mechanik*, New York: Springer.

[119] Hamilton, W. R. [1834] On a general method in dynamics, Part I, *Philos. Trans. R. Soc. Lond.*, 247–308.

[120] Hamilton, W. R. [1835] On a general method in dynamics, Part II, *Philos. Trans. R. Soc. Lond.*, 95–144.

[121] Hamilton, W. R. [1862] On the existence of a symbolic and biquadratic equation, which is satisfied by the symbol of linear or distributive operation on a quaternion, *Philos. Mag. Ser. 4* **24**, 127–128.

[122] Hamilton, W. R. [1864] On a new and general method of inverting a linear and quaternion function of a quaternion, *Proc. R. Ir. Acad.* **8**, 182–183.

[123] Hamilton, W. R. [1864] On the existence of a symbolic and biquadratic equation, which is satisfied by the symbol of linear operation in quaternions, *Proc. R. Ir. Acad.* **8**, 190–191.

[124] Hart, J. B., R. E. Miller and R. L. Mills [1987] A simple geometric model for visualizing the motion of a Foucault pendulum, *Am. J. Phys.* **55**, 67–70.

[125] Hartman, P. [1960] A lemma in the theory of structural stability of differential equations, *Proc. Am. Math. Soc.* **11**, 610–620.

[126] Hartman, P. [2002] *Ordinary Differential Equations*, Second Edition, Philadelphia: SIAM.

[127] Heisenberg, W. [1927] Über den anschaulichen Inhalt der quantentheoretischen Kinematik und Mechanik, *Z. Phys.* **43**(3–4), 172–198; translated in Wheeler, J. A. and H. Zurek (eds.) (1983) *Quantum Theory and Measurement*, Princeton NJ: Princeton University Press, 62–84.

[128] Hénon, M. [1969] Numerical exploration of the restricted problem. V. Hill's case: periodic orbits and their stability, *Astron. Astrophys.* **1**, 223–238.

[129] Hénon, M. [2003] New families of periodic orbits in Hill's problem of three bodies, *Celest. Mech. Dyn. Astron.* **85**(3), 223–246.

[130] Hénon, M. [2005] Families of asymmetric periodic orbits in Hill's problem of three bodies, *Celest. Mech. Dyn. Astron.* **93**, 87–100.

[131] Hénon, M. and C. Heiles [1964] The applicability of the third integral of motion: some numerical experiments, *Astrophys. J.* **69**(1), 73–79.

[132] Hermans, J. [1995] A symmetric sphere rolling on a surface, *Nonlinearity* **8**, 1–23.

[133] Hero [c. 50] *Catoptrics*, Alexandria: np.

[134] Herstein, I. N. [1975] *Topics in Algebra*, New York: Wiley.

[135] Hertz, H. R. [1894] *Gessamelte Werke, Band III: Der Prinzipien der Mechanik in neuem Zussamenhange dargestellt*, Leipzig: Barth.

[136] Hill, G. W. [1886] On the part of the motion of the lunar perigee which is a function of the mean motions of the sun and moon, *Acta Math.* **8**, 1–36.

[137] Hirsch, M. W. and S. Smale [1974] *Differential Equations, Dynamical Systems and Linear Algebra*, New York: Academic Press.

[138] Hirsch, M. W., S. Smale and R. L. Devaney [2004] *Differential Equations, Dynamical Systems, and an Introductions to Chaos*, Second Edition, New York: Elsevier Academic Press.

[139] Hodgkin, A. and A. Huxley [1952] A quantitative description of membrane current and its application to conduction and excitation in nerve, *J. Physiol.* **117**, 500–544.

[140] Hofer, H. and E. Zehnder [1990] A new capacity for symplectic manifolds; in Rabinowitz, P. and E. Zehnder (eds.), *Analysis, Et Cetera: Research Papers Published in Honor of Jurgen Moser's 60th Birthday*, New York: Academic Press, 405–428.

[141] Hofer, H. and E. Zehnder [1994] *Symplectic Invariants and Hamiltonian Dynamics*, Boston: Birkhäuser.

[142] Hohmann, W. [1925] *Die Erreichbarkeit der Himmelskörper*, Munich: Oldenbourg.

[143] Hölder, O. [1896] Über die principien von hamilton und maupertuis, *Nachr. Akad. Wiss. Göttingen Math.-Phys. Kl. II* **2**, 122–157.

[144] Holm, D. D. [2008] *Geometric Mechanics Part I: Dynamics and Symmetry*, London: Imperial College.

[145] Holm, D. D. [2011] *Geometric Mechanics Part II: Rotating, Translating, and Rolling*, Second Edition, London: Imperial College.

[146] Holmes, P. J. [1977] Bifurcations to divergence and flutter in flow-induced oscillations: a finite-dimensional analysis, *J. Sound Vib.* **53**(4), 471–503.

[147] Holmes, P. J. and J. E. Marsden [1978] Bifurcations to divergence and flutter in flow-induced oscillations: a finite-dimensional analysis, *Automatica* **14**(4), 367–384.

[148] Hopf, E. [1942] Abzweigung einer periodischen Lösung von einer stationären Lösung eines Differentialsystems, *Ber. Math.-Phys. Kl. Sächs. Akad. Wiss. Leipzig* **94**, 1–22.

[149] Howell, K. C. [1984] Three-dimensional, periodic, 'halo' orbits, *Celest. Mech.* **32**, 53–71.

[150] Hsiao, F. Y. and D. J. Scheeres [2007] Fundamental constraints on uncertainty evolution in Hamiltonian systems, *IEEE Trans. Autom. Control* **52**(4), 686–691.

[151] Ibn al-Haytham (Alhazen) [1021] *The Book of Optics* (in Arabic), Cairo: n.p.; translated in Smith, A. M. (trans.) (2001) *Trans. Am. Philos. Soc.* **91**(4–5), 1–819 and Smith, A. M. (trans.) (2006) *Trans. Am. Philos. Soc.* **96**(2–3), 1–697.

[152] Ince, E. L. [1926] *Ordinary Differential Equations*, London: Longmans; reprinted in (1944) New York: Dover.

[153] Jackson, J. [1913] Retrograde satellite orbits, *Mon. Not. R. Astron. Soc.* **74**, 62–82.

[154] Jordan, D. W. and P. Smith [1987] *Nonlinear Ordinary Differential Equations*, Second Edition, Oxford: Oxford University Press.

[155] Kalman, R. E. [1960] Contributions to the theory of optimal control, *Bol. Soc. Mat. Mexicana* **5**, 102–119.

[156] Kalman, R. E. [1960] On the general theory of control systems, in: *Proc. 1st IFAC World Congress*, 481–493.

[157] Kalman, R. E., Y. C. Ho, and K. S. Narendra [1963] Controllability of linear dynamical systems, *Contrib. Differ. Equ.* **1**, 189–213.

[158] Kirk, D. E. [1970] *Optimal Control Theory: An Introduction*, Englewood Cliffs NJ: Prentice Hall; reprinted in (2004), New York: Dover.

[159] Kobayashi, S. and K. Nomizu [1996] *Foundations of Differential Geometry: Volume I*, New York: Wiley.

[160] Kobayashi, S. and K. Nomizu [1996] *Foundations of Differential Geometry: Volume II*, New York: Wiley.

[161] Koiller, J. [1992] Reduction of some classical nonholonomic systems with symmetry, *Arch. Ration. Mech. Anal.* **118**, 113–148.

[162] Koon, W. S. and J. E. Marsden [1997] Optimal control for holonomic and nonholonomic mechanical systems with symmetry and lagrangian reduction, *SIAM J. Control Optim.* **35**(3), 901–929.

[163] Koon, W. S., M. W. Lo, J. E. Marsden, S. D. Ross [2008] *Dynamical Systems, the Three Body Problem, and Space Mission Design*, Marsden Books.

[164] Kozlov, V. V. [1982] Dynamics of systems with nonintegrable constraints I (in Russian), *Vestn. Mosk. Univ. Ser. I Mat. Mekh.* **37**(3), 92–100.

[165] Kozlov, V. V. [1982] Dynamics of systems with nonintegrable constraints II (in Russian), *Vestn. Mosk. Univ. Ser. I Mat. Mekh.* **37**(4), 70–76.

[166] Kozlov, V. V. [1983] Dynamics of systems with nonintegrable constraints III (in Russian), *Vestn. Mosk. Univ. Ser. I Mat. Mekh.* **38**(3), 102–111.

[167] Kozlov, V. V. [1983] Realization of nonintegrable constraints in classical mechanics (in Russian), *Dokl. Akad. Nauk SSSR* **272**, 550–554; translated in *Sov. Phys. Dokl.* **28**, 735–737.

[168] Kozlov, V. V. [1987] Dynamics of systems with nonintegrable constraints IV. Integral priciples (in Russian), *Vestn. Mosk. Univ. Ser. I Mat. Mekh.* **42**(5), 76–83.

[169] Kozlov, V. V. [1988] Dynamics of systems with nonintegrable constraints V. The freedom principle and ideal constraint condition (in Russian), *Vestn. Mosk. Univ. Ser. I Mat. Mekh.* **43**(6), 51–54.

[170] Krenner, A. J. [1977] The higher order maximal principle and its application to singular extremals, *SIAM J. Control Optim.* **15**, 256–293.

[171] Kreyszig, E. [1978] *Introductory Functional Analysis with Applications*, New York: Wiley.

[172] Lagrange, J. L. [1788] *Mechanique Analytique*, Chez la Veuve Desaint.

[173] Lanczos, C. [1986] *The Variational Principles of Mechanics*, New York: Dover.

[174] Landau, L. D. and E. M. Lifshitz [1976] *Mechanics*, Third Edition, Oxford: Pergamon.

[175] Lawden, D. F. [1962] Impulsive transfer between elliptic oribts; in Leitmann, G. (ed.), *Optimization Techniques*, New York: Academic Press, 323–351.

[176] Lawden, D. F. [1963] *Optimal Trajectories for Space Navigation*, London: Butterworths.

[177] Lawden, D. F. [1975] *Analytical Methods of Optimization*, Edinburgh: Scottish Academic Press; reprinted in (2006) New York: Dover.

[178] Lawson, H. B. [1977] *The Qualitative Theory of Foliations*, Providence RI: AMS.

[179] Lee, T., M. Leok, and N. H. McClamroch [2007] Lie group variational integrators for the full body problem in orbital mechanics, *Celest. Mech. Dyn. Astron.* **98**, 121–144.

[180] Lee, T., M. Leok, and N. H. McClamroch [2007] Lie group variational integrators for the full body problem, *Comput. Methods Appl. Mech. Eng.* **196**, 2907–2924.

[181] Leimkuhler, B. and S. Reich [2004] *Simulating Hamiltonian Dynamics*, Cambridge UK: Cambridge University Press.

[182] Lenz, F., F. K. Diakonos, and P. Schmelcher [2007] Classical dynamics of the time-dependent elliptical billiard, *Phys. Rev. E* **76**, 066213.

[183] Lenz, F., F. K. Diakonos, and P. Schmelcher [2008] Tunable Fermi acceleration and the driven elliptical billiard, *Phys. Rev. Lett.* **100**, 014103.

[184] Leproux, P., S. Fevrier, V. Doya, P. Roy, and D. Pagnoux [2003] Modeling and optimization of double-clad fiber amplifiers using chaotic propagation of pump, *Opt. Fiber Technol.* **7**(4), 324–339.

[185] Levi-Civita, T. [1917] Nozione di parallelismo in una varietà qualunque, *Rend. Circ. Mat. Palermo* **42**, 173–205.

[186] Levi-Civita, T. and U. Amaldi [1951] *A Course in Theoretical Mechanics (in Italian)*, Bologna: Zanichelli.

[187] Lewis, A. D. and R. M. Murray [1995] Variational principles in constrained systems: theory and experiment, *Int. J. Non-Linear Mech.* **30**, 793–815.

[188] Lichtenberg, A. J. and M. A. Lieberman [1992] *Regular and Chaotic Dynamics*, New York: Springer.

[189] Lindelf, E. [1894] Sur l'application de la methode des approximations successives aux quations diffrentielles ordinaires du premier ordre, *C. R. Hebd. Séances Acad. Sci.* **114**, 454–457; available from http://gallica.Bnf.Fr/ark:/12148/bpt6k3074r/f454.Table.

[190] Liouville, J. [1838] Note sur la Théorie de la Variation des constantes arbitraires, *J. Math. Pures Appl., Sér. 1*, **3**, 342–349.

[191] Lorenz, E. N. [1963] Deterministic nonperiodic flow, *J. Atmos. Sci.* **20**(2), 130–141.

[192] Lyapunov, A. M. [1892] *The General Problem of the Stability of Motion* (in Russian), Ph.D. Thesis, Moscow University; reprinted in Fuller, A. T. (ed., trans.) (1992) *Int. J. Control* **55**(3), 531–772.

[193] Magnus, W. [1954] On the exponential solution of differential equations for a linear operator, *Commun. Pure Appl. Math.* **7**, 649–673.

[194] Magnus, W. and S. Winkler [1966] *Hill's Equation*, New York: Wiley Interscience; corrected version reprinted in (1979) New York: Dover.

[195] Marsden, J. E. and M. McCracken [1976] *The Hopf Bifurcation and Its Applications*, New York: Springer.

[196] Marsden, J. E. and T. S. Ratiu [1994] *Introduction to Mechanics and Symmetry*, New York: Springer.

[197] Marsden, J. E. and M. West [2001] Discrete mechanics and variational integrators, *Acta Numer.* **10**, 357–514.

[198] Marsden, J. E., T. S. Ratiu, and A. Weinstein [1983] Semi-direct products and reduction in mechanics, *Trans. Am. Math. Soc.* **281**, 147–177.

[199] Marsden, J. E., T. S. Ratiu, and A. Weinstein [1984] Reduction and Hamiltonian structures on duals of semidirect product Lie algebras, *Contemp. Math.* **28**, 55–100.

[200] Marsden, J. E., T. S. Ratiu, and J. Scheurle [2000] Reduction theory and the Lagrange–Routh equations, *J. Math. Phys.* **41**(6), 3307–4237.

[201] Martinez, S., J. Cortes, and M. de Leon [2001] Symmetries in vakonomic dynamics: applications to optimal control, *J. Geom. Phys.* **38**, 343–365.

[202] Maruskin, J. M. [2008] *On the Dynamical Propagation of Subvolumes and on the Geometry and Variational Principles of Nonholonomic Systems*, Ph.D. Thesis, Ann Arbor: The University of Michigan.

[203] Maruskin, J. M. and A. M. Bloch [2007] The Boltzmann–Hamel equations for optimal control, *Proc. 46th IEEE Conf. Decis. Control*, 554–559.

[204] Maruskin, J. M. and A. M. Bloch [2011] The Boltzmann–Hamel equations for the optimal control of mechanical systems with nonholonomic constraints, *Int. J. Robust Nonlinear Control* **21**(4), 373–386.

[205] Maruskin, J. M., D. J. Scheeres, and A. M. Bloch [2009] Dynamics of symplectic subvolumes, *SIAM J. Appl. Dyn. Syst.* **8**(1), 180–201.

[206] Maruskin, J. M., A. M. Bloch, J. E. Marsden, and D. Zenkov [2012] A fiber bundle approach to the transpositional relations in nonholonomic systems, *J. Nonlinear Sci.*, **22**(4): 431–461.

[207] Matukuma, T. [1933] Periodic orbits in Hill's case: third paper (periodic ejectional orbits), *Proc. Imp. Acad.* **9**(8), 364–366.

[208] Matukuma, T. [1957] *Sendai Astron. Rap.*, **51**.

[209] Maupertuis, P. L. M. [1740] Lois du repos des corps, *Acad. R. Sci.*

[210] McDuff, D. and D. Salamon [1999] *Introduction to Symplectic Topology*, Oxford: Oxford University Press.

[211] Meiss, J. D. [2006] *Differential Dynamical Systems*, Philadelphia: SIAM.

[212] Misner, C. W., Thorne, K. S., and Wheeler, J. A. [1973] *Gravitation*, San Francisco: Freeman.

[213] Moore, W. [1813] *A Treatise on the Motion of Rockets*, Royal Military Academy at Woolwich; reprinted with commentary in Johnson, W. (1995) Contents and commentary on William Moore's "A treatise on the motion of rockets and an essay on naval gunnery," *In. J. Impact Eng.* **16**(3): 499–521.

[214] Morita, S. [2001] *Geometry of Differential Forms*, Providence RI: AMS.

[215] Moser, J. K. [1965] On the volume elements of manifolds, *Trans. Am. Math. Soc.* **120**, 280–296.

[216] Munkres, J. R. [2000] *Topology*, Second Edition, Englewood Cliffs NJ: Prentice Hall.

[217] Munthe-Kaas, H. [1995] Lie Butcher theory for Runge–Kutta methods, *BIT* **35**, 572–587.

[218] Munthe-Kaas, H. [1998] Runge–Kutta methods on Lie groups, *BIT* **38**, 92–111.

[219] Munthe-Kaas, H. [1999] Higher order Runge–Kutta methods on manifolds, *J. Appl. Numer. Math.* **29**, 115–127.

[220] Nagumo, J., S. Arimoto, and S. Yoshizawa [1962] An active pulse transmission line simulating nerve axon, *Proc. Inst. Radio Eng.* **50**, 2061–2070.

[221] Neimark, J. I. and N. A. Fufaev [1960] Transpositional relations in the analytical mechanics of nonholonomic systems, *J. Appl. Math. Mech.* **24**, 1541–1548.

[222] Neimark, J. I. and N. A. Fufaev [1972] *Dynamics of Nonholonomic Systems*, Providence RI: AMS.

[223] Nelson, E. [1967] *Tensor Analysis*, Princeton NJ: Princeton University Press.

[224] Newton, I. [1687] *Philosophia Naturalis Principia Mathematica*, Josephi and Streater.

[225] Nijmeijer, H. and A. J. van der Schaft [1990] *Nonlinear Dynamical Control Systems*, New York: Springer.

[226] Nöther, E. [1918] Invariante Variationsprobleme, *Nachr. d. König. Gesellsch. d. Wiss. zu Göttingen, Math-phys*, 235–257.

[227] Olsen, L. F. and H. Degn [1985] Chaos in biological systems, *Q. Rev. Biophys.* **18**(2), 165–225.
[228] O'Reilly, O. M. [1996] The Dynamics of rolling disks and sliding disks, *Nonlinear Dyn.* **10**, 287–305.
[229] Oseledec, V. I. [1968] A multiplicative ergodic theorem: Liapunov characteristic numbers for dynamical systems, *Trans. Mosc. Math. Soc.* **19**(2), 197–231.
[230] Ott, E. [2002] *Chaos in Dynamical Systems*, Second Edition, Cambridge UK: Cambridge University Press.
[231] Palais Jr., J. and W. de Melo [1981] *Geometric Theory of Dynamical Systems*, New York: Springer.
[232] Papastavridis, J. G. [2002] *Analytical Mechanics*, Oxford: Oxford University Press.
[233] Peano, G. [1888] Intégration par séries des équations différentielles linéaires, *Math. Ann.* **32**, 450–456.
[234] Perko, L. [1982] Families of symmetric periodic solutions of Hill's problem i: first species periodic solutions for $C \ll -1$, *Am. J. Math.* **104**(2), 321–352.
[235] Perko, L. [2001] *Differential Equations and Dynamical Systems*, New York: Springer.
[236] Picard, C. E. [1890] Memoire sur la théorie des équations aux dérivées partielles et la méthode des approximations successives, *J. Math. Pures Appl., Sér. 4*, T6, 145–210; available from http://gallica.Bnf.Fr/ark:/12148/bpt6k107454k.
[237] Poincaré, H. [1881] Mémoire sur les courbes définies par une equation différentielle, *J. Math.* **7**, 375–422.
[238] Poincaré, H. [1892] *Les Méthodes Nouvelles de la Mécanique Céleste, Tome I: Solutions Périodiques. Non-Existence des Intégrales Uniformes. Solutions Asymptotiques*, Paris: Gauthier-Villars.
[239] Poincaré, H. [1893] *Les Méthodes Nouvelles de la Mécanique Céleste, Tome II: Méthodes de MM. Newcomb, Gyldén, Lindstedt et Bohlin*, Paris: Gauthier-Villars.
[240] Poincaré, H. [1899] *Les Méthodes Nouvelles de la Mécanique Céleste, Tome III: Invariants Intégraux. Solutions Périodiques du Deuxiéme Genre. Solutions Doublement Asymptotiques*, Paris: Gauthier-Villars.
[241] Poincaré, H. [1901] Sur la stabilité de l'équilibre des figures piriformes affectées par une masse fluide en rotation, *Philos. Trans. R. Soc. Lond. Ser. A* **198**, 333–373.
[242] Poinsot, L. [1851] *Théorie Nouvelle de la Rotation des Corps*, Paris: Bachelier; available from http://www.archive.org/details/thorienouvelled00poingoog.
[243] Polderman, J. W. and J. C. Willems [1998] *Introduction to Mathematical Systems Theory: A Behavioral Approach*, New York: Springer.
[244] Pontryagin, L. S. [1959] Optimal regulation processes (in Russian), *Usp. Mat. Nauk* **14**: 3–20.
[245] Pontryagin, L. S., V. G. Boltyanskii, R. V. Gamkrelidze, and E. F. Mishchenko [1962] *The Mathematical Theory of Optimal Processes* (in Russian); translated in Neustadt, L. W. (ed.) and K. N. Trirogoff (trans.), *L. S. Pontryagin: Selected Works* **IV**, New York: Wiley Interscience.
[246] Prussing, J. E. [1967] *Optimal Multiple-Impulse Orbital Rendezvous*, Sc.D. Thesis, MIT.
[247] Prussing, J. E. [1991] Simple proof of the global optimality of the Hohmann transfer, *J. Guid. Control Dyn.* **15**, 1037–1038.
[248] Rangarajan, G., S. Habib, and R. D. Ryne [1998] Lyapunov exponents without rescaling and reorthogonalization, *Phys. Rev. Lett.* **80**(17), 3747–3750.
[249] Rasband, S. N. [1990] *Chaotic Dynamics of Nonlinear Systems*, New York: Wiley.
[250] Rinzel, J. and G. B. Ermentrout [1989] Analysis of neural excitability and oscillations; in Koch, C. and I. Segev (eds.), *Methods in Neuronal Modeling: From Synapses to Networks*, Cambridge MA: MIT Press.
[251] Rojo, A. and A. Bloch [2018] *The Prinicple of Least Action: History and Physics*, Cambridge University Press.

[252] Ross, K. A. [1980] *Elementary Analysis: The Theory of Calculus*, New York: Springer.
[253] Rössler, O. E. [1972] An equation for continuous chaos, *Phys. Lett. A* **57**(5), 397–398.
[254] Routh, E. J. [1882] *Dynamics of a System of Rigid Bodies*, London: Macmillan.
[255] Routh, E. J. [1884] *The Advanced Part of a Treatise on the Dynamics of a System of Rigid Bodies*, London: Macmillan.
[256] Roy, A. E. [2005] *Orbital Motion*, Fourth Edition, Bristol: Institute of Physics.
[257] Rudin, W. [1991] *Functional Analysis*, Second Edition, New York: McGraw-Hill.
[258] Rumyantsev, V. V. [1978] On Hamilton's principle for nonholonomic systems, *J. Appl. Math. Mech.* **42**, 407–419.
[259] Rumyantsev, V. V. [2001] Routh's equations and variational principles, *J. Appl. Math. Mech.* **65**, 529–713.
[260] Scheeres, D. J. [2009] Minimum energy asteroid reconfigurations and catastrophic disruptions, *Planet. Space Sci.* **57**(2), 154–164.
[261] Scheeres, D. J. [2012] *Orbital Motion in Strongly Perturbed Environments*, New York: Springer Praxis.
[262] Scheeres, D. J., B. G. Williams, J. K. Miller [2000] Evaluation of the dynamic environment of an asteroid: applications to 433 Eros, *J. Guid. Control Dyn.* **23**(3), 466–475.
[263] Scheeres, D. J., F. Y. Hsiao, R. S. Park, B. F. Villac, and J. M. Maruskin [2005] Fundamental limits on spacecraft orbit uncertainty and distribution propagation, *J. Astronaut. Sci.* **54**(3–4), 505–523.
[264] Scheeres, D. J., M. A. de Gosson, and J. M. Maruskin [2012] Application of symplectic topology to orbit uncertainty and spacecraft navigation, *J. Astronaut. Sci.* **59**(1–2) 63–83.
[265] Schutz, B. F. [1980] *Geometric Method of Mathematical Physics*, Cambridge UK: Cambridge University Press.
[266] Schwartz, A. J. [1963] A generalization of a Poincaré–Bendixson theorem to closed two-dimensional manifolds, *Am. J. Math.* **85**, 453–458; errata, *Am. J. Math.* **85**, 753.
[267] Sel'kov, E. E. [1968] Self-oscillations in glycolysis: a simple kinetic model, *Eur. J. Biochem.* **4**, 79.
[268] Serovaiski, S.Ya. [2003] *Counterexamples in Optimal Control Theory*, Utrecht: VSP.
[269] Shi, D., Y. Berchenko-Kogan, D. Zenkov and A. Bloch [2017] Hamel's formalism for infinite-dimensional nonholonomic systems, *J. Nonlinear Sci.* **27**, 241–283.
[270] Sinai, Ya. G. [1963] On the foundations of the ergodic hypothesis for a dynamical system of statistical mechanics (in Russian), *Dokl. Akad. Nauk SSSR* **153**(6), 1261–1264; translated in *Sov. Math., Dokl.* **4**, 1818–1822.
[271] Smart, E. H. [1951] *Advanced Dynamics, Volume II*, London: Macmillan.
[272] Sontag, E. D. [1998] *Mathematical Control Theory: Deterministic Finite Dimensional Systems*, Second Edition, New York: Springer.
[273] Steenrod, N. E. [1951] *The Topology of Fibre Bundles*, Princeton NJ: Princeton University Press.
[274] Strogatz, S. H. [1994] *Nonlinear Dynamics and Chaos*, Westview.
[275] Suslov, G. K. [1901] On a particular variant of d'Alembert's principle (in Russian), *Mat. Sb.* **22**, 687–691.
[276] Suslov, G. K. [1946] *Theoretical Mechanics (in Russian)*, Moscow: Gostekhizdat.
[277] Sussman, H. J. [1998] An introduction to the coordinate-free maximum principle; in Jackubczyk, B. and W. Respondek (eds.), *Geometry of Feedback and Optimal Control*, Dekker, 463–557.
[278] Sussman, H. J. [1998] Geometry and optimal control; in Baillieul, J. and J. C. Williams (eds.), *Mathematical Control Theory*, New York: Springer, 140–198.
[279] Synge, J. L. [1928] Geodesics in nonholonomic geometry, *Math. Ann.* **99**, 738–751.

[280] Tabor, M. [1990] *Chaos and Integrability in Nonlinear Dynamics: An Introduction*, New York: Wiley.
[281] Terrell, W. [1999] Some fundamental control theory I: controllability, observability, and duality, *Am. Math. Mon.* **106**, 705–719.
[282] Thomas, L. H. [1926] The motion of the spinning electron, *Nature* **117**(2945), 514.
[283] Thompson, W. (Lord Kelvin) [1892] *Collected Papers IV*, 516.
[284] Tsiolkovsky, K. E. [1903] The exploration of cosmic space by means of reaction devices (in Russian), *Sci. Rev.* **5**.
[285] Tu, L. W. [2008] *An Introduction to Manifolds*, New York: Springer.
[286] van der Pol, B. [1927] On relaxation-oscillations, *London Edinburgh Dublin Philos. Mag. J. Sci.* **2**(7), 978–992.
[287] Vierkandt, A. [1892] Über gleitende und rollende Bewegung, *Monatshefte Math. Phys.* **III**, 31–54.
[288] Voronets, P. [1901] On the equations of motion for nonholonomic systems (in Russian), *Mat. Sb.* **22**, 659–686.
[289] Warner, F. W. [1989] *Foundations of Differential Manifolds and Lie Groups*, Second Edition, New York: Springer.
[290] Weinberg, S. [1972] *Gravitation and Cosmology: Principles and Applications of the General Theory of Relativity*, New York: Wiley.
[291] Weinstein, A. [1971] Symplectic manifolds and their Lagrangian submanifolds, *Adv. Math.* **6**, 329–346.
[292] Weyl, H. [1938] Cartan on groups and differential geometry, *Bull. Am. Math. Soc.* **44**(9), 598–601.
[293] Whittaker, E. T. [1937] *A Treatise on the Analytical Dynamics of Particles and Rigid Bodies*, Cambridge UK: Cambridge University Press.
[294] Wiggins, S. [2000] *Introduction to Applied Nonlinear Dynamical Systems and Chaos*, New York: Springer.
[295] Wirtinger, W. [1936] Eine Determinantenidentität und ihre Anwendung auf analytische Gebilde in euklidischer und Hermitescher Maßbestimmung, *Monatshefte Math.* **44**(1), 343–365.
[296] Zadunaisky, P. E. [1960] The orbit of satellite 1958 Alpha (Explorer 1) during the first 10500 revolutions, *Smithson. Astrophys. Obs.*, Special Report #50.
[297] Zenkov, D. V. [1995] The geometry of the Routh problem, *J. Nonlinear Sci.* **5**(6), 503–519.
[298] Zenkov, D. V., A. M. Bloch, and J. E. Marsden [1998] The energy momentum method for the stability of nonholonomic systems, *Dyn. Stab. Syst.* **13**, 123–166.

Index

https://doi.org/10.1515/9783110597806-013

De Gruyter Studies in Mathematical Physics

Volume 47
Eugene Stefanovich
Elementary Particle Theory: Volume 3: Relativistic Quantum Dynamics, 2018
ISBN 978-3-11-049090-9, e-ISBN (PDF) 978-3-11-049322-1,
e-ISBN (EPUB) 978-3-11-049139-5

Volume 46
Eugene Stefanovich
Elementary Particle Theory: Volume 2: Quantum Electrodynamics, 2018
ISBN 978-3-11-049089-3, e-ISBN (PDF) 978-3-11-049320-7,
e-ISBN (EPUB) 978-3-11-049143-2

Volume 45
Eugene Stefanovich
Elementary Particle Theory: Volume 1: Quantum Mechanics, 2018
ISBN 978-3-11-049088-6, e-ISBN (PDF) 978-3-11-049213-2,
e-ISBN (EPUB) 978-3-11-049103-6

Volume 44
Vladimir V. Kiselev
Collective Effects in Condensed Matter Physics, 2018
ISBN 978-3-11-058509-4, e-ISBN (PDF) 978-3-11-058618-3,
e-ISBN (EPUB) 978-3-11-058513-1

Volume 43
Robert F. Snider
Irreducible Cartesian Tensors, 2017
ISBN 978-3-11-056363-4, e-ISBN (PDF) 978-3-11-056486-0,
e-ISBN (EPUB) 978-3-11-056373-3

Volume 42
Javier Roa
Regularization in Orbital Mechanics: Theory and Practice, 2017
ISBN 978-3-11-055855-5, e-ISBN (PDF) 978-3-11-055912-5,
e-ISBN (EPUB) 978-3-11-055862-3

Volume 41
Esra Russell, Oktay K. Pashaev
Oscillatory Models in General Relativity, 2017
ISBN 978-3-11-051495-7, e-ISBN (PDF) 978-3-11-051536-7, e-ISBN (EPUB) 978-3-11-051522-0

www.degruyter.com

Bei Fragen zur Produktsicherheit wenden Sie sich bitte an:
If you have any questions regarding product safety,
please contact:

Walter de Gruyter GmbH
Genthiner Straße 13
10785 Berlin
productsafety@degruyterbrill.com